INFECTIOUS DISEASE AND HOST–PATHOGEN EVOLUTION

It has long been recognized that an important factor in human evolution is the struggle against infectious disease, and more recently, it has been revealed that complex genetic polymorphisms are the direct result of that struggle. As molecular biological techniques become more sophisticated, a number of breakthroughs in the area of host–pathogen evolution have led to an increased interest in this field.

From the historical beginnings of J. B. S. Haldane's original hypothesis to current research, this book strives to evaluate infectious diseases from an evolutionary perspective. It provides a survey of the latest information regarding host–pathogen evolution related to major infectious diseases and parasitic infections, including malaria, influenza, and leishmaniasis.

Written by leading authorities in the field, and edited by a former pupil of Haldane, *Infectious Disease and Host–Pathogen Evolution* will be a valuable reference for those working in related areas of microbiology, parasitology, immunology, and infectious disease medicine, as well as genetics, evolutionary biology, and epidemiology.

Krishna R. Dronamraju is currently the President of the Foundation for Genetic Research in Houston, Texas. He has authored or edited 13 books and has authored over 200 research papers in the field of genetics. A former student of J. B. S. Haldane, whose pioneering work in epidemiologic research he cites as the basis for this book, Dronamraju holds several distinguished honorary positions in both the United States and Europe.

T0222431

Krishna R. Dronamraju at the U.K. Genome Center (courtesy of Michele Wambaugh).

OTHER BOOKS BY KRISHNA R. DRONAMRAJU

Biological Wealth and Other Essays (World Scientific, 2002)

Biological and Social Issues in Biotechnology Sharing (Ashgate Publishing, 1998)

Science and Society (University Press of America, 1998)

Haldane's Daedalus Revisited (Oxford University Press, 1995)

If I Am Remembered: The Life and Work of Julian Huxley (World Scientific, 1993)

The History and Development of Human Genetics: Progress in Different Countries (World Scientific, 1992)

Selected Genetic Papers of JBS Haldane (Garland Publishing Company, 1990)

The Foundations of Human Genetics (C. C. Thomas, 1989)

Cleft Lip and Palate: Aspects of Reproductive Biology (C. C. Thomas, 1986)

Haldane: The Life and Work of JBS Haldane with Special Reference to India (Aberdeen University Press, 1985)

Haldane and Modern Biology (The Johns Hopkins University Press, 1968)

Infectious Disease and Host–Pathogen Evolution

Edited by
Krishna R. Dronamraju
Foundation for Genetic Research, Houston, TX

CAMBRIDGE
UNIVERSITY PRESS

CAMBRIDGE UNIVERSITY PRESS
Cambridge, New York, Melbourne, Madrid, Cape Town, Singapore,
São Paulo, Delhi, Dubai, Tokyo

Cambridge University Press
The Edinburgh Building, Cambridge CB2 8RU, UK

Published in the United States of America by Cambridge University Press, New York

www.cambridge.org
Information on this title: www.cambridge.org/9780521126557

First published 2004
This digitally printed version 2009

A catalogue record for this publication is available from the British Library

Library of Congress Cataloguing in Publication data
Infectious diseases and host–pathogen evolution / edited by Krishna R. Dronamraju.

    p. cm.

    Includes bibliographical references and index.

    ISBN 0-521-82066-9

    1. Microorganisms – Evolution.   2. Host-parasite relationship.   3. Evolutionary genetics.
4. Human evolution.   I. Dronamraju, Krishna R.

    QR13.I546   2004
    616.9'041 – dc21   2003055194

ISBN 978-0-521-82066-0 Hardback
ISBN 978-0-521-12655-7 Paperback

Additional resources for this publication at www.cambridge.org/9780521126557

To my friend Sir Arthur C. Clarke

"Likewise it absorbed a whole microcosmos of living creatures – the bacteria and viruses which, upon an older planet, had evolved into a thousand deadly strains.

"Even as the *Morning Star* set course for her distant home, Venus was dying. The films and photographs and specimens that Hutchins was carrying in triumph were more precious even than he knew. They were the only record that would ever exist of life's third attempt to gain a foothold in the solar system. Beneath the clouds of Venus, the story of Creation was ended."

From Before Eden, by Arthur C. Clarke (1961)

Contents

Contributors

L. Abel
Laboratory of Human Genetics of
    Infectious Diseases
INSERM, U. 550
Paris
France

L. Argiro
Immunology and Genetics of
    Parasitic Diseases
INSERM, U. 399
Marseille
France

Francisco J. Ayala
University of California, Irvine
Department of Ecology and
    Evolutionary Biology
321 Steinhaus Hall
Irvine, CA 92697-2525

B. Bucheton
Immunology and Genetics of
    Parasitic Diseases
INSERM, U. 399
Marseille
France

Robin M. Bush
Department of Ecology and
    Evolutionary Biology
University of California, Irvine
Irvine, CA 92697

Luca Cavalli-Sforza
Genetics Department, M346
Stanford University Medical School
Stanford, CA 94305-5120

C. Chevillard
Immunology and Genetics of
    Parasitic Diseases
INSERM, U. 399
Marseille
France

Kyle D. Cochran
57 Robbs Hill Road
Lunenberg, MA 01462

Gregory M. Cochran
6708 Loftus Street, N.E.
Albuquerque, NM 87109

Rita R. Colwell
Center of Marine Biotechnology
University of Maryland
  Biotechnology Institute
701 East Pratt Street
Baltimore, MD 21202

Lindsay G. Cowell
Department of Immunology
Campus Box 3010
Duke University Medical Center
Durham, NC 27701

Nancy J. Cox
Influenza Branch
Centers for Disease Control &
  Prevention
Atlanta, GA 30333

James F. Crow
Genetics Department
University of Wisconsin
Madison, WI 53706

A. Dessein
Immunology and Genetics of
  Parasitic Diseases
INSERM, U. 399
Marseille
France

Helia Dessein
Immunology and Genetics of
  Parasitic Diseases
INSERM, U. 399
Marseille
France

Krishna R. Dronamraju
Foundation for Genetic Research
PO Box 27701-0
Houston, TX 77227

Dieter Ebert
Université de Fribourg
Departement de Biologie
Ecologie et Evolution, Chemin du
  Musée 10
1700 Fribourg
Switzerland

S.H. El-Safi
Institute for Tropical Medicine
PO Box 1304
Khartoum
Sudan

N.M.A. Elwali
Institute of Nuclear Medicine and
  Molecular Biology
University of Gezira
Wad Medani
Sudan

Ananias A. Escalante
Division of Parasitic Diseases
National Center for Infectious
  Diseases
Centers for Disease Control and
  Prevention
Public Health Service
US Department of Health and
  Human Services
Atlanta, Georgia 30333

Shah M. Faruque
International Centre for Diarrhoeal
  Disease, Bangladesh
Dhaka
Bangladesh

Sylvain Gandon
School of Biological Sciences
University of Edinburgh
Edinburgh EH9 3JT
UK

Altaf A. Lal
U.S. Embassy
New Delhi, 110021
India

Tom Little
Institute for Cell, Animal and
  Population Biology
University of Edinburgh, Kings
  Buildings
West Mains Road
Edinburgh EH9 3JT
UK

Margaret J. Mackinnon
School of Biological Sciences
University of Edinburgh
Edinburgh EH9 3JT
UK

S. Marquet
Immunology and Genetics of
  Parasitic Diseases
INSERM, U. 399
Marseille
France

N. Avrion Mitchison
Department of Immunology
Windeyer Institute of Medical
  Science
46 Cleveland Street
London W1T 4JF
UK

Newton E. Morton
Human Genetics Division
University of Southampton
Duthie Building (MP 808)
Southampton General Hospital
Tremona Road
Southampton SO16 6YD
UK

Brigitte Muller
Deutsches Rheuma Zentrum Berlin
Shumannstrasse 21/22
10117 Berlin-Mitte
Germany

G. Balakrish Nair
International Centre for Diarrhoeal
  Disease, Bangladesh
Dhaka
Bangladesh

Sean Nee
School of Biological Sciences
University of Edinburgh
Edinburgh EH9 3JT
UK

Andrew F. Read
School of Biological Sciences
University of Edinburgh
Edinburgh EH9 3JT
UK

Stephen M. Rich
Division of Infectious Disease
Tufts University School of
  Veterinary Medicine
200 Westboro Road
North Grafton, MA 01536

V. Rodrigues
Laboratory of Immunology
University of Medicine
Triangulo Miniero
Uberaba
Brazil

Michel Tibayrenc
Unité de Recherche "génétique des
   maladies infectieuses"
Unité Mixte de Recherche
   no. 9926 Centre National
   de la Recherche Scientifique/
   Institut de Recherche pour le
   Développement
911 avenue Agropolis, BP 64501
34394 Montpellier Cedex 5
France

Sarah A. Tishkoff
Department of Biology
Biology/Psychology Building
University of Maryland
College Park, MD 20742

Brian C. Verrelli
Department of Biology
Biology/Psychology Building
University of Maryland
College Park, MD 20742

D.J. Weatherall
Weatherall Institute of Molecular
   Medicine
University of Oxford, UK

Peter A. Zimmerman
The Center for Global Health and
   Diseases
Case Western Reserve University
School of Medicine, W147D
2109 Adelbert Road
Cleveland, OH

Introduction

Krishna R. Dronamraju

The subject of infectious disease has never before been so intricately linked to our daily lives as it is today. Recent political and social events have underscored the importance of understanding the forces that have shaped, and continue to shape, the evolution of infectious diseases and their underlying genetic basis. The following chapters discuss infectious disease from an evolutionary perspective within the context of the occurrence of genetic polymorphisms in human populations. It was my mentor, J. B. S. Haldane (1949a,b), who first drew attention to the significant role infectious diseases have played in shaping our own evolution.

Haldane's idea was the culmination of long years of investigation into the fundamental nature of evolutionary forces that have shaped the biology of all species on this planet (Haldane 1924, 1932). He was one of the great trio of founding fathers of population genetics, the two others being R. A. Fisher and S. Wright. Haldane and Fisher were also pioneers in human genetics. Haldane contributed quite profoundly to pedigree analysis and gene mapping, as well as to the estimation of mutation rates, selection effects and other branches of human population genetics, and the impact of genetic knowledge on our ethical outlook (Dronamraju 1990, 1995).

In his often-quoted paper, entitled "Disease and Evolution," Haldane (1949b) wrote: "... the struggle against disease, and particularly infectious disease, has been a very important evolutionary agent, and that some of its results have been rather unlike those of the struggle against natural forces, hunger, and predators, or with members of the same species." Haldane further pointed out that a disease may be an advantage in certain instances, suggesting that "Europeans have used their genetic resistance

1

to such viruses as that of measles as a weapon against primitive peoples as effective as fire-arms." He suggested further that, in certain circumstances, parasitism will be a factor promoting polymorphism and may even tend to encourage speciation. Haldane (1949b) has been often credited with the idea that the very high frequency of thalassemia that is found in the Mediterranean region might reflect heterozygote advantage against malaria. However, Haldane's (1949b) reference to the greater resistance of thalassemia heterozygotes to malaria does not appear in the text of that paper but in the discussion footnotes where the Italian biologist Giuseppe Montalenti acknowledges (in Italian) Haldane's idea as a personal communication. Consequently, some English-language readers appear to be confused about the precise nature of Haldane's contribution to this important subject.

In his contribution, Weatherall (p. 19) has drawn attention to a slightly earlier publication of Haldane's (1949a), in which Haldane stated his idea clearly and explicitly. Haldane (1949a) wrote: "... the possibility that the heterozygote is fitter than the normal must be seriously considered... The corpuscles of the anaemic heterozygotes are smaller than normal and more resistant to hypotonic solutions, ... more resistant to attacks by the sporozoa which cause malaria, a disease prevalent in Italy, Sicily and Greece, where the gene is frequent." Haldane's comments were made while discussing Neel and Valentine's (1947) estimate for the mutation rate for thalassemia major (Cooley's anemia). Haldane (1949a) wrote: "Neel and Valentine believe that the heterozygote is less fit than normal, and think that the mutation rate is above 4×10^{-4} rather than below it. I believe that the possibility that the heterozygote is fitter than the normal must be seriously considered. Such increased fitness is found in the case of several lethal and sublethal genes in *Drosophila* and *Zea*." Haldane (1949a) noted that if the heterozygote had an increased fitness of only 2%, this would account for the incidence without invoking any mutation at all. Earlier, in his classic, *The Causes of Evolution*, Haldane (1932) wrote that a study of the causes of death in man, animals, and plants clearly indicates that one of the principal agents of survival during the course of evolution is immunity to disease.

In a review of Haldane's paper of 1949b,[1] Lederberg (1999) cited previous work on host resistance genes against rust fungi in wheat. However, no previous author before Haldane (1949a,b) had suggested the evolutionary

---

[1] Although Haldane's (1949b) paper in *Ricerca Scientifica* is often quoted, his hypothesis about thalassemia and malaria was more directly and clearly stated in his address to the Eighth International Congress of Genetics in Stockholm, in July 1948 (Haldane 1949a).

role of resistance against infectious disease with special reference to heterozygosity in human populations. By drawing attention to Haldane's (1949a) earlier paper, where Haldane made a direct reference to the greater fitness of thalassemia heterozygotes, Weatherall (2004) has cleared up the matter of priority on this subject. Crow has briefly reviewed Haldane's ideas with special reference to disease and evolution in his introductory essay.

INFECTIOUS DISEASES

Much of the general information about infectious diseases can be easily accessed from such sources as the World Health Organization (WHO) and Centers for Disease Control (CDC) websites on the internet. Infectious diseases are responsible for almost half of the mortality in developing countries. AIDS, malaria, and tuberculosis cause much of the mortality. Population movements and lack of even the most basic measures of public hygiene complicate the picture. The emergence of drug resistance and the absence of effective vaccines are worsening the situation. Current scientific and medical tools are far from satisfactory, partly because the methodologies employed are cumbersome and expensive and are not designed to yield immediate results. Meanwhile, pathogens are evolving ever more rapidly. Research has been compartmentalized to such an extent that it has hampered an integrated approach to the study of the coevolution of host–pathogen vectors (Tibayrenc 2001, 2004). Research described in the following pages emphasizes the need to study host–pathogen coevolution.

MALARIA

Several papers are devoted to malaria. Almost three million malaria-related deaths are estimated to occur each year, largely in sub-Saharan Africa. More than 40% of the world's population lives in countries where the disease is endemic. In view of the enormity of this problem, it is encouraging that the genome sequences of the major mosquito vector *Anopheles gambiae* and of the parasite *Plasmodium falciparum* were recently published. The availability of these sequence data and that of the human host will accelerate further research aimed at solving a number of puzzles, such as the identification of potential vaccine antigens and the manipulation of genomes to block transmission of disease. Future progress will depend on close collaboration among many participants worldwide, especially those in malaria-endemic countries.

Haldane's idea opened up a whole new field of evolutionary epidemiology. This subject is reviewed by Weatherall. The first evidence in support of Haldane's hypothesis came in the mid-1960's, when it was found that heterozygotes for the sickle cell gene might be protected against *Plasmodium falciparum* malaria. More recently it has been found that the heterozygous state for Hb S offers approximately 80% protection against cerebral malaria or the profound anemia of malaria.

Genetic diversity of the malarial parasite *Plasmodium falciparum* has rightly received the most attention. However, the development of antimalarial vaccines has been hampered by the extensive polymorphism observed in Plasmodium's proteins. Among the several studies that have taken place, Escalante et al. (1998) studied the genetic polymorphism at ten *Plasmodium falciparum* loci that are considered potential targets for specific antimalarial vaccines. They concluded that at five of the loci there is definite evidence for positive selection and that even moderate or low host immune activity generates sufficient selective pressure to be detectable in the parasite's polymorphism. The study of genetic diversity of *P. falciparum* is hampered by certain limitations. The number of gene loci studied remains small. Malaria research has focused upon immunologically relevant protein parts. The polymorphism of the gene cannot be assessed completely. In the present volume, Escalante and Lal review the current status of molecular evolutionary and population genetic research of malarial parasites, emphasizing the need for longitudinal studies that link population-based information with clinical end points. Another aspect, molecular variation at the G6PD locus in relation to resistance against malarial infection, is reviewed by Tishkoff and Verelli in this volume.

The antiquity of these polymorphisms is of much interest. Rich and Ayala (2000) have proposed that *P. falciparum* is derived from a single parasite during the past 5,000 to 50,000 years. In the following pages, Rich and Ayala have suggested that the extant world populations of *P. falciparum* have evolved from a single strain within the past several thousand years. Their estimate is in sharp contrast with that of Hughes and colleagues (Hughes 1993; Hughes and Hughes 1995; Hughes and Verra 2001), who proposed a model that requires that variants of genes encoding *P. falciparum* surface proteins may be older than the species itself. Their original estimates of the ages of the most divergent alleles (of *Msp-1* and *Csp*) are 35 million and 2.1 million years, respectively. However, on the basis of an analysis of introns, Volkman et al. (2001) have estimated the age of the most recent common ancestor of all extant *P. falciparum* to be between 3,200 and 7,700 years. Population genetic studies of *P. falciparum* have produced conflicting results. Some have suggested that today's population includes multiple ancient lineages predating human speciation, whereas others suggest it includes one or a small number of these ancient lineages.

Zimmerman investigated the relationship between Duffy blood group and resistance to *Plasmodium vivax*. Two central issues are: Does *P. vivax* impose a selective burden on human survival? Did the heterozygous Duffy-negative condition increase fitness sufficiently to explain the dramatic fixation of this trait in human populations living in malaria-endemic Africa?

Evolutionary considerations of other parasites include the dynamics of *Daphnia* and their microparasites by Little and Ebert, susceptibility to visceral leishmaniasis and to schistosomiasis by Dessein et al., and the evolution of influenza viruses by Bush and Cox.

GENETIC AND EVOLUTIONARY CONSIDERATIONS

In their paper "Infection and the Diversity of Regulatory DNA," Cowell et al. pose a number of intriguing questions, such as "Can regulatory DNA be identified from its sequence properties?," and "Can we detect a pattern in the appearance of nucleotide substitutions in the promoter region, or do they occur at random?," etc. Whether it is regulatory DNA discussed by these authors, or linkage disequilibrium discussed by Morton, or Cavalli-Sforza on cultural evolution, these authors are concerned with aspects of infectious disease that have so far not received adequate attention. A different perspective is presented by Cochran and Cochran who argue in favor of the infectious etiology of diabetes.

VIRULENCE

Read et al. discuss the evolution of pathogen virulence in response to public health intervention, utilizing a model with the following assumptions: there is genetic variation in parasite virulence; vaccination is imperfect; there is a positive genetic correlation between virulence and transmission; and the cost of virulence is host death.

The evolution of virulence can be affected in rapidly evolving pathogens when only a few individual pathogens are transmitted from one infected host to another in the process of initiating a new infection. Bergstrom et al. (1999) observed that such bottlenecks are likely to drive down the virulence of a pathogen because of stochastic loss of the most virulent pathogens, through a process analogous to Muller's ratchet. These authors argued that the patterns of accumulation of deleterious mutation may explain differing levels of virulence in vertically and horizontally transmitted diseases. Furthermore, it has been suggested that virulence should be low in vertically transmitted diseases, because high virulence reduces the number of offspring produced, resulting in a reduction of the available pool of new infections (Ewald 1994). On the other hand, virulence can be higher in horizontally transmitted pathogens, because their survival is not solely dependent on the host's descendants. There is reason to believe that under horizontal transfer more virulent strains are able to transfer their abundance to other hosts and that natural selection will maintain high virulence in horizontally transmitted pathogens.

FURTHER RESEARCH

These are rapidly changing fields. Advances within each of these disciplines may soon render the accounts presented here obsolete. Meanwhile, I hope these accounts will have served a purpose as stepping stones in aiding that eventuality. Some key areas for further research on the evolutionary aspects of the host–pathogen relationship are outlined in the following summary.

(1) Causes and sources of infectious diseases:
    (a) genetic variation and structure of pathogen populations and the genetic relationships between pathogenic members of closely related taxa;
    (b) population analyses of the contributions and sources of transfer of genes and accessory elements coding for virulence determinants, host range and specificity, and drug resistance; and
    (c) genetic factors (pathogen, host, or vector) responsible for geographic and temporal variations in disease frequency and severity.
(2) Biology of pathogens:
    (a) genetic and evolutionary history of host–pathogen range;
    (b) model systems exploring the molecular biology of host barriers encountered by pathogens; and
    (c) exploration of molecular and pathogen models involving the extension of the host range of pathogens.
(3) Interactions between hosts and pathogens:
    (a) contribution of population dynamic and evolutionary processes to the pathogenesis and virulence of infecting organisms; and
    (b) exploration of model systems to investigate the relationship between the evolution of pathogenic organisms and factors affecting host susceptibility.
(4) Consequences of intervention:
    (a) in the host, reversion to virulence of live vaccines and evolution of resistance following vaccination;
    (b) exploration of model systems to predict the genetic and evolutionary consequences of vaccination, antimicrobial drug therapy, and other intervention strategies on host, pathogen, and vector populations;

(c) mechanisms and consequences of antibiotic action;

(d) investigation of new resistance mechanisms and evolutionary strategies to aid drug development; and

(e) exploration of genetic, physiological, and environmental variables that are involved in generating and maintaining variation in pathogen, host, and vector populations.

ACKNOWLEDGMENTS

The idea for this book was first suggested by some colleagues who took part in a conference at the U.K. Genome Centre Campus at Hinxton. I was responsible for organizing that conference with the kind support of the Wellcome Trust. Several individuals provided advice and support, although only some of those who took part in the conference contributed to this book. Others who were not able to attend the conference also contributed some chapters. In particular, I would like to thank David Weatherall, Adrian Hill, Debbie Carly, and Pat Goodwin for help with the program. I am grateful to Katrina Halliday and Michael Shelley at Cambridge University Press for editorial help with the publication.

REFERENCES

Bergstrom, C. T., McElhany, P., and Real, L. A. (1999). Transmission bottlenecks as determinants of virulence in rapidly evolving pathogens. *Proc. Natl. Acad. Sci. USA, 96*, 5095–100.

Dronamraju, K. R., Ed. (1990). *Selected Genetic Papers of J. B. S. Haldane*. Garland Publishing Co., New York.

Dronamraju, K. R., Ed. (1995). *Haldane's Daedalus Revisited*. Oxford University Press, Oxford.

Escalante, A. A., Freeland, D. E., Collins, W. E., and Lal, A. A. (1998). The evolution of primate malaria parasites based on the gene encoding cytochrome b from the linear mitochondrial genome. *Proc. Natl. Acad. Sci. USA, 95*, 8124–9.

Ewald, P. W. (1994). *Evolution of Infectious Diseases*. Oxford University Press, Oxford.

Haldane, J. B. S. (1924). A mathematical theory of natural and artificial selection. Part I. *Trans. Camb. Phil. Soc., 23*, 19–41.

Haldane, J. B. S. (1932). *The Causes of Evolution*. Longmans, Green, London.

Haldane, J. B. S. (1949a). The rate of mutation of human genes. In *Proceedings of the Eighth International Congress of Genetics, Hereditas, 35*, 267–73.

Haldane, J. B. S. (1949b). Disease and evolution. *La Ricerca Scientifica, 19*, 2–11.

Hughes, A. L. (1993). Coevolution of immunogenic proteins of *Plasmodium falciparum* and the host's immune system. In *Mechanisms of Molecular Evolution* (N. Takahata and A. G. Clark, Eds.), pp. 109–27. Sinauer Assoc., Sunderland, MA.

Hughes, A. L., and Hughes, M. K. (1995). Natural selection in *Plasmodium* surface proteins. *Mol. Biochem. Parasitol., 71*, 99–113.

Hughes, A. L., and Verra, F. (2001). Very large long-term effective population size in the virulent human malaria parasite *Plasmodium falciparum*. *Proc. R. Soc. Lond., B, 268*, 1855–60.

Lederberg, J. (1999). J. B. S. Haldane (1949) on infectious disease and evolution. *Genetics, 153*, 1–6.

Neel, J. V., and Valentine, W. N. (1947). Further studies on the genetics of thalassaemia. *Genetics, 32*, 38–63.

Rich, S. M., and Ayala, F. J. (2000). Population structure and recent evolution of *Plasmodium falciparum*. *Proc. Natl. Acad. Sci. USA, 97*, 6994–7001.

Siddiqui, M. R., and colleagues (2001). A major susceptibility locus for leprosy in India maps to chromosome 10p13. *Nat. Genet., 27*, 439–41.

Tibayrenc, M. (2001). The golden age of genetics and the dark age of infectious diseases. *Infect., Genet., Evol., 1*(1), 1–2.

Tibayrenc, M. (2004). The impact of human genetic diversity in the transmission and severity of infectious diseases. In *Infectious Disease and Host–Pathogen Evolution* (K. R. Dronamraju, Ed.), p. 315, 2004. Cambridge University Press, Cambridge, UK (this volume).

Volkman, S. K., Barry, A. E., Lyons, E. J., Nielsen, K. M., Thomas, S. M., Choi, M., Thakore, S. S., Day, K. P., Wirth, D. F., and Hartl, D. L. (2001). Recent origin of *Plasmodium falciparum* from a single progenitor. *Science, 293*, 482–4.

Weatherall, D. J. (2004). J. B. S. Haldane and the malaria hypothesis. In *Infectious Disease and Host–Pathogen Evolution* (K. R. Dronamraju, Ed.), p. 18, 2004. Cambridge University Press, Cambridge, UK (this volume).

PART ONE

J. B. S. HALDANE

J. B. S. Haldane (1892–1964) arriving in India, 1957 (photo courtesy of Indian Statistical Institute).

ONE

Haldane's Ideas in Biology with Special Reference to Disease and Evolution

James F. Crow

I don't know who first had the idea that disease is a very potent agent of natural selection. I do know, however, that one of the first was J. B. S. Haldane. In *The Causes of Evolution* (Haldane 1932), he wrote: "A study of the causes of death in man, animals, and plants leaves no doubt that one of the principal characters possessing survival value is immunity to disease. Unfortunately, this is not a very permanent acquisition, because the agents of disease also evolve, and on the whole more rapidly than their victims." Insightful as it was, this statement seems to have had little impact on evolutionary thought at the time (Sarkar 1992).

A further development of the subject came a few years later in a now-famous paper "Disease and evolution" (Haldane 1949b). Recognizing that density is often a limiting factor in population growth, he argued that the most important density-dependent limiting factor is a parasite. Over-crowding favors the parasite by, among other things, weakening the hosts and facilitating transmission among them. As an example, he discussed the detailed causes of death at various times in the life cycle of the gall insect, *Urophora jaceana*, and its various parasitoids. He noted that genetic diversity is much greater for disease resistance than, for example, resistance to predators. When a variety of wheat has been exposed to rust, a new resistant strain soon appears, followed by a new parasite strain that attacks the resistant host, and so on. Again he emphasized that the parasite can change faster than the host. He speculated that genes conferring resistance to specific strains of parasites might be especially mutable. He noted that variability could also be induced by recombination, although it would be years before such mechanisms, both germinal and somatic, were demonstrated in immunoglobins.

Haldane suggested that sometimes heterosis is a diversifying factor, but he thought that more often diversification is brought about by a rare type being at a selective advantage. Either mechanism leads to a stable polymorphism. I might note that it is not possible from gene frequency studies alone to distinguish between heterosis and rarity-advantage (Denniston and Crow 1990). Haldane (1949b) went on to suggest that selection of rare biochemical genotypes might also be an important agent for speciation. In this article, he developed a mathematical model for simultaneous host–parasite evolution. He noted that evolution of resistance is an antisocial agency, because diseases spread more readily in large, dense populations. There were many other ideas: the paper was far ahead of its time.

It is now generally known that Haldane was responsible for the idea that superior resistance to malaria in the heterozygote is the reason for the high incidence of thalassemia and sickle cell anemia. The beginning of this theory was remarkably inauspicious. The high incidence of the trait despite the fact that homozygotes have a fitness of essentially zero was a widely recognized puzzle. Neel and Valentine (1947) thought that a high mutation rate was the reason. They noted that the frequency of thalassemia homozyotes is about 4×10^{-4}. Assuming, naively, that there is no selection in heterozygotes, they estimated the mutation rate to be as high as 4×10^{-4}. Haldane (1949a) regarded this as improbably high and noted that if heterozygotes had an increased fitness of only 2 percent, this would account for the incidence without invoking any mutation at all. Haldane went on to suggest a specific mechanism for the greater fitness of heterozygotes. In his words, "The corpuscles of the anaemic heterozygotes are smaller than normal, and more resistant to hypotonic solutions. It is at least conceivable that they are also more resistant to attacks by the sporozoa which cause malaria, a disease prevalent in Italy, Sicily, and Greece, where the gene is frequent." Haldane went on to say that the fitness of heterozygotes might be quite different in countries such as the United States, where malaria is relatively rare.

Curiously, the "Disease and Evolution" paper (Haldane 1949b) has nothing about malaria in the main text, but in the later discussion Montalenti refers to the fact that Haldane had pointed out in a personal communication that thalassemia heterozygotes may be more resistant to malaria. Haldane, always one to consider more than one possibility, noted that these heterozygotes might also have an advantage in an iron-deficient diet. He suggested an analogy to a *vermilion*-eyed Drosophila. In a tryptophan-deficient diet, by not using tryptophan for eye pigment, the mutant fly conserves this scarce commodity for more vital uses.

The actual demonstration of malaria resistance came later. Allison (1954) found that young children with the sickle cell trait had significantly fewer *Plasmodium falciparum* parasites than normal homozygotes. He also found correlations between the frequency of hemoglobin S and the incidence of malaria, the incidence of heterozygotes being as high as 40 percent in some highly malarious areas. At the time, Allison did not know of Haldane's earlier suggestion regarding thalassemia, but they eventually got together (Allison 1968; Lederberg 1999). The interest of population geneticists was aroused by Allison's paper at the Cold Spring Harbor Symposium of 1955. Sickle cell anemia has become the standard example of heterozygote advantage and appears in numerous textbooks. Its very popularity attests to the rarity of other good examples.

Although Haldane wrote little about this subject in later years, it was never absent from his thoughts. One example is a clever paper (Haldane and Jayakar 1963) showing that a stable polymorphism can arise when, for example, a genotype is favored most of the time but is susceptible to a rare epidemic disease. The condition for stability is that the arithmetic mean is greater than unity while the geometric mean is less than unity. This example shows not only Haldane's interest in disease but also his knack for finding clever selection schemes.

Haldane did not take the next step, to ask the source of the variation on which disease-resistance mechanisms depend (Sarkar 1992). This field was developed mainly by Hamilton (1990), who argued that resistance to parasites was the major reason for the maintenance of sexual reproduction. Hamilton devoted many years to the baffling question of why sexual reproduction is so common when it is obviously not very efficient *qua* reproduction and typically entails a two-fold fitness cost. His search for an explanation and the development of the idea that parasitism is a major factor are discussed in his recent book (Hamilton 2001). This volume not only reprints all of his papers on this subject but also includes introductory comments for each one. In these, Hamilton recounts in a charmingly personal and informal way the circumstances under which each article was written. Among other difficulties, he had more than his share of problems with publishers, which is surprising in view of his high standing in the field. This should be comforting to other young authors confronting similar problems.

A further idea in the realm of genetics and disease is found in the work of Cochran, Ewald, and Cochran (2000). They argue that any harmful trait that is too frequent to be easily explained by recurrent mutation or balanced polymorphism is probably due to infection. They cite peptic ulcer as an example. This was traditionally attributed to gastric acidity, smoking, alcohol, stress, and of course genetic susceptibility. Recently *Helicobacter pylori* has been shown to be a major causative factor. Cochran et al. believe that there may be many more instances to be found. Not all their examples are convincing, but the idea is certainly worth a much more careful exploration than it has had.

Haldane's writing on disease and evolution is but a tiny fraction of his enormous output. He wrote 23 books, some 70 book reviews in *The Journal of Genetics*, and more than 400 scientific papers. He not only did great science, but he was one of the best popular science writers. He could simplify without introducing distortions. Among his "popular" writings are 345 articles for the *Daily Worker* during his communist period.

This is not the place to attempt a summary of Haldane's many accomplishments. Their sheer number would likely thwart any such attempt. But to give a bit of a flavor of his work, here are some of the subjects that he wrote about. Many of these are taken from my foreword to Dronamraju's (1990) collection of Haldane articles and a centennial appreciation (Crow 1992).

Haldane's early work on regulation of blood alkalinity is now textbook material. He pioneered the theory of enzyme kinetics and wrote a classical book on the subject. He discovered the first case of genetic linkage in mammals. He derived the first gene-mapping function and coined the terms "morgan" and "centimorgan" for recombination units. He instigated pioneer work on the biochemistry of flower pigments. He was the first to measure the mutation rate of a human gene, in this case hemophilia, and he showed that it is an order of magnitude higher in males than in females. He derived the equilibrium relationship between mutation and selection and used this, among other things, to measure mutation rates. He pointed out that the Rhesus blood group factor is not in stable equilibrium and suggested that this implies a hybrid origin for the European population. He was the first to estimate the selective advantage of a gene in a natural population, the change in melanin pigmentation in the peppered moth in smoke-polluted industrial England. He was the first to compute the probability of fixation of a new, selectively favored mutation, a formula refined by Kimura (1983) and used extensively in developing the neutral theory of molecular evolution. Haldane introduced the ideas of genetic load and the cost of a gene substitution and found a surprising simplification in each case. He found a new theory for the origin of the universe. He developed a number of statistical techniques. One example is an unbiased estimate of the frequency of rare types, such as eosinophil white blood cells. He suggested sampling until a fixed number of eosinophils have been counted, then estimating the proportion by omitting the last observation. He developed the inventive idea of partial sex linkage and ways this could be used for chromosome mapping (alas, it has not stood up). He produced his own theory for the origin of dominance, generally more acceptable than that of Fisher. He constructed the first gene map in humans; it involved X-linked loci. He coined the unit, the "darwin" as a measure of morphological evolution rate. He tried several experiments in non-violent animal research, in the spirit of his newly adopted country, India. As early as 1924 he propounded the heterotrophic theory of the origin of life, now fully accepted. He pioneered in the theory of polyploid segregation. During World War II, experimenting on himself and colleagues, he developed methods for rescuing people from sunken submarines. His best known work is a series of papers in the 1920s and early 1930s on the mathematical theory of natural and artificial selection. Together with Sewall Wright and R. A. Fisher he founded the mathematical theory of evolution in Mendelian populations. And I could go on.

I will mention two books among many that I found of interest. One is *New Paths in Genetics* (Haldane 1941). In this he first introduced the words *cis* and *trans* in a genetic context. He gave a coherent account of the ornithine cycle and the genes involved, based on human recessive diseases. This book made an enormous impression on me at a time when I was first beginning an academic career.

The other book, *What is Life* (Haldane 1947), is an example of Haldane at his best and worst. This is a series of essays written for the *Daily Worker*, often scribbled while he was riding on a commuter train. Haldane's prodigious memory meant that he did not need to have references at hand. These are excellent popular science writings with characteristic Haldane simplicity and clarity, and they often display his vast erudition. At the same time, they are annoying in that they almost always include a Marxist sermon, although I couldn't help admiring his inventiveness in finding such a twist to almost any scientific finding.

According to M. J. D. White (1965), Haldane was "probably the most erudite biologist of his generation, and perhaps of the century." As I have already emphasized, his contributions were legion. Yet his name is now regularly associated with only two phenomena. One is "Haldane's Rule," the observation that in hybrids, sterile or inviable individuals are usually the heterogametic sex, a topic of much recent research. The other is the Briggs–Haldane equation for enzyme kinetics, which gave realism and specificity to the Michaelis–Menton equation. I think Haldane's name is not associated with other theories or observations because of his eclecticism and open-mindedness. He was generous toward the work of others, with which he was sure to be acquainted because he read so widely. Being egotistical and supremely self-assured, he felt no need to worry about rivals.

Several years ago, I expressed the hope that some publisher would reprint Haldane's entire output. Henry Bennett (1971–4) has collected and reprinted all of Fisher's papers, and the collection of five large volumes is a godsend. Reprinting all of Haldane's writings would be a herculean task, both because they are so numerous and because they were published in all sorts of places. But if reprinting is impractical, putting them on a web site would not be so daunting and I very much hope it happens.

REFERENCES

Allison, A. C., 1954. Protection afforded by the sickle-cell trait against subtertian malarial infection. *Brit. Med. J.* I:290–2.

Allison, A. C., 1955. Aspects of polymorphism in man. *Cold Spring Harbor Symp. Quant. Biol.* 20:239–55.

Allison, A. C., 1968. Genetics and infectious disease, pp. 179–201, in *Haldane and Modern Biology*, edited by K. R. Dronamraju. Johns Hopkins Press, Baltimore.

Bennett, J. H., Ed., 1971–74. *Collected Papers of R. A. Fisher.* 5 volumes. University of Adelaide, Adelaide, Australia.

Cochran, G. M., P. W. Ewald, and K. D. Cochran, 2000. Infectious causation of disease: an evolutionary perspective. *Perspect. Biol. Med.* 43:406–48.

Crow, J. F., 1992. Centennial: J. B. S. Haldane, 1892–1964. *Genetics* 130:1–6.

Denniston, C., and J. F. Crow, 1990. Alternative fitness models with the same allele frequency dynamics. *Genetics* 125:201–5.

Dronamraju, K. R., Ed., 1990. *Selected Genetic Papers of J. B. S. Haldane.* Garland Publishing, New York/London.

Haldane, J. B. S., 1932. *The Causes of Evolution.* Longmans, Green & Co., London. Reprinted 1990 with Introduction and Afterword by E. G. Leigh. Princeton University Press, Princeton, NJ.

Haldane, J. B. S., 1941. *New Paths in Genetics.* Allen and Unwin, London.

Haldane, J. B. S., 1947. *What is Life?* Boni and Gaer, New York.

Haldane, J. B. S., 1949a. The rate of mutation of human genes. *Proceedings of the 8th International Congress on Genetics*, pp. 267–73.

Haldane, J. B. S., 1949b. Disease and evolution. *Ricera Sci. Suppl.* A 19:68–76. Reprinted in 1990. *Selected Genetic Papers of J. B. S. Haldane*, edited by K. R. Dronamraju. Garland Publishing, New York/London.

Haldane, J. B. S., and S. D. Jayakar, 1963. Polymorphism due to selection of varying direction. *J. Genet.* 58:318–23.

Hamilton, W. D., 1990. Sexual reproduction as an adaptation to resist parasites (a review). *Proc. Natl. Acad. Sci. USA* 87:3566–73.

Hamilton, W. D., 2001. *Narrow Roads of Gene Land*, Vol. 2: *Evolution of Sex.* Oxford University Press, Oxford.

Kimura, M., 1983. *The Neutral Theory of Molecular Evolution.* Cambridge University Press, Cambridge, UK.

Lederberg, J., 1999. J. B. S. Haldane (1949) on infectious disease and evolution. *Genetics* 153:1–3.

Neel, J. V., and W. N. Valentine, 1947. Further studies on the genetics of thalassemia. *Genetics* 32:38–63.

Sarkar, S., 1992. Sex, disease, and evolution – variations on a theme from J. B. S. Haldane. *BioScience* 142:448–54.

White, M. J. D., 1965. J. B. S. Haldane. *Genetics* 52:1–7.

TWO

J. B. S. Haldane and the Malaria Hypothesis

D. J. Weatherall

In 1948, J. B. S. Haldane proposed that the selective agent for maintaining the high frequency of thalassemia in the Mediterranean races might be malaria. Thus was born what, in the human hemoglobin field at least, later became known as the malaria hypothesis. In this short essay, I examine the origins of this hypothesis, ask how it has stood the test of time, and summarize the broader field of research which it has spawned and which attempts to understand the genetic basis of variability in individual susceptibility to infection in general. Many of these issues have been reviewed in detail over recent years and are only summarized here.

THE MALARIA HYPOTHESIS

Although the first clinical descriptions of the thalassemias appeared in the 1920's, it only became apparent that these conditions follow a Mendelian recessive or co-dominant pattern of inheritance in the period immediately preceding World War II. At this time relatively simple methods for carrier screening became available and hence it was possible to carry out population studies to attempt to determine gene frequencies.

The extraordinarily high frequency of the genes for thalassemia puzzled population geneticists, particularly those who had become interested in mutation rates as a result of their studies of the survivors of the atomic bombs that had been dropped on Hiroshima and Nagasaki. Extensive studies of gene frequencies of thalassemia were carried out quite independently in the 1940's by workers in the United States and Italy (Neel and Valentine, 1947; Silvestroni and Bianco, 1947), but because of lack of communication during the war and its aftermath it was not until the early 1950's that this broad body of work could be integrated and assessed. Assuming that the fitness of a thalassemia homozygote is zero and that the heterozygote is selectively neutral, Neel and Valentine calculated a mutation rate for thalassemia of 1 in 2,500. An even higher rate was proposed by Silvestroni (1949), who also suggested that some form of positive selection could be operating to maintain the high frequency of thalassemia heterozygotes.

These issues were discussed at the 8th International Congress of Genetics in Stockholm in 1948 and Haldane spoke as follows (Haldane, 1949a):

> Neel and Valentine believed that the thalassemia heterozygote is less fit than normal, and think that the mutation rate is above 4×10^{-4} rather than below it. I believe that the possibility that the heterozygote is fitter than normal must be seriously considered. Such increased fitness is found in the case of several lethal and sub-lethal genes in *Drosophila* and *Zea*. A possible mechanism is as follows. The corpuscles of the anaemic heterozygotes are smaller than normal, and are more resistant to hypotonic solutions. It is at least conceivable that they are also more resistant to attacks by the sporozoa which cause malaria, a disease prevalent in Italy, Sicily and Greece where the gene is frequent... Until more is known about the physiology of this gene in various environments I doubt if we can accept the hypothesis that it arises very frequently by mutation in a small section of the human species.

Considering that at the time Haldane proposed this hypothesis it was thought that thalassemia was restricted to Mediterranean populations, and virtually nothing was known about the heterogeneity of the thalassemias or that there were other common inherited disorders of hemoglobin, Haldane's insights seem all the more remarkable. This view was not held by Lederberg in a recent review of Haldane's contribution however (Lederberg, 1999). He suggested that the concept of genetic resistance to infection was already known by the time Haldane proposed this mechanism for the high frequency of thalassemia. In this discussion he does not refer to the paper just quoted but to a paper published in the same year based on a lecture given by Haldane at the Symposium on Ecological and Genetic Factors in Speciation among Animals, held in Milan (Haldane, 1949b). Haldane certainly does not mention thalassemia or malaria in this paper and it is possible that the geneticist Montalenti could have brought his attention to the work of Silvestroni and colleagues in Italy at this meeting. However, in a footnote recording the discussion, Montalenti acknowledges an earlier verbal communication from Haldane to the effect that carriers of thalassemia may be more resistant to malaria. Haldane is also recorded as suggesting that they may also be advantaged in an environment in which iron deficiency is common. Since the Milan meeting actually followed the one in Stockholm, it is clear that Haldane had already formulated his ideas on heterozygote advantage before these discussions. Certainly his paper in Stockholm was the first clear exposition of this concept – one that was to become central to our understanding of the population genetics of the hemoglobin disorders. But for a variety of reasons which will become apparent in the sections that follow, it has taken close to 50 years to put the Haldane hypothesis onto a solid experimental footing. In fact, he has turned out to be right about the general mechanisms involved in maintaining the high frequency of the thalassemia mutations, although he was almost certainly wrong in his suggestion of the specific mechanisms involved!

INHERITED DISORDERS OF HEMOGLOBIN

Since Haldane's short sorti into the human hemoglobin field, a great deal has been learnt about the inherited disorders of hemoglobin which, collectively, are the commonest monogenic diseases in man. They comprise the structural hemoglobin variants and the thalassemias, a heterogeneous collection of inherited defects in the synthesis of the alpha or beta chains of human adult hemoglobin.

Although many hundreds of structural hemoglobin (Hb) variants have been identified, only three – Hbs S, C, and E – reach polymorphic frequencies (Weatherall and Clegg, 2001a, 2001b). The gene for Hb S is distributed broadly throughout sub-Saharan Africa, the Middle East, and parts of the Indian subcontinent, where heterozygote frequencies range from 5–40% or more of the population. Hemoglobin C is restricted to West and North Africa, where it occurs at slightly lower frequencies than Hb S. Hemoglobin E, the commonest hemoglobin variant, is found in the eastern half of the Indian subcontinent and throughout Southeast Asia, where, in some parts, heterozygote rates may exceed 60–70% of the population.

Thalassemias occur at a high frequency in a broad band extending from parts of sub-Saharan Africa and the Mediterranean basin, throughout the Middle East, the Indian subcontinent, and Southeast Asia, to Melanesia and the Pacific Islands. The gene frequency for the thalassemias is extremely patchy, even within closely related geographical regions (see Weatherall and Clegg, 2001a, 2001b). Overall, the carrier frequencies for β thalassemia in these areas range from 1–20%, though rarely higher, whereas those for the milder forms of α thalassemia range from 10–20% in parts of sub-Saharan Africa, through 40% or more in some Middle Eastern and Indian populations, to as high as 70% or more in northern Papua New Guinea and isolated populations in northern India.

Studies of the hemoglobin disorders at the molecular level have provided invaluable information about their population genetics and heterogeneity (Weatherall and Clegg, 2001a). Analyses of globin–gene haplotypes, that is, the patterns of restriction-fragment–length polymorphisms (RFLPs) in the α or β globin gene clusters associated with these conditions have been of particular importance in studies of their evolution. For example, they indicate that the sickle-cell mutation may have occurred at least twice, once in Africa and once in either the Middle East or India. Similar data have been interpreted as suggesting multiple origins of the genes for Hb S and Hb E in Africa and Asia, respectively. These latter conclusions have to be interpreted with caution, however. A more convincing explanation for a great deal of the haplotype diversity found with these variants is that it reflects redistribution on different backgrounds by gene conversion and recombination (Flint *et al.*, 1998).

Over 200 different mutations have been found to underlie β thalassemia and each high-frequency population has its own particular mutations. The situation is more complex in the case of α thalassemia, largely because the α globin genes are duplicated. There are two main groups of α thalassemias: α^{o} thalassemias, in which both α globin genes are deleted or otherwise inactivated, and α^{+} thalassemias, in which one of the pair of linked genes is deleted or inactivated to a variable degree by a point mutation. The homozygous states for the deletion forms of α thalassemia are represented as --/-- and $-\alpha/-\alpha$. These conditions are extremely heterogeneous at the molecular level and many different sized deletions have been found to cause both α^{+} and α^{o} thalassemia (Weatherall and Clegg, 2001b). Similarly, the α globin genes may also be inactivated by a wide range of different point mutations. And, as in the case of β thalassemia, all the high-frequency areas for α thalassemia have different sets of mutations.

EARLY ATTEMPTS TO TEST THE MALARIA HYPOTHESIS

Interestingly, it was work in the sickle-cell-anemia field rather than thalassemia which first suggested that Haldane's view that protection against malaria might be a factor in maintaining the high frequency of hemoglobin disorders was shown to be more or less correct. The extensive evidence that indicated that the sickle-cell trait offers protection against *Plasmodium falciparum* malaria has been reviewed extensively elsewhere (Allison, 1965). But although this work provided reasonably solid evidence in support of the malaria hypothesis, studies in the thalassemia field turned out to be less conclusive.

The early attempts to relate the distribution of thalassemia to the frequency of malaria, either at the present time or in the past, are reviewed in detail elsewhere (Weatherall and Clegg, 2001b). Although suggestive evidence for the association between β thalassemia and endemic malaria was obtained from population studies in Sardinia, when such correlations were sought in other parts of the world they were simply not found. In retrospect, perhaps this should not surprise us. These studies were often carried out using inefficient carrier-detection systems and before the extraordinary heterogeneity of the thalassemias was appreciated, against a background of some futile, if colorful, hypotheses about how the thalassemia genes had arisen and how they had become distributed among the world's populations. But without knowledge of their molecular pathology it was almost impossible to distinguish between selection, drift, migration, and founder effects as the basis for the population distribution of the thalassemias. It was only with the tools of the molecular era that it became possible to tackle some of these difficult questions.

RECENT INVESTIGATIONS OF THE MALARIA HYPOTHESIS

Over recent years, case control studies in Africa using strict World Health Organization (WHO) criteria for the severity of malaria have provided stronger numerical data regarding the protective effect of the sickle-cell trait against *P. falciparum* malaria. It turns out that sickle-cell carriers enjoy almost 80% protection against the severe complications of malaria, notably cerebral malaria and profound anemia (Hill *et al.*, 1991). Work in West Africa suggests also that the relatively high frequencies of Hb C have been maintained by resistance to *P. falciparum* (Modiano *et al.*, 2001). In this case there is evidence for both heterozygote and homozygote protection and the authors suggest that, unlike the sickle-cell mutation, this could be an example of a transient polymorphism, based, presumably, on the perceived lack of clinical or hematological changes in Hb C homozygotes. However, if this is so, it is hard to understand why the frequency of Hb C is not higher in African populations. Furthermore, it is not absolutely clear whether Hb C homozygotes are completely unaffected by the condition, and further work is required to pursue this interesting suggestion.

Work carried out over recent years in the southwest Pacific has provided strong evidence that the high frequency of the milder varieties of α thalassemia, the α^+ thalassemias, is related to protection against *P. falciparum* malaria. The work on which this conclusion is based, which includes both population and case control studies, has been reviewed in detail recently (Weatherall and Clegg, 2002). In short, the frequency of α^+ thalassemia in this region follows a clinal distribution from north/west to south/east, with the highest frequencies in the north coastal region of Papua New Guinea and the lowest in New Caledonia. This distribution is strongly correlated with malaria endemicity, but there is no similar geographical correlation with other polymorphic markers (Flint et al., 1986). The possibility that the α thalassemias had been introduced from mainland populations of Southeast Asia was excluded by finding that the molecular forms of α thalassemia in Melanesia and Papua New Guinea are different from those of the mainland and are set in different α globin gene haplotypes. Further studies along these lines have provided strong evidence that the occurrence of α thalassemia in other parts of this region, where malaria has never been recorded, is the result of population migrations (O'Shaughnessy et al., 1990; Flint et al., 1993). These findings were strengthened by a prospective case-control study which provided strong evidence of protection of both α^+ thalassemia homozygotes and heterozygotes against P. falciparum; the risk of contracting severe malaria as defined by the strictest WHO guidelines was 0.4 for α^+ thalassemia homozygotes and 0.66 for α^+ thalassemia heterozygotes (Allen et al., 1997).

Although they are less extensive, molecular analyses of the β globin genes in thalassemic and non-thalassemic persons in different populations have provided some indirect evidence that the high-frequency of β thalassemia also reflects heterozygote protection against malaria. Every high-frequency population that has been studied has a completely different set of β thalassemia mutations. The arrangement of RFLPs in the β globin gene complex, or haplotype, is divided into two regions, the 3' and 5' sub-haplotypes, which are separated by a recombination hotspot (Chakravarti et al., 1984). Surprisingly perhaps, particular β thalassemia mutations are always closely associated with specific β globin gene haplotypes, most strongly with the 3' sub-haplotype which contains the β globin gene but also with the 5' sub-haplotype, despite the presence of the hotspot (Weatherall and Clegg, 2001b). These findings suggest that a recent, that is in evolutionary terms, cause is responsible for the expansion of the β thalassemia mutations. It appears that migration has not yet had sufficient time to disperse them unlike the normal β globin gene background haplotypes, nor has recombination yet had time to disrupt these linkages. As judged by the population distribution of the β thalassemias, it seems likely that the various genes involved have been amplified by protection against P. falciparum malaria fairly recently so that none of the other population forces – migrations, recombination, and drift, for example – have had sufficient time to bring them into genetic equilibrium with the haplotype backgrounds.

However, although these observations provide strong circumstantial evidence that the high frequency of β thalassemia reflects heterozygote advantage against malaria, this conclusion will only be proved to be correct beyond a doubt by the application of the kind of case control studies that have been so successful in the case of the sickle-cell and α thalassemia genes.

HOW DO THE HEMOGLOBIN DISORDERS PROTECT AGAINST MALARIA INFECTION?

Despite all the evidence that Haldane's original proposal was correct in principle, when we come to consider the mechanisms of protection it is becoming increasingly clear that Haldane's proposal was overly simplistic. Considering the state of knowledge at the time, it was not unreasonable for Haldane to suggest that it might be the smaller, less hemoglobinized red cells of thalassemia heterozygotes, and the less friendly environment that these cells might offer malarial parasites, that is the basic mechanism for protection. Although it has been extremely difficult to test this part of the hypothesis, the bulk of the evidence now suggests that this is not the case.

Once it became possible to carry out *in vitro* culture of malarial parasites, many groups started to analyze the rates of invasion and growth of parasites in normal as compared with thalassemic red cells. At the same time, similar studies were initiated in the red cells of carriers of hemoglobins S and C. The results of these studies have been summarized recently (Nagel, 2001; Weatherall and Clegg, 2001b) and are only outlined here.

One of the main problems in work of this type is comparing like with like. For although it has been known for some time that *P. vivax* only invades very young red cells, or reticulocytes, it was thought for a long time that *P. falciparum* was less selective. However, once it became possible to carry out *in vitro* invasion studies it was found that *P. falciparum* undoubtedly has a predilection for younger red cells (Pasvol *et al.*, 1980). Because there are subtle differences in the turnover rates even between the red cells of those with the thalassemia trait and normal individuals, these observations have to be taken into account when assessing the results of *in vitro* invasion assays. Overall, some abnormalities of invasion and growth have been found in the red cells in the more severe forms of thalassemia. But in the milder forms, which would have to have come under selection to maintain high gene frequencies, no consistent abnormalities have been reported. Attempts to monitor parasite growth over a number of cycles have also given inconsistent results. More reproducible findings have been found in the case of red cells from individuals with the sickle-cell trait. Here there appears to be a consistent reduction in parasite development provided these cells are subjected to hypoxic conditions (Friedman, 1978; Pasvol *et al.*, 1978; Roth *et al.*, 1978).

There is increasing evidence that the mechanisms of protection, at least in the case of α thalassemia, against severe malaria may be much more subtle. A number of studies have implicated specific changes in the *P. falciparum*–infected thalassemic red blood cell membrane in this process. For example, after incubation in malaria-immune serum, α thalassemic red cells have been found to bind significantly more antibody than control cells, suggesting that they may be more easily recognized immunologically, thus leading to slower disease progression and to protection against more severe forms of malaria (Luzzi *et al.*, 1991; Williams *et al.*, 2002). Similarly, infected α thalassemic red cells are more susceptible to phagocytosis *in vitro* (Bunyaratvej *et al.*, 1986; Yuthavong *et al.*, 1988) and are less able than normal to form rosettes, phenomenon that have been correlated with severe malaria (Udomsangpetch *et al.*, 1993; Carlson *et al.*, 1994). Recent studies hint that some of these phenomena may be related to altered red-cell-membrane band 3 protein which may be a target for enhanced antibody binding in α thalassemic red cells and hence could be involved in protection against malaria (Williams *et al.*, 2002).

These *in vitro* observations are backed up to some degree by *in vivo* findings. For example, in a cohort study in an island with holoendemic malaria in Vanuatu it was found that the incidence of uncomplicated malaria and enlargement of the spleen, an index of malaria infection, were significantly higher in young children with α thalassemia than in normal children. The effect was most marked in the youngest children and, surprisingly, with the non-lethal parasite *P. vivax*. It was suggested that early susceptibility to *P. vivax* may be acting as a natural vaccine by inducing cross-species protection against *P. falciparum* (Williams *et al.*, 1996).

Surprisingly, in the case control studies that showed clear evidence that α thalassemia protects against severe malaria, it was found that protection is also mediated against other infectious diseases (Allen *et al.*, 1997). Whether this reflects a secondary effect mediated through improving the general health of children in highly malarious areas by protecting them against malaria, or whether it results from a more specific mechanism is not yet clear. But these observations underline the remarkable effects on the health of populations, particularly children, that have been and are being mediated by the hemoglobin disorders in modifying susceptibility to infective agents.

There are, therefore, now extensive data in support of Haldane's hypothesis, although it is quite clear that the mechanisms involved are far more complex than those that he proposed. It is now clear that this is not simply a reflection of the properties of small red cells but, rather, a much more complicated train of events, at least some of which may have an immunological basis.

EXTENDING THE MALARIA HYPOTHESIS

The hemoglobin variants are not the only red cell polymorphisms that have been maintained by exposure of human populations to malaria. For example, the high frequency of persons in parts of Africa who do not carry the Duffy blood group antigen reflects the protective effect of this genotype against infection with *P. vivax* (Miller *et al.*, 1976). It is now known that this effect is mediated because this particular genetic variant disrupts the Duffy antigen/chemokine receptor (DARC promoter) and so alters a GATA-1-binding site, which inhibits DARC expression on red cells and therefore prevents DARC-mediated entry of *P. vivax* (Tournamille *et al.*, 1995). A variety of other associations between blood group antigens and susceptibility to malaria have been reported.

The extraordinarily high frequency of glucose-6-phosphate-dehydrogenase (G6PD) among tropical populations, an X-linked disorder that causes hemolytic reactions in response to certain drugs or other oxidants, is also a reflection of protection against *P. falciparum*. As in the case of thalassemia, several hundred different mutations are responsible for their condition, and their pattern varies between different populations (Luzzatto *et al.*, 2001). Both hemizygous males and heterozygous females appear to be protected against malaria, in both East and West Africa (Ruwende *et al.*, 1995). Distribution studies in Vanuatu have shown a strong correlation with malaria (Ganczakowski *et al.*, 1995). The most likely protective mechanisms appear to be impaired parasite growth or more efficient phagocytosis of parasitized red cells at early stages of parasite maturation.

Another important example of malaria-related balanced polymorphism involves mutations in the red-cell membrane protein, band 3, that cause the Melanesian form of ovalocytosis, that is, oval-shaped red cells (Mgone *et al.*, 1996). The homozygous state for this condition appears to be lethal, but it is clear that heterozygotes are protected against malaria though in a particularly interesting and unusual fashion. They appear to be fully susceptible to malaria infection and yet have almost complete protection against the development of cerebral malaria (Genton *et al.*, 1995; Allen *et al.*, 1999). This finding suggests that the defect in the red cell membrane must also alter the infected red cell's interaction with vascular endothelium, the nature of which remains to be characterized.

But the relationship between varying susceptibility to malaria and genetic polymorphisms is not confined to the red cell. There is now a broad body of evidence for associations between both HLA Class I and II alleles of the human histocompatibility complex and susceptibility to malaria (Hill *et al.*, 1992). It appears that certain cytokines have similar properties. Tumor necrosis factor-α (TNF-α), a cytokine that is produced by white blood cells and has widespread effects on the immune system, has been analyzed in a number of studies. Several different polymorphisms in the promoter regions of the gene for TNF-α have been identified and have been associated with particularly severe forms of malaria, at least one of which may cause increased expression of TNF-α (McGuire *et al.*, 1994; Wilson *et al.*, 1997; Knight *et al.*, 1999). Polymorphisms of proteins that are involved in the adhesion of parasitized red cells to the vascular endothelium have also been implicated. For example, CD36, an important molecule of this kind, is quite polymorphic, and certain polymorphisms have been found to be associated with growth susceptibility and resistance to severe forms of malaria (Aitman *et al.*, 2000; Pain *et al.*, 2001). Similarly, a variant of the intracellular adhesion molecule 1 (ICAM-1) has been found more commonly in Kenyan children with severe malaria (Fernandez-Reyes *et al.*, 1997), although it is not associated with severe disease in West Africans (Bellamy *et al.*, 1998).

There are other data derived from population genetics studies that suggest that there may be so far unidentified immune mechanisms responsible for variations in individual response to *P. falciparum* malaria. For example, analyses of sympatric ethnic groups with very similar exposure to malaria have shown remarkable differences in infection rates, the severity of malaria, and the prevalence and levels of antibodies to a variety of malaria antigens. Associated investigations of these populations showed no differences in the use of protective measures or any other socio-cultural or environmental factors which might be involved in these modified responses (Modiano *et al.*, 1996).

Most recently, following the partial sequencing of the human genome and considerable progress towards completing the sequence of the mouse genome, several "whole-genome" studies have been carried out in an attempt to define additional factors that may modify the clinical course of malaria. There is growing evidence that analyses of murine malaria of this type may be of considerable value in the future for clarifying the genes involved in human response to the disease (Foote *et al.*, 1997; Fortin *et al.*, 1997).

Although no human malaria resistance genes have been found by genome searches, there have already been successes derived from similar studies of murine malaria models, notably *P. chabaudi*. Using this approach a number of chromosomal regions have been pinpointed as being likely to contain resistance loci for murine malaria. Once these have been defined further and the particular genes involved have been isolated, this approach could be used to search for possible associations of the corresponding human syntenic chromosomal regions with susceptibility to and/or severity of disease in areas endemic for malaria.

BEYOND MALARIA

After the successes in defining increasing numbers of malaria-related polymorphisms it is not surprising that the attention of this field was directed towards other infectious agents. There is now a rapidly expanding literature covering the associations between a wide range of viruses, bacteria, and other parasites and resistance or susceptibility to infections associated with these agents and different genetic polymorphisms. It is beyond the scope of this essay to cover this field in detail and it has been the subject of several recent reviews (Cooke and Hill, 2001; Weatherall and Clegg, 2002).

Not surprisingly, the human HLA/DR gene complex was one of the first to be explored. There is now clear evidence that different alleles of this complex are involved in variable susceptibility or resistance to a wide range of infectious diseases, including tuberculosis, HIV/AIDS, hepatitis B and C, typhoid fever, and leprosy. Similarly, there is a growing list of polymorphisms involving cytokines and immune effectors that are involved in variable susceptibility to infectious agents.

Another important principle that has been raised by these studies is the value of analyzing potential polymorphisms of receptors for particular organisms in relationship to susceptibility to infection. For example, there is a variant of the promoter of the chemokine receptor 5 (CCR5) that confers protection in homozygotes against HIV infection (Dean *et al.*, 1996). Furthermore, heterozygotes have a delay in the progression to AIDS, an observation that has been confirmed in several studies. A variety of other common CCR5 polymorphisms have also been analyzed and a number of other susceptibility associations for HIV/AIDS have been demonstrated.

These examples of variation in genetic response to infection have stemmed mainly from studies of particular candidate genes. As in the case of malaria there is increasing interest in searching for susceptibility genes by using whole-genome linkage analysis. Already, promising results have been obtained in the cases of leprosy, tuberculosis, and chronic hepatitis. Similarly, using unrelated sib-pair analysis, a susceptibility locus for *Schistosoma mansoni* has been defined on chromosome 5 in a region that contains a number of immune-related candidate genes, including those for several cytokines (Marquet *et al.*, 1999).

EVOLUTIONARY IMPLICATIONS

The concept that the genetic make-up of human populations may have been shaped by exposure to a wide variety of infectious agents in the past is certainly not new. It was an explanation commonly proposed as a mechanism for the diversity of human blood group antigens in different populations, although there was little evidence at the time to support it. But the introduction of the technology of molecular biology has started to provide more definitive answers to some of these questions.

It is clear that a recurrent pattern has emerged regarding the distribution and molecular pathology of the human malaria-related polymorphisms. Despite the high level of protection afforded to heterozygotes for the sickle-cell gene against malaria, and the very high frequency of this variant throughout Africa, the Middle East, and India, remarkably it has never been found further east than India. Similarly, although Hb E reaches extremely high frequencies throughout Southeast Asia it is not seen further west than the eastern side of the Indian subcontinent (Weatherall and Clegg, 2001a). Analyses of the β globin gene haplotypes associated with a particular sickle-cell mutation in Africa have suggested that it is recent (45–70 generations) in origin (Currat *et al.*, 2002).

The α and β thalassemias occur throughout the tropical climes, with the exception of Central and South America, yet in each of the high-frequency populations there are different sets of mutations. The fact that there has been no homogenization of mutations within human populations, together with their relationships to the haplotypes of the β globin gene discussed above, suggests that, as in the case of the sickle mutation, the expansion of the β thalassemia must have occurred fairly recently.

Studies of haplotype diversity and linkage disequilibrium at the human G6PD locus provide further evidence of the recent origin of alleles that confirm malarial resistance. For example, in an analysis of two G6PD haplotypes it was found that two common variants appeared to have evolved independently between 3,000 and 11,000 years ago (Tishkoff *et al.*, 2001).

These observations on the fairly recent, at least in evolutionary terms, appearance of genetic polymorphisms that confer resistance to malaria are in keeping with estimations of the spread of *P. falciparum* derived from studies of polymorphic systems of the parasite. For example, in one analysis, involving 25 intron sequences and encompassing both general metabolic and housekeeping genes, there were very few nucleotide polymorphisms, suggesting that the parasite originated in something like its present form between 9,000 and 20,000 years ago (Volkman *et al.*, 2001). These data are in general agreement with other studies of polymorphic genes of *P. falciparum* (Rich and Ayala, 2000). This time scale is in keeping with the idea that it was the development of agriculture somewhere between 5,000 and 10,000 years ago that provided the conditions necessary for the effective spread of malaria in human populations. Of course, *P. falciparum* may be very much older than this, and other genetic studies suggest that this is the case and that today's population includes multiple ancient lineages pre-dating human speciation (Hartl *et al.*, 2002). What its natural host might have been is, of course, unknown, but from the human and genetic data summarized earlier, it seems likely that it became the unstable parasite that it is today in human populations quite recently.

The picture that has emerged, therefore, is that during our relatively short exposure to severe forms of malaria we have utilized a wide range of different genetic polymorphisms in order to modify our response to this lethal disease. Undoubtedly, this has had a profound effect on the genetic constitution of human populations. And, as evidenced by the hemoglobin disorders and G6PD deficiency, it has left in its wake some extremely common monogenic diseases. On the other hand, it may well have changed the capacity of populations to respond to other infections, often in a protective way. Clearly, we have only uncovered a small fraction of the remarkable genetic variation that exposure to malaria has left behind.

It is almost certain that other parasitic, bacterial, and viral infections have, depending on their virulence and length of exposure to human populations, played a major role in modifying our genetic make-up. We have already seen how this may be happening in the case of HIV in sub-Saharan Africa. Of course, given the changing virulence of different infectious organisms with time it may not always be possible to relate current patterns of polymorphisms directly to varying genetic response to infectious agents. For example, there are some puzzling features about the almost total absence of the Duffy group antigen in certain African populations. As mentioned above, this genotype is associated with almost complete protection against *P. vivax* malaria. However, this is a milder form of malaria, at least at the present time, and unless it was more severe in the past there may be another explanation for the high prevalence of those who do not carry the Duffy antigen in African populations.

CONCLUSION

Although there has been rapid progress towards a better understanding of the genetic mechanisms involved in the varying susceptibility to infectious agents, it is likely that we have only seen the tip of the iceberg so far. Furthermore, despite having identified many of the genes involved, the precise cellular mechanisms that mediate these variable responses have, overall, remained elusive. As well as their being of considerable biological interest, the further exploration of these mechanisms may have practical importance. For example, the elucidation of the mechanism whereby variation in the HLA/DR alleles provides protection against *P. falciparum* malaria in some African populations has already led to a promising approach to developing malaria vaccines. A better understanding of our inherent mechanisms for combating infection may provide other avenues for the control of infectious disease.

Research in this field may have other important practical implications. For example, as we move towards the development of vaccines for malaria and other communicable diseases, and if the aim is attenuation rather than complete protection, it is vital to know ahead of time whether certain individuals in particular populations in which the efficacy of such vaccines is being explored already have some resistance to particular pathogens. If malaria vaccines were to be tested in African populations with a high frequency of the sickle-cell trait, or in Papuan populations with an equally high frequency of α thalassemia, large numbers of the population would already have a very considerable degree of inherent protection against the parasite. Lack of knowledge of this likelihood ahead of time will make the design of vaccine trials extremely difficult.

It is doubtful whether Haldane could have had any idea of the large industry that his short paper of 1948 has spawned. Indeed, it is only in the past few years that the genetics of susceptibility to infection has become a topic of major interest to human geneticists. Although there is a long way to go, enough has been learnt to suggest that the malaria hypothesis is no longer a hypothesis and that the further exploration of this field is likely to yield extraordinarily important information, not only about human evolution and the reasons for the genetic variability of different ethnic groups, but also about how to approach the ever present problem of infectious disease in human communities.

ACKNOWLEDGMENTS

I am very grateful to Dr. Ida Bianco and her colleagues for helpful discussions on the early screening programs for thalassemia in Italy at a meeting held in her honor in Rome in 1999. This essay is based on two more extensive reviews of this field (Weatherall and Clegg, 2001b; Weatherall and Clegg, 2002). I thank Liz Rose for her help in typing it and the Leverhulme Trust for support.

REFERENCES

Aitman, T. J., Cooper, L. D., Norsworthy, P. J., Wahid, F. N., Gray, J. K., Curtis, B. R., McKeigue, P. M., Kwiatkowski, D., Greenwood, B. M., & Snow, R. W. (2000). Malaria susceptibility and CD36 mutation. *Nature*, **405**, 1015–16.

Allen, S. J., O'Donnell, A., Alexander, N. D. E., Alpers, M. P., Peto, T. E. A., Clegg, J. B., & Weatherall, D. J. (1997). α^+-thalassemia protects children against disease due to malaria *and* other infections. *Proceedings of the National Academy of Sciences, USA*, **94**, 14,736–41.

Allen, S. J., O'Donnell, A., Alexander, N. D. E., Mgone, C. S., Peto, T. E. A., Clegg, J. B., Alpers, M. P., & Weatherall, D. J. (1999). Prevention of cerebral malaria in children in Papua New Guinea by Southeast Asian ovalocytosis band 3. *American Journal of Tropical Medicine and Hygiene*, **60**, 1056–60.

Allison, A. C. (1965). Population genetics of abnormal hemoglobins and glucose-6-phosphate dehydrogenase deficiency. In *Abnormal Hemoglobins in Africa* (ed. by J. H. P. Jonxis), p. 365. Blackwell Scientific Publications, Oxford.

Bellamy, R., Kwiatkowski, D., & Hill, A. V. (1998). Absence of an association between intercellular adhesion molecule 1, complement receptor 1 and interleukin 1 receptor antagonist gene polymorphisms and severe malaria in a West African population. *Transaction of the Royal Society of Tropical Medicine and Hygiene*, **92**, 312–16.

Bunyaratvej, A., Butthep, P., Yuthavong, Y., Fucharoen, S., Khusmith, S., Yoksan, S., & Wasi, P. (1986). Increased phagocytosis of *Plasmodium falciparum*-infected erythrocytes with hemoglobin E by peripheral blood monocytes. *Acta Haematologica*, **76**, 155–8.

Carlson, J., Nash, G. B., Gabutti, V., Al-Yaman, F., & Wahlgren, M. (1994). Natural protection against severe *Plasmodium falciparum* malaria due to impaired rosette formation. *Blood*, **84**, 3909–14.

Chakravarti, A., Buetow, K. H., Antonarakis, S. E., Waber, P. G., Boehm, C. D., & Kazazian, H. H. (1984). Nonuniform recombination within the human β-globin gene cluster. *American Journal of Human Genetics*, **36**, 1239–58.

Cooke, G. S., & Hill, A. V. S. (2001). Genetics of susceptibility to human infectious disease. *Nature Reviews Genetics*, **2**, 967–77.

Currat, M., Trabuchet, G., Rees, D., Perrin, P., Harding, R. M., Clegg, J. B., Langaney, A., & Excoffier, L. (2002). Molecular analysis of the β-globin gene cluster in the Niokholo Mandenka population reveals a recent origin of the β^S Senegal mutation. *American Journal of Human Genetics*, **70**, 207–23.

Dean, M., Carrington, M., Winkler, C., Huntley, G. A., Smith, M. W., Allikmets, R., Goedert, J. J., Buchbinder, S. P., Vittinghoff, E., and Gomperts, E. (1996). Genetic restriction of HIV-1 infection and progression to AIDS by a deletion allele of the *CKR5* structural gene. Hemophilia Growth and Development Study, Multicenter Hemophilia Cohort Study, San Francisco City Cohort, ALIVE Study. *Science*, **273**, 1856–62.

Fernandez-Reyes, D., Craig, A. G., Kyes, S. A., Peshu, N., Snow, R. W., Berendt, A. R., Marsh, K., & Newbold, C. I. (1997). A high frequency African coding polymorphism in the N-terminal domain of *ICAM-1* predisposing to cerebral malaria in Kenya. *Human Molecular Genetics*, **6**, 1357–60.

Flint, J., Harding, R. M., Boyce, A. J., & Clegg, J. B. (1998). The population genetics of the hemoglobinopathies. In *Baillière's Clinical Haematology; 'Hemoglobinopathies'* (ed. by D. R. Higgs, & D. J. Weatherall), pp. 1–51. Baillière Tindall and W. B. Saunders, London.

Flint, J., Harding, R. M., Clegg, J. B., & Boyce, A. J. (1993). Why are some genetic diseases common? Distinguishing selection from other processes by molecular analysis of globin gene variants. *Human Genetics*, **91**, 91–117.

Flint, J., Hill, A. V. S., Bowden, D. K., Oppenheimer, S. J., Sill, P. R., Serjeantson, S. W., Bana-Koiri, J., Bhatia, K., Alpers, M. P., & Boyce, A. J. (1986). High frequencies of α thalassemia are the result of natural selection by malaria. *Nature*, **321**, 744–9.

Foote, S. J., Burt, R. A., Baldwin, S. M., & 5 colleagues (1997). Mouse loci for malaria-induced mortality and the control of parasitaemia. *Nature Genetics*, **17**, 380–1.

Fortin, A., Belouchi, A., Tam, M. F., & 4 colleagues (1997). Genetic control of blood parasitaemia in mouse malaria maps to chromosome 8. *Nature Genetics*, **17**, 382–3.

Friedman, M. J. (1978). Erythrocytic mechanism of sickle cell resistance to malaria. *Proceedings of the National Academy of Sciences, USA*, **75**, 1994.

Ganczakowski, M., Town, M., Bowden, D. K., Vulliamy, T. J., Kaneko, A., Clegg, J. B., Weatherall, D. J., & Luzzatto, L. (1995). Multiple glucose 6-phosphate dehydrogenase-deficient variants correlate with malaria endemicity in the Vanuatu archipelago (Southwestern Pacific). *American Journal of Human Genetics*, **56**, 294–301.

Genton, B., Al-Yaman, F., Mgone, C. S., Alexander, N., Paniu, M. M., & Alpers, M. P. (1995). Ovalocytosis and cerebral malaria. *Nature*, **378**, 564–5.

Haldane, J. B. S. (1949a). The rate of mutation of human genes. *Proceedings of the VIII International Congress of Genetics Hereditas*, **35**, 267–73.

Haldane, J. B. S. (1949b). Disease and evolution. *Ricera Sci.*, **19**, 2.

Hartl, D. L., Volkman, S. K., Nielsen, K. M., Barry, A. E., Day, K. P., Wirth, D. F., & Winzeler, E. A. (2002). The paradoxical population genetics of *Plasmodium falciparum*. *Trends in Parasitology*, **18**, 266–71.

Hill, A. V. S., Allsopp, C. E. M., Kwiatkowski, D., Anstey, N. M., Twunmasi, P., Rowe, P. A., Bennett, S., Brewster, D., McMichael, A. J., & Greenwood, B. M. (1991). Common west African HLA antigens are associated with protection from severe malaria. *Nature*, **352**, 595–600.

Hill, A. V. S., Elvin, J., Willis, A., Aidoo, M., Allsopp, C. E. M., Gotch, F. M., Gao, X. M., Takiguchi, M., Greenwood, B. M., & Townsend, A. R. M. (1992). Molecular analysis of the asscoiation of HLA-B53 and resistance to severe malaria. *Nature*, **360**, 434–9.

Knight, J. C., Udalova, J., Hill, A. V., Greenwood, B. M., Peshu, N., Marsh, K., & Kwiatkowski, D. (1999). A polymorphism that affects OCT-1 binding to the TNF promoter region is associated with severe malaria. *Nature Genetics*, **22**, 145–50.

Lederberg, J. (1999). J. B. S. Haldane (1949) on infectious disease and evolution. *Genetics*, **153**, 1–3.

Luzzatto, L., Mehta, A., & Vulliamy, T. (2001). Glucose 6-phosphate dehydrogenase. In *The Metabolic and Molecular Basis of Inherited Disease* (ed. by C. R. Scriver, A. L. Beaudet, W. S. Sly, D. Valle, B. Childs, & B. Vogelstein), pp. 3367–98. McGraw Hill, New York.

Luzzi, G. A., Merry, A. H., Newbold, C. I., Marsh, K., Pasvol, G., & Weatherall, D. J. (1991). Surface antigen expression on *Plasmodium falciparum*-infected erythrocytes is modified in α- and β-thalassemia. *Journal of Experimental Medicine*, **173**, 785–91.

Marquet, S., Abel, L., Hillaire, D., & Dessein, A. (1999). Full results of the genome-wide scan which localises a locus controlling the intensity of infection by *Schistosoma mansoni* on chromosome 5q31–q33. *European Journal of Human Genetics*, **7**, 88–97.

McGuire, W., Hill, A. V. S., Allsopp, C. E. M., Greenwood, B. M., & Kwiatkowski, D. (1994). Variation in the TNF-α promoter region is associated with susceptibility to cerebral malaria. *Nature*, **371**, 508–11.

Mgone, C. S., Koki, G., Paniu, M. M., Kono, J., Bhatia, K. K., Genton, B., Alexander, N. D. E., & Alpers, M. P. (1996). Occurrence of the erythrocyte band 3 (AE1) gene deletion in relation to malaria endemicity in Papua New Guinea. *Transaction of the Royal Society of Tropical Medicine and Hygiene*, **90**, 228–31.

Miller, L. H., Mason, S. J., Clyde, D. F., & McGinniss, M. H. (1976). The resistance factor to *Plasmodium vivax* in blacks. *New England Journal of Medicine*, **295**, 302–4.

Modiano, D., Luoni, G., Sirima, B. S., Simporé, J., Verra, F., Konaté, A., Rastrelli, E., Olivieri, A., Calissano, C., & Paganotti, G. M. (2001). Hemoglobin C protects against clinical *Plasmodium falciparum* malaria. *Nature*, **414**, 305–8.

Modiano, D., Petrarca, V., Sirima, B. S., Nebié, I., Diallo, D., Esposito, F., & Coluzzi, M. (1996). Different response to *Plasmodium falciparum* malaria in West African sympatric ethnic groups. *Proceedings of the National Academy of Sciences, USA*, **93**, 13,206–11.

Nagel, R. L. (2001). Malaria and hemoglobinopathies. In *Disorders of Hemoglobin* (ed. by M. H. Steinberg, B. G. Forget, D. R. Higgs, & R. L. Nagel), pp. 832–60. Cambridge University Press, Cambridge, UK.

Neel, J. V., & Valentine, W. N. (1947). Further studies on the genetics of thalassemia. *Genetics*, **32**, 38–63.

O'Shaughnessy, D. F., Hill, A. V. S., Bowden, D. K., Weatherall, D. J., Clegg, J. B., *et al.* (1990). Globin genes in Micronesia: origins and affinities of Pacific Island peoples. *American Journal of Human Genetics*, **46**, 144–55.

Pain, A., Urban, B. C., Kai, O., Casals-Pascual, C., Shafi, J., Marsh, K., & Roberts, D. J. (2001). A non-sense mutation in the *Cd36* gene is associated with protection from severe malaria. *Lancet*, **357**, 1502–3.

Pasvol, G., Weatherall, D. J., & Wilson, R. J. (1980). The increased susceptibility of young red cells to invasion by the malarial parasite *Plasmodium falciparum*. *British Journal of Haematology*, **45**, 285–95.

Pasvol, G., Weatherall, D. J., & Wilson, R. J. M. (1978). A mechanism for the protective effect of hemoglobin S against *P. falciparum*. *Nature*, **274**, 701.

Rich, S. M., & Ayala, F. J. (2000). Population structure and recent evolution of *Plasmodium falciparum*. *Proceedings of the National Academy of Sciences, USA*, **97**, 6994–7001.

Roth, E. F., Jr., Friedman, M., Ueda, Y., Tellez, L., Trager, W., & Nagel, R. L. (1978). Sickling rates of human AS red cells infected *in vitro* with *Plasmodium falciparum* malaria. *Science*, **202**, 650–2.

Ruwende, C., Khoo, S. C., Snow, R. W., Yates, S. N. R., Kwiatkowski, D., Gupta, S., Warn, P., Allsopp, C. E. M., Gilbert, S. C., & Peschu, N. (1995). Natural selection of hemi- and heterozygotes for G6PD deficiency in Africa by resistance to severe malaria. *Nature*, **376**, 246–9.

Silvestroni, E. (1949). Microcitemia e malattie a substrato microcitemico; falcemia e malattie falcemiche. 50° Congresso della Societa di Medicina Internationale. Roma: 1949.

Silvestroni, E., & Bianco, I. (1947). Sulla frequenza dei porta tori di malatia di morbo di Codey e primi observazioni sulla frequenza dei portatore di microcitemia nel Ferrarese e inakune regioni limitrofe. *Bollettinoe Atti della Accademia Medica di Roma*, **72**, 32.

Tishkoff, S. A., Varkonyi, R., Cahinhinan, N., Abbes, S., Argyropoulos, G., Destro-Bisol, G., Drousiotou, A., Dangerfield, B., Lefranc, G., & Loiselet, J. (2001). Haplotype diversity and linkage disequilibrium at human *G6PD*: recent origin of alleles that confer malarial resistance. *Science*, **293**, 455–62.

Tournamille, C., Colin, Y., Cartron, J. P., & Le Van Kim, C. (1995). Disruption of a GATA motif in the *Duffy* gene promoter abolishes erythroid gene expression in Duffy-negative individuals. *Nature Genetics*, **10**, 224–8.

Udomsangpetch, R., Sueblinvong, T., Pattanapanyasat, K., Dharmkrong-at, A., Kittilayawong, A., & Webster, H. K. (1993). Alteration in cytoadherence and rosetting of *Plasmodium falciparum*-infected thalassemic red blood cells. *Blood*, **82**, 3,752–9.

Volkman, S. K., Barry, A. E., Lyons, E. J., Nielsen, K. M., Thomas, S. M., Choi, M., Thakore, S. S., Day, K. P., Wirth, D. F., & Hartl, D. L. (2001). Recent origin of *Plasmodium falciparum* from a single progenitor. *Science*, **293**, 482–4.

Weatherall, D. J., & Clegg, J. B. (2001a). Inherited hemoglobin disorders: an increasing global health problem. *Bulletin of the World Health Organization*, **79**, 704–12.

Weatherall, D. J., & Clegg, J. B. (eds.) (2001b). *The Thalassemia Syndromes*. (4 ed.) Blackwell Science, Oxford.

Weatherall, D. J., & Clegg, J. B. (2002). Genetic variability in response to infection. In Malaria and after, *Paeds and Immunity*, **3**, 331–7.

Williams, T. N., Maitland, K., Bennett, S., Ganczakowski, M., Peto, T. E. A., Newbold, C. I., Bowden, D. K., Weatherall, D. J., & Clegg, J. B. (1996). High incidence of malaria in α-thalassemic children. *Nature*, **383**, 522–5.

Williams, T. N., Weatherall, D. J., & Newbold, C. I. (2002). The membrane characteristics of *Plasmodium falciparum*-infected and -uninfected heterozygous α^0 thalassemic erythrocytes. *British Journal of Haematology*, **118**, 663–70.

Wilson, A. G., Symons, J. A., McDowell, T. L., McDevitt, H. O., & Duff, G. W. (1997). Effects of a polymorphism in the human tumor necrosis factor α promoter on transcriptional activation. *Proceedings of the National Academy of Sciences, USA,* **94,** 3195–9.

Yuthavong, Y., Butthep, P., Bunyaratvej, A., Fucharoen, S., & Khurmith, S. (1988). Impaired parasite growth and increased susceptibility to phagocytosis of *Plasmodium falciparum* infected alpha-thalassemia and hemoglobin Constant Spring red blood cells. *American Journal of Clinical Pathology,* **89,** 521–5.

PART TWO

MALARIAL PARASITES

Evolutionary Genetics of *Plasmodium falciparum*, the Agent of Malignant Malaria

Stephen M. Rich and Francisco J. Ayala

We have investigated the population structure of *P. falciparum* by analyzing several genes and conclude that the extant world populations of this parasite have evolved from a single strain within the past several thousand years. The evidence is based on a lack of synonymous polymorphisms among nuclear antigenic and nonantigenic genes; among introns; and among mitochondrial loci. Coalescence calculations for silent nucleotide variation converge in each case into a single ancestral allele. The extensive polymorphisms observed in the highly repetitive central region of the *Csp* gene, as well as the apparently very divergent two classes of alleles at the *Msp-1* gene, are consistent with this conclusion.

Understanding the population structure and evolution of *Plasmodium* has important implications for the control of human malaria.

The human toll of malaria is stunning, perhaps the greatest of all human afflictions (Sherman, 1998). Malaria is caused by species of *Plasmodium*, a parasitic protozoan. Four species of *Plasmodium* are parasitic to humans: *P. falciparum, P. malariae, P. ovale,* and *P. vivax. P. falciparum* is the most pervasive and malignant human malarial parasite. It causes yearly 300 million to 500 million cases of clinical illness and 1.5 million to 2.7 million deaths in sub-Saharan Africa, plus 5–20 million clinical cases and 100,000 deaths elsewhere in the world, 80% of them in Asia (Trigg and Kondrachine, 1998).

The genus *Plasmodium* consists of nearly 200 named species that parasitize reptiles, birds, and mammals. *Plasmodium* belongs to the Apicomplexa, a large and complex phylum with about 5,000 known species and as many as 60,000 yet to be described (Levine, 1988, pp. 1–21; Vivier and Desportes, 1989; Corliss, 1994; Escalante and Ayala, 1995). The Apicomplexa

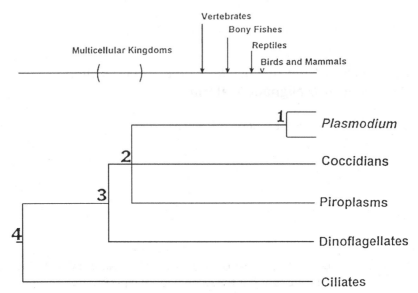

Figure 3.1. Evolutionary relationships of the phyla Apicomplexa, Dinozoa, and Ciliophora. The age of the Apicomplexa phylum (node 2) is thought to be about 1,000 million years, or nearly as old as the origin of the multicellular phyla of plants, animals, and fungi. The radiation of the *Plasmodium* genus (node 1) is likely to be older than 100 million years.

are all parasites, characterized by the eponym structure, the apical complex. The taxonomy and phylogeny of the phylum have been the subject of controversy and frequent revision. One issue is whether *Plasmodium* evolved directly from monogenetic parasites of the ancient marine invertebrates from which the chordates evolved or whether they originated by lateral transfer from other, already digenetic, vertebrate parasites (Huff, 1938; Manwell, 1955; Mattingly, 1965; Garnham, 1966; Barta, 1989). There is no fossil record of apicomplexans (Margulis *et al.*, 1993), but molecular investigations indicate that the phylum is very ancient, perhaps as old as the multicellular kingdoms of plants, fungi, and animals, and thus somewhat older than one billion years (Escalante and Ayala, 1995; Ayala *et al.*, 1998) (Figure 3.1).

The origin of the genus *Plasmodium* is dated to more than 100 million years ago, and it may be as old as 500 million years (Ayala *et al.*, 1999). The phylogeny of the genus has been considerably elucidated through molecular studies (e.g., Barta *et al.*, 1991; Van de Peer *et al.*, 1996; Morrison and Ellis, 1997), including our own investigations of three genes, the small

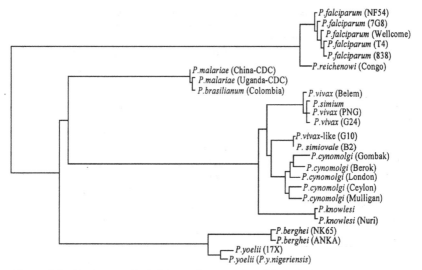

Figure 3.2. Phylogeny of 12 *Plasmodium* species (32 strains) inferred from *Csp* gene sequences. When several strains have identical sequences, only one is represented. (Modified from Ayala *et al.*, 1998.)

subunit ribosomal RNA gene (*SSUrRNA*; Escalante and Ayala, 1994, 1995) and those coding for the circumsporozoite protein (*Csp*; Escalante *et al.*, 1995) and the mitochondrial cytochrome b (Escalante *et al.*, 1998a). Figure 3.2 is a phylogenetic tree of *Plasmodium* species derived from *Csp* gene sequences. Table 3.1 lists the *Plasmodium* species and provides information about their geographic distributions and hosts.

Three general conclusions can be derived from the *Csp* phylogenetic results:

1. The human parasites *P. falciparum*, *P. malariae*, and *P. vivax* are remotely related to one another, so that the evolutionary divergence of these three human parasites greatly predates the origin of the hominids and even the primates. It follows that their parasitic associations with humans are phylogenetically independent; that is, at least three of these species have been laterally transmitted to the human ancestral lineage from other, nonprimate hosts. This conclusion is consistent with the diversity of physiological and epidemiological characteristics of the four *Plasmodium* species (Coatney, 1976; López-Antuñano and Schumunis, 1993). Phylogenies obtained with other genes, such as the small subunit ribosomal RNA gene (*SSUrRNA*; Escalante and

Table 3.1. *Plasmodium* species investigated with the *Csp* sequences

Species	Number of Strains	Host	Geographic Distribution
P. falciparum	8	human	tropics worldwide
P. malariae	2	human	tropics worldwide
P. vivax	4	human	tropics worldwide
P. vivax-like	2	human	tropics worldwide
P. reichenowi	1	chimpanzee	African tropics
P. brasilianum	1	monkey	New World tropics
P. simiovale	1	monkey	tropics worldwide
P. cynomolgi	5	monkey	Asian tropics
P. simium	2	monkey	New World tropics
P. knowlesi	2	monkey	Asian tropics
P. berghei	2	rodent	African tropics
P. yoelii	2	rodent	Africa

Ayala, 1994, 1995) and the mitochondrial gene coding from cyto-chrome B (Escalante *et al.*, 1998a), also show that evolutionary divergence of the four human parasites greatly predates the origin of the hominids. Host-shifts appear to have been a common occurrence among avian and reptilian malaria parasites (Bensch *et al.*, 2000; Ricklefs and Fallon, 2002).

2. *Plasmodium falciparum* is more closely related to *P. reichenowi*, the chimpanzee parasite, than to any other *Plasmodium* species. The time of divergence between these two *Plasmodium* species is estimated at 8–11 million years (My) ago, which is roughly consistent with the time of divergence between the two host species, human and chimpanzee (Escalante and Ayala, 1994, 1995). A parsimonious interpretation of this state of affairs is that *P. falciparum* is an ancient human parasite, associated with our ancestors at least since the divergence of the hominids from the great apes, and that the divergence of *P. falciparum* and *P. reichenowi* is associated with the divergence of their host species, humans and chimps. This conclusion is also corroborated by the phylogenies derived from *SSUrRNA* gene sequences (Escalante and Ayala, 1994, 1995).

Some authors (McCutchan *et al.*, 1996) failed to separate unambiguously *P. falciparum* and *P. reichenowi* when they analyzed amino acid rather than nucleotide sequences. This ambiguity can be attributed to the difficulty of aligning, for several *Plasmodium* species, amino acid sequences that are quite different and variable in length

(Escalante *et al.*, 1995; see also Rich *et al.*, 1997), with the consequence that only the more conserved amino acids can be reliably aligned. When the comparison is made between DNA sequences of *P. reichenowi* and *P. falciparum*, the difference between the two species is unambiguous (see below for the distinct composition of the central *Csp* repeat region).

3. The highly divergent human parasites *P. malariae* and *P. vivax* are genetically indistinguishable from two primate counterparts, the New World monkey parasites *P. brasilianum* and *P. simium*, respectively. We infer that a lateral transfer between hosts has occurred in recent times, either from monkeys to humans or vice versa. The *Csp* genetic distance between *P. malariae* and *P. brasilianum* is 0.002 ± 0.002, not greater than the distance among the various isolates of *P. malariae* ($n = 2$), *P. vivax* ($n = 4$), or *P. falciparum* ($n = 8$) (Table 3.2; Ayala *et al.*, 1998). This suggests that *P. malariae* (isolated from humans) and *P. brasilianum* (isolated from New World monkeys) have not long been divergent from one another and might be considered a single species exhibiting "host polymorphism" (Escalante and Ayala, 1994), i.e., the ability to parasitize more than one host species. A similar hypothesis might be put forth with respect to *P. vivax* and *P. simium*, because these two are also genetically indistinguishable (genetic distance, 0.004 ± 0.001; Table 3.2).

Whether or not the two species in each human–primate parasite pair (*P. vivax–P. simium* and *P. malariae–P. brasilianum*) should be considered the same or distinct species is merely a matter of taxonomy and nomenclatural convenience and hence is not biologically substantive. What is more important is the conclusion that two of the four known human malaria parasites have nearly identical platyrrhine (New World monkey) parasite relatives. This is a strong indication that a host-switch has occurred in recent times (or even continues to occur). A host-switch is defined as a *horizontal* shift of a parasite from one host species to another distantly related host species. This is in stark contrast to the observed relationship of *P. falciparum* and *P. reichenowi* that has evolved *vertically* (i.e., from a common ancestor), in parallel with their respective human and chimpanzee host lineages.

Determining the direction of the host-switch between human and platyrrhine – either from monkey to human, or human to monkey – holds great biological relevance to understanding the evolution of the genus and the origin of disease. Humans and platyrrhine monkeys are distantly related and have only been geographically associated at a time after the

Table 3.2. Average genetic distance within and between various *Plasmodium* species, based on the *Csp* gene

Species	Number of Strains	Intraspecific	*malariae*	*vivax*	*simium*	*brasilianum*
					Interspecific	
P. falciparum	8	.009 ± .001	.697 ± .003	.581 ± .003	.837 ± .002	.687 ± .004
P. malariae	2	.004 ± .003		.517 ± .006	.513 ± .004	.002 ± .002
P. vivax	4	.004 ± .001			.004 ± .001	.517 ± .000
P. simium	2	.000 ± .000				.508 ± .187

first human colonization of the Americas, which occurred within the past 15,000 years. Indeed, the host-switch may have occurred following the second influx of humans, when Europeans began to colonize America in the sixteenth century. Whether 500 or 15,000 years have passed since the host-switch, either would be a mere moment in evolutionary time and so it is not surprising that the human and platyrrhine parasites are genetically so little diverged. Both *P. simium* and *P. brasilianum* are known to be infectious to humans (Gilles and Warrell, 1993). Epidemiological serosurveys of humans and monkeys in French Guiana indicate that platyrrhines may actually serve as zoonotic reservoirs for human disease (Fandeur *et al.*, 2000), thus lending support to the host-polymorphism hypothesis.

Unlike the Old World primate parasite, *P. reichenowi*, which thrives exclusively in chimpanzees, the platyrrhine malaria parasites are quite capacious in their host preference, and so these New World parasites appear quite susceptible to host-switches. *P. simium* infects at least three, and *P. brasilianum* has been identified in as many as 26 species of New World monkeys (Gysin, 1998). We have argued in the past, on the grounds of evolutionary parsimony, that the host-switches observed in *vivax/simium* and *malariae/brasilianum* were most likely to have occurred from primates to humans (Escalante and Ayala, 1995; Ayala *et al.*, 1998). Based on the observed host distribution, the alternative explanation of a switch from humans to primates seems less plausible because it would require that multiple independent switches have occurred. For example, in the case of the *malariae–brasilianum* host-switch, it is improbable that no less than 26 human to platyrrhine (or platyrrhine to platyrrhine) host-switches would have occurred in the 15,000-year history of human habitation of South America.

P. vivax and *P. malariae* have widespread global distributions, whereas the complementary *P. simium* and *P. brasilianum* are restricted to South America. This is not inconsistent with a host-switch of the platyrrhine parasites to human hosts because humans have been remarkably vagile in the past several centuries and could have carried their parasites wherever they may have traveled. For example, a survey of ribosomal DNA sequences has revealed the occurrence of a *P. brasilianum* isolate in Myanmar (Kawamoto *et al.*, 2002). The most plausible explanation for New World monkey malaria in Southeast Asia is that an infected human carried it there. This pattern of host transmission may become more evident as additional molecular genotypes of malaria isolates collected from around the globe become available.

Geographical distribution records are somewhat ambiguous with respect to determining the direction of host-switch in the case of *P. vivax* and

P. simium. *P. vivax* is the most cosmopolitan of the human malarias, but it is notably absent from sub-Saharan Africa. Absence of *vivax*-malaria in the region has been attributed to the widespread occurrence of a genetic mutation of the Duffy blood group proteins in indigenous sub-Saharan peoples. Duffy proteins are expressed on the surface of erythrocytes and are necessary for receptor–ligand mediated invasion of *P. vivax* into red blood cells (Miller *et al.*, 1976). Duffy receptors do not play a role in infection by the main African malaria parasite *P. falciparum*. Individuals with an Fy^{a-b-} mutation do not express the Duffy receptor, suggesting an adaptive response for resistance to *P. vivax*. This adaptation is found primarily in particular parts of Africa, and its occurrence is inconsistent with a recent introduction of *vivax*-malaria to humans, suggesting that the occurrence of this human mutation was in response to an ancient exposure to *P. vivax*, which has since been nearly extirpated from the continent. The possibility that the Duffy mutation may have arisen in response to some other selection pressure cannot be eliminated. Indeed, Duffy-negative individuals are resistant to *P. knowlesi* (Mason *et al.*, 1977), which is an Old World monkey parasite and hence one with which human ancestors have shared a common geographical range for millions of years. The Duffy mutations may have reached a high frequency in sub-Saharan Africa as a counter response to risk of exposure to this zoonotic malaria known to be infectious to humans lacking the Duffy-negative genotype.

Historical documentation of nonmalignant (i.e., non-*P. falciparum*) malaria in humans is similarly equivocal. Firstly, there is no record of quartan malaria in South America prior to European colonization. This would be consistent with the interpretation that *P. vivax* (as well as *P. malariae* and *P. falciparum*) was introduced to the New World by the European colonizers and their African slaves. The weakness of this argument is that it relies on negative evidence, particularly unreliable when there are few records or studies that would have likely manifested the presence of malaria in the New World before the year 1,500, even if it had indeed been present.

In the Old World, historical records are more complete. Chinese medical writings (dated 2,700 B.C.), cuneiform clay tablets from Mesopotamia (~2,000 B.C.), the Ebere Egyptian Papyrus (ca. 1,570 B.C.), and Vedic-period Indian writings (1,500–800 B.C.) mention severe periodic fevers, spleen enlargement, and other symptoms suggestive of malaria (Sherman, 1998). Spleen enlargement and the malaria antigen have been detected in Egyptian mummies more than 3,000 years old (Miller *et al.*, 1994; Sherman, 1998). Hippocrates' (460–370 B.C.) discussion of tertian and quartan fevers "leaves little doubt that by the fifth century B.C. *Plasmodium malariae* and *P. vivax* were present in Greece" (Sherman, 1998, p. 3). If this interpretation is correct, the association of *malariae* and *vivax* with humans could not be attributed to a host-switch from monkeys to humans that would have occurred after the European colonization of the Americas. This would be definitive evidence, so long as one accepts the interpretation that the fevers described by Hippocrates were indeed caused by the two particular species *P. vivax* and *P. malariae* (rather than, say, *P. ovale*).

The matter will be resolved by comparing the genetic diversity of the human and primate parasites. Genetic diversity will be greater in the donor host than in the recipient host of the switch. If the transfer has been from human to monkeys, the amount of genetic diversity, particularly at silent nucleotide sites and other neutral polymorphisms, will be much greater in *P. vivax* than in *P. simium*, and in *P. malariae* than in *P. brasilianum* (including in each comparison the polymorphisms present in the various monkey host species). A transfer from monkey to humans should yield much lower polymorphism in the human than in the monkey parasites. Due to their lesser role in human mortality and morbidity, these malaria species have not garnered the attention that has been lavished on *P. falciparum*, and so very little genetic diversity data are available for *P. vivax* and *P. malariae* and far, far less for *P. brasilianum* and *P. simium*. Acquisition of this kind of data will be of great benefit in evaluating the origins of human malaria and in determining whether animals may serve as disease reservoirs.

Another chasm in our understanding of malaria parasite evolution is in regard to its shared evolution with the definitive host, the mosquito. Not only are the diverse human malaria parasites all associated with human disease, they also share the commonality of having an *Anopheles* mosquito as their primary vector. A great deal of information about these mosquitoes has been collected in recent years, including the now complete sequence of the *Anopheles gambiae* (Holt *et al.*, 2002). Chromosomal inversions, allozymes, and microsatellites have all proven useful in determining phylogeny and population structure of several anopheline species (Powell *et al.*, 1999; Walton *et al.*, 2000; Coluzzi *et al.*, 2002; della Torre *et al.*, 2002; Krzywinski and Besansky, 2002; Sharakhov *et al.*, 2002). Certainly, the phylogeny and population structure of malaria parasites will be linked to that of their vectors, but to date little work has been done in this regard. A complete understanding of the evolution of malaria will require that this gap be filled.

GENETIC POLYMORPHISMS IN *P. FALCIPARUM*: RECENT
GEOGRAPHIC EXPANSION OF MALIGNANT MALARIA

P. falciparum is the most consequential of the human malarias. We have established that *P. falciparum* (or, more precisely, its immediate ancestors) has been a parasite of the human lineage since before the divergence of humans and chimpanzees. Now, we want to examine the extant genetic variation in the global *P. falciparum*. It is well known that *P. falciparum* populations are highly polymorphic with respect to antigenic determinants, drug resistance, allozymes, and chromosome sizes (e.g., Sinnis and Wellems, 1988; Creasey *et al.*, 1990; Kemp and Cowman, 1990; McConkey *et al.*, 1990; Babiker and Walliker, 1997). Investigation of DNA sequence variation has focused on genes coding for antigenic determinants, where amino acid polymorphisms (nonsynonymous nucleotide substitutions) are common (Hughes, 1992; Hughes and Hughes, 1995). Antigenic and drug resistance polymorphisms respond to natural selection, which is most effective in large populations – millions of humans are infected by *P. falciparum* and one single patient may harbor 10^{10} parasites (McConkey *et al.*, 1990). The replacement of one allele by another, or the rise of a polymorphism with two or more alleles at high frequency, may occur even in one generation. If the selection pressure is strong enough, all individuals exposed to the selective agent may die, except those carrying a resistant mutation. With populations as large as those of *P. falciparum*, any particular mutation is expected to arise in any one generation; and the same mutation may arise – and rise to high frequency – independently in separate populations.

On the contrary, silent (i.e., synonymous) nucleotide polymorphisms are often adaptively neutral (or very nearly so) and not directly subject to natural selection. Thus, silent nucleotide polymorphisms reflect the mutation rate and the time elapsed since their divergence from a common ancestor. The population structure of *P. falciparum* is, consequently, best investigated by examining the incidence of synonymous polymorphisms. The comparison between synonymous and nonsynonymous polymorphisms may provide, in addition, insights into the population dynamics of the parasite. Table 3.3 summarizes the relevant data for 10 genes for which several sequences are available (Rich *et al.*, 1998). The gene sequences analyzed derive from isolates of *P. falciparum* representative of the global malaria endemic regions (see Table 3.1 in Rich *et al.*, 1998; for the *Csp* gene, see Rich *et al.*, 1997).

Five possible hypotheses that can account for the absence of silent polymorphisms in *P. falciparum* are (Ayala *et al.*, 1998):

Table 3.3. Polymorphisms in 10 loci of *P. falciparum*

Gene	Chromosome Location	Length (bp)	n_i^*	$D_n^\dagger$	$D_s^\dagger$	Number of Synonymous Sites 4-fold	2-fold
Dhfr	4	609	32	4	0	2,144	4,128
Ts	4	1,215	10	0	0	1,250	2,640
Dhps	8	1,269	12	5	0	1,536	2,724
Mdr1	5	4,758	3	1	0	1,350	2,088
Rap1	–	2,349	9	8	0	1,092	1,668
Calm	14	441	7	0	0	364	602
G6pd	14	2,205	3	9	0	726	1,404
Hsp86	7	2,241	2	0	0	532	910
Tpi	–	597	2	0	0	180	262
Csp1 5'end	3	387	25	7	0	688	2,010
Csp1 3'end	3	378	25	17	0	1,050	1,625
Total	–	–	–	51	0	10,912	20,061

* n_i is the number of sequences.
† D_n and D_s are the observed number of nonsynonymous and synonymous polymorphisms, respectively.

(a) persistent low effective population size,

(b) low rates of spontaneous mutation,

(c) strong selective constraints on silent variation,

(d) one or more recent selective sweeps affecting the genome as a whole, or most of it, and

(e) a demographic sweep, i.e., a recent population bottleneck, so that extant world populations of *P. falciparum* would have recently derived from a single common ancestor.

Hypothesis (*a*) can readily be excluded for the present, given that *P. falciparum* occurs in many millions of infected humans. If the effective worldwide population of *P. falciparum* had been very small (tens or at most hundreds of individuals) for very many generations until not long ago, this would effectively amount to a population bottleneck (as in hypothesis *e*).

There seems to be no reason to suspect that spontaneous mutation rates are exceptional in *P. falciparum* (hypothesis *b*), and there are two arguments against it. One is the high incidence of polymorphisms at antigenic and drug-sensitivity sites, both in worldwide samples (Kemp *et al.*, 1987; Bickle *et al.*, 1993; Qari *et al.*, 1994; Escalante *et al.*, 1998b) and in laboratory selection experiments with mice (Cowman and Lew, 1989). The other argument is that there is divergence, in synonymous as well as nonsynonymous sites, between *P. falciparum* and other *Plasmodium* species (Hughes, 1993; Escalante and Ayala, 1994; Escalante *et al.*, 1995).

Similarly, hypothesis (c), namely selective constraints due to codon bias and high AT content, cannot account for the total absence of silent variation in *P. falciparum* (Escalante *et al.*, 1998b; Rich and Ayala, 1998, 1999). Two lines of evidence suggest that the paucity of synonymous polymorphisms cannot be attributed entirely to codon bias. Firstly, in the case of 4-fold redundant codons, the bias is for codons with either A or T in the third position (at the cost of G and C) (see Table 3.5 in Ayala *et al.*, 1999). This bias reflects the overall 71.6% AT richness of the *falciparum* genome (61.1, 70.1, and 83.5 for first, second, and third positions, respectively). The fact that the mean ratio of A/T in the third position of 4-fold codons (Leu, Ile, Val, Ser, Pro, Thr, Ala, Arg, and Gly) is 1.1, suggests that, while A/T ↔ C/C changes may be restricted, A ↔ T changes seem not to be, because there is no evidence of one base being favored over the other. Secondly, levels of codon bias found in several species of *Plasmodium* are similar to those in *falciparum* (see Table 3.6 in Ayala *et al.*, 1999). Presumably, these species would, therefore, have the same constraint on synonymous substitutions in 4-fold sites as *P. falciparum*. Yet synonymous polymorphisms occur in these other species, as well as among them and between them and *falciparum*, suggesting that high codon bias does not preclude synonymous substitutions. It is interesting to note that in *P. vivax*, the first, second, and third position GC content (47.2, 60.1, and 47.8%, respectively) is more homogenous than it is in *P. falciparum*.

Comparisons between *P. falciparum* and *P. reichenowi*, at each of five genes for which data are available in both species, indicate high numbers of synonymous substitutions (average $K_s = 0.072$ and $K_n = 0.046$, for synonymous and nonsynonymous substitutions, respectively, calculated from Escalante *et al.*, 1998a, Table 3.7). Synonymous substitutions have accumulated between these two lineages over the 8 million years since their divergence, AT richness notwithstanding.

Hypothesis (d) proposes that natural selection may account for the rapid spread of a favored genotype throughout populations, particularly when the population is large and/or the selection is strong. The repeated appearance throughout global malaria endemic regions of drug-resistant phenotypes, determined by nonsynonymous substitutions at the *Dhfr*, *Dhps*, and other loci, is most likely due to natural selection. Selection sweeps are known in other organisms, such as *Drosophila melanogaster* (Hudson *et al.*, 1997; Sáez *et al.*, 2003). Natural selection can account for

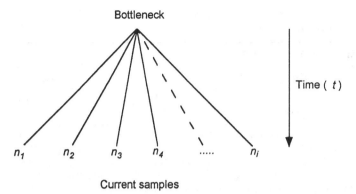

Figure 3.3. Schematic representation of a star phylogeny. t represents the time elapsed between the population bottleneck (cenancestor) and the present.

the absence of synonymous variation at any one of the 10 loci shown in Table 3.3, if the particular gene itself (or a gene with which it is linked) has been subject to a recent worldwide selective sweep, without sufficient time for the accumulation of new synonymous mutations. However, the 10 genes are located on, at least, 6 different chromosomes, and thus 6 independent selective sweeps would need to have occurred more or less concurrently, which seems unlikely. A selective sweep simultaneously affecting all chromosomes could happen if the population structure of *P. falciparum* were predominantly clonal, rather than sexual (see Rich *et al.*, 1997).

We shall now consider hypothesis (*e*), namely that the absence of silent polymorphism is a consequence of a recent population bottleneck (or demographic sweep) so that the extant world populations of *P. falciparum* can be traced to a single recent common ancestor. If a population grows to a large size after a bottleneck, it is reasonable to assume that the genealogy of a sample of multiple strains collected from widely distributed localities would be a star-like phylogeny with their last common ancestor ("cenancestor") at the vertex of the star (Figure 3.3; Slatkin and Hudson, 1991). Under this assumption, and ignoring the possibility of multiple hits at individual sites, the number of neutral polymorphisms that we observe in a sample of multiple strains will be Poisson-distributed with a mean that depends on the neutral mutation rate, the time elapsed, and the number of lineages examined. The expected number of polymorphisms is $\lambda = \mu_a t \sum n_i l_i + \mu_b t \sum n_i m_i$, where μ_a and μ_b are the neutral mutation rates at the third position of 4-fold and 2-fold degenerate codons, respectively; t is the time since the bottleneck; n_i is the number of lineages sampled at the *i*th locus; and l_i and m_i are, respectively, the number of 4-fold and

Table 3.4. Estimated upper-boundary times (t_{95} and t_{50}, in years) to the cenancestor of the world populations of *P. falciparum*

Estimated Mutation Rate $\times$ 10^{-9}			
(μ_a)	(μ_b)	t_{95}	t_{50}
7.12	2.22	24,511	5,670
3.03	0.95	57,481	13,296

Note: The t_{95} and t_{50} are the upper boundaries of the confidence intervals. Thus, in the first row the cenancestor lived less than 24,511 years ago with a 95% probability, and less than 5,670 years ago with a 50% probability. The mutation rates in the top row assume that the *Plasmodium* radiation occurred 55 Myr ago; in the bottom row, 129 Myr ago. Based on these two assumptions, μ_a and μ_b are the estimated neutral mutation rates of 4-fold and 2-fold degenerate codons, respectively.
Source: Table adapted from Ayala *et al.*, 1998.

2-fold synonymous sites examined at the *i*th locus. This expression suggests an estimator of the time of the bottleneck, obtained by solving for *t* and replacing λ (the *expected* number of polymorphisms) by *S* (the *observed* number of polymorphisms):

$$\hat{t} = \frac{S}{\mu_a \sum n_i l_i + \mu_b \sum n_i m_i}.$$

In the present sample, $S = 0$, so $t = 0$. Because *S* has a Poisson distribution, we can estimate confidence intervals by using appropriate values of the distribution in place of *S*. Estimates of the neutral mutation rates, μ_a and μ_b, may be obtained by comparing species for which the time of divergence is known: the number of neutral substitutions between species divided by the time elapsed is an estimator of the mutation rate. We have obtained four estimates of neutral mutation rates, based on two comparisons: *P. falciparum* with *P. berghei* and *P. falciparum* with *P. reichenowi* (Rich *et al.*, 1998). A summary of the results is shown in Table 3.4. We estimate the 95% confidence interval for the *falciparum* bottleneck as 0–24,511 or 0–57,481 years. We also give in Table 3.4 the t_{50} values, which represent the time such that, if the bottleneck had been older, there is a probability greater than 50% that we would have observed greater neutral variation than has actually been observed (which is zero). We have referred to the conclusion that the extant world populations of *P. falciparum* are of recent origin as the Malaria's Eve hypothesis.

In the few years following our first proposal of Malaria's Eve (Rich *et al.*, 1998), the issue has been the subject of contentious debate. In 1998, the amount of sequence data available for the species was rather limited, but since that time this data set has grown substantially, culminating in the complete genome sequence of *P. falciparum* published in 2002 (Gardner *et al.*, 2002). Other investigators have, accordingly, sought to carefully scrutinize the Malaria's Eve hypothesis.

One of these studies entailed a large-scale sequencing survey of 25 introns located on the second chromosome in eight *P. falciparum* isolates collected from global sites (Volkman *et al.*, 2001). The findings of this study confirmed our previous result: there is an extreme scarcity of silent site polymorphism among extant populations of *P. falciparum*. Among some 32,000 nucleotide sites examined, Volkman *et al.* (2001) found only three silent single nucleotide polymorphisms (SNPs) and concluded that the age of Malaria's Eve was somewhere between 3,200 and 7,700 years.

Conway *et al.* (2000) have presented further evidence in support of Malaria's Eve based on analysis of the *P. falciparum* mitochondrial genome. They examined the entire mitochondrial DNA (mtDNA) sequence of *P. falciparum* isolates originating from Africa (NF54), Brazil (7G8), and Thailand (K1 and T9/96), as well as the chimpanzee parasite *P. reichenowi*. Alignment of the four complete mtDNA sequences (5,965 bp) showed that 139 sites contain fixed differences between *falciparum* and *reichenowi*, whereas only 4 sites were polymorphic within *falciparum*. The corresponding estimates of divergence (*K*, between *P. reichenowi* and *P. falciparum*) and diversity (π, within *P. falciparum* strains) are 0.1201 and 0.0004, respectively. In short, divergence in *mtDNA* sequence between the two species is 300-fold greater than the diversity within the global *P. falciparum* population. If we use the *rDNA*-derived estimate of 8 million years as the divergence time between *P. falciparum* and *P. reichenowi*, then the estimated origin of the *P. falciparum* mtDNA lineages is 26,667 years (i.e., 8 million/300), which corresponds quite well with our estimate based on 10 nuclear genes (Rich *et al.*, 1998). In a subsequent survey of a total of 104 isolates from Africa ($n = 73$), Southeast Asia ($n = 11$), and South America ($n = 20$), Conway *et al.* (2000) determined that the extant global population of *P. falciparum* is derived from three mitochondrial lineages that started in Africa and migrated subsequently (and independently) to South America and Southeast Asia. Each mitochondrial lineage is identified by a unique arrangement of the four polymorphic *mtDNA* nucleotide sites.

Arguments against the Malaria's Eve hypothesis come in two forms. The first argument is that the loci chosen in the studies described above are a biased sample and do not reflect the levels of polymorphism in the genome as a whole. The second counterargument concedes that nucleotide polymorphisms are scarce, however this is not attributable to

recent origin, but rather to strong selection pressure against the occurrence of synonymous nucleotide substitutions.

One study that reports an "ancient" origin of a *P. falciparum* is based on
a survey of sequences available from the GenBank database (Hughes and
Verra, 2001). As with the data in our original paper (Rich *et al.*, 1998), these
GenBank sequences are compiled from a variety of sources and many of the
entries may contain sequencing errors associated with *taq* misincorpora-
tion during the PCR amplification of alleles. Moreover, some sequences included in the Hughes and Verra (2001) paper were not carefully examined,
and the comparisons include multiple nucleotide sequences from a single clone derived in different laboratories. For example, GenBank entries
AF239801 and AF282975 are both *falcipain-2* sequences from *P. falciparum*
clone W2. Regardless of possible errors, the overwhelming message from
their compiled data is that there is indeed a dearth of polymorphism. In
fact, among the 23 loci examined, which comprised over 10,000 codons,
only six contained synonymous substitutions in 4-fold degenerate codons.
Nonetheless, Hughes and Verra (2001) concluded that the time to the most
recent common ancestry of *P. falciparum* must be 300,000–400,000 years.

A most ambitious effort to quantify polymorphism in *P. falciparum* involved a survey of >200 kb from the completely sequenced chromosome 3
(Mu *et al.*, 2002). The authors reported 31 and 62 polymorphisms among
80,415 noncoding and 192,400 synonymous nucleotide sites, respectively.
Using the equation and mutation rates from our paper (Rich *et al.*, 1998),
Mu *et al.* (2002) estimated the common ancestor to be between 102,000
and 177,000 years old. At this level of polymorphism, i.e., 62 of 192,400 (or
0.03%), the possible error rate (in PCR amplification and sequencing) becomes relevant and bears great impact on estimates of recent ancestry. Mu
et al. (2002) reamplified and resequenced 56 of the regions containing single nucleotide polymorphisms (both synonymous and nonsynonymous)
and in this second pass found that two of the polymorphisms were in
error (an error rate of ~4.0%). It is because of a possibly high error rate
that the previously described paper by Volkman *et al.* (2001) incorporated
a highly redundant approach to assure integrity of the data. Their methods
involved meticulous bidirectional sequencing of three clones from each of
three independent DNA amplifications, or an 18-fold redundancy (Hartl
et al., 2002).

Another concern about calculation of the age of Malaria's Eve pertains
to the estimation of mutation rates. The estimates used by Mu *et al.* (2002)
are from a comparison of a very small number of nucleotides (708 bp) between the rhoptry-associated protein gene of *P. falciparum* and *P. reichenowi*

(Rich *et al.*, 1998). The neutral mutation rate may vary among chromosomal regions, and its estimation is subject to sampling error. Even slight
perturbations in its calculation will have exponential effects on estimation of age of the common ancestor. Reliable estimates of the mean age of
Malaria's Eve are in the range of 4,000 to 180,000 years. Although at first
glance this range of nearly two orders of magnitude appears unsatisfac-
tory, the differences are in fact quite small in light of the 6–8 million year
age of the species, dating back to its split from the chimpanzee parasite.
This means that the global, extant distribution of *P. falciparum*, with its
abundant diversity of antigens and drug resistance factors, originated in
only a small fraction (at most ~3%) of the time since the origin of the
species. This finding contrasts greatly with the previous estimates of some
antigenic variation as being 35 million years old (Hughes and Hughes,
1995; see Rich *et al.*, 2000).

Despite discrepancies in the estimation of age of the Malaria's Eve
common ancestry, it is clear that nucleotide polymorphisms are scarce
in many portions of the *P. falciparum* genome (Conway and Baum, 2002;
Hartl *et al.*, 2002). A second criticism of the recent origin hypothesis concedes the paucity of synonymous site polymorphism but attributes this
to constraints on the genome itself. One proposition is that the extreme
AT content of the *P. falciparum* genome suggests that some constraint is
acting upon mutations that lead to unfavorable codon sequences (Saul
and Battistutta, 1988; Arnot, 1991; Saul, 1999). As we have argued above,
this does not seem to be the case, because in spite of AT content as high as
84% in third positions, there appears to be an equal proportion of A and
T nucleotides in third positions of 4-fold degenerate codons (Rich and
Ayala, 1999, 2000). Moreover, the fact that synonymous substitutions
are in evidence in the divergence between *P. falciparum* and *P. reichenowi*
(which has a similarly extreme AT content) indicates that mutations can
and do occur (Rich and Ayala, 1999).

Hartl *et al.* (2002) have pointed out that genomic constraints seem
unlikely given the variability of microsatellite markers among introns, in-
tergenic regions and, in some cases, coding sequences (Su and Wellems,
1996; Anderson *et al.*, 1999, 2000; Volkman *et al.*, 2001). Nonetheless, Forsdyke (2002) has argued that the extreme conditions of the *P. falciparum*
genome present a situation where selection for genomic composition exceeds the selection on the proteins encoded by these genes. The argument
is leveled not so much against the Malaria's Eve hypothesis in particular,
but rather the author attempts to refute the notion that neutral evolution
is even possible. This warrants further discussion.

In an attempt to assign adaptive significance to the occurrence of a simple-repetitive sequence element (the Epstein–Barr nuclear antigen-1, *EBNA-1*) in the genome of the Epstein–Barr virus (EBV), Forsdyke (2002) argues that the selective pressure for particular genomic content and/or arrangement supersedes the selection acting on encoded proteins (phenotype). The *EBNA-1* can be removed from the genome without any loss of function in the virus. Because EBV, like most viruses, tends to lose extraneous genetic elements nonessential to its survival, Forsdyke (2002) maintains that the *EBNA-1* must have a function other than that typically assigned to genes, i.e., to encode messages. To establish this fact, he has developed several descriptive parameters that are based on the nucleotide composition and secondary-folding potential of nucleotide sequences. These parameters are termed potential "pressures" acting on the genome to maintain a particular configuration and/or composition. Forsdyke (2002) tested whether the region in question has extraordinary values for the pressure parameters and found that in the *EBNA-1* region there is an excessive skew in purine content (A and G), which would limit the potential for folding of the molecule and hence reduce recombination. The potential benefit of this situation is not explained and its biological relevance remains unclear.

The analysis of the EBV provided the analytical basis of Forsdyke's claim that *P. falciparum* is under pressure for reduced nucleotide polymorphism. He chose to examine the individual sequence content of two *P. falciparum* genes coding for surface antigens, *Csp* and *Msp-2* (merozoite surface protein-2). As with the *EBNA-1*, he found that there was a high bias toward purines (primarily A in this case) and a strong potential for secondary folding within the repetitive regions of both *Msp-2* and *Csp*. The only conclusion drawn from this was that the high folding potential might enhance recombination in the repeat regions of both genes. The model is neither predictive nor explanatory and does not even offer much in the way of descriptive value. If it were demonstrated that these extraordinary pressure regions had significantly less (or greater) synonymous site polymorphism and that pressure was predictive of this polymorphism, the author's claim might bear some relevancy. However, neither of these claims can be made, particularly because the author chose to examine two of the most highly polymorphic loci known in *P. falciparum*. What is clear is that silent-site polymorphisms are in evidence in non-*falciparum* malaria species and that synonymous substitutions have occurred in the evolution of *P. falciparum* and *P. reichenowi*. On this basis, we maintain that although substitutions may be constrained due to nucleotide composition and/or codon usage bias, these constraints do not explain the paucity of

P. falciparum synonymous-site variation. Therefore, the Malaria's Eve hypothesis remains the most likely explanation for this state of affairs.

In addition to the analyses of genetic polymorphism data, there is independent information in support of the Malaria's Eve hypothesis. Sherman (1998) notes the late introduction and low incidence of *falciparum* malaria in the Mediterranean region. Hippocrates (460–370 B.C.) describes quartan and tertian fevers, but there is no mention of severe malignant tertian fevers, which suggests that *P. falciparum* infections had not yet occurred in classical Greece, as recently as 2,400 years ago. Interestingly, Tishkoff *et al.* (2001) traced the origin of malaria-resistant *G-6pd* genotypes in humans to the spread of agricultural societies some 5,000 years ago. The recent origin of this mutation in humans suggests a similarly recent association with widespread exposure to the malaria parasite.

How can we account for a recent demographic sweep of *P. falciparum* across the globe, given its long-term association with the hominid lineage? One likely hypothesis is that human parasitism by *P. falciparum* has long been highly restricted geographically and has dispersed throughout the Old World continents only within the past several thousand years, perhaps within the past 10,000 years, after the Neolithic revolution (Coluzzi, 1994, 1997, 1999). Three possible scenarios may explain this historically recent dispersion:

(1) changes in human societies,
(2) genetic changes in the host–parasite–vector association that have altered their compatibility, and
(3) climatic changes that entailed demographic changes (migration, density, etc.) in the human host, the mosquito vectors, and/or the parasite.

One factor that may have impacted the widespread distribution of *P. falciparum* in human populations from a limited original focus, probably in tropical Africa, is changes in human living patterns, particularly the development of agricultural societies and urban centers that increased human population density (Livingston, 1958; Weisenfeld, 1967; de Zulueta *et al.*, 1973; de Zulueta, 1994; Coluzzi, 1997, 1999; Sherman, 1998). Genetic changes that have increased the affinity within the parasite–vector–host system are also a possible explanation for a recent expansion, not mutually exclusive with the previous one. Coluzzi (1997, 1999) has cogently argued that the worldwide distribution of *P. falciparum* is recent and has come about, in part, as a consequence of a recent dramatic rise in vectorial capacity due to repeated speciation events in Africa of the most anthropophilic members of the species complexes of the *Anopheles gambiae* and *A. funestus* mosquito vectors. Biological processes implied by this account may have been associated with, and even dependent on, the onset of agricultural societies in Africa (scenario 1) and climatic changes (scenario 3), specifically, a gradual increase in ambient temperatures after the Würm glaciation, so that about 6,000 years ago climatic conditions in the Mediterranean region and the Middle East made the spread of *P. falciparum* and its vectors beyond tropical Africa possible (de Zulueta *et al.*, 1973; de Zulueta, 1994; Coluzzi, 1997, 1999). The three scenarios are likely interrelated. Once demographic and climatic conditions became suitable for propagation of *P. falciparum*, natural selection would have facilitated evolution of *Anopheles* species that were highly anthropophilic and effective *falciparum* vectors (de Zulueta *et al.*, 1973; Coluzzi, 1997, 1999).

NATURAL SELECTION AND THE EVOLUTION OF ANTIGENIC GENES

There is an apparent contradiction between the paucity of synonymous polymorphisms and the abundance of replacement changes observed in antigenic loci. Natural selection may offer an explanation because strong positive selection, particularly where immune evasion is at stake, can very likely fix even rare mutations. However, Hughes and colleagues proposed a model that requires that variants of genes encoding *P. falciparum* surface proteins be as old or much older than the species itself (Hughes *et al.*, 1983; Hughes, 1993; Hughes and Hughes, 1995). They originally estimated that the ages of the most divergent alleles of *Msp-1* and *Csp* alleles are 35 and 2.1 million years, respectively. It must first be noted that the polymorphisms in these antigenic genes, whether or not they are of ancient origin, do not contradict the recent origin of *P. falciparum* current world populations. As with the misinterpretation of the Mitochondrial Eve model of human origins, it should be noted that the hypothetical Malarial Eve does not represent a single ancestral individual but rather current populations derive from few individuals at some point in the past. Ancient polymorphisms at certain loci under strong balancing (diversifying) selection can be maintained through a severe constriction in population numbers, or even through a number of generations with small populations that would lead to the virtual complete elimination of neutral allelic polymorphisms, as noted above to account for the scarcity of silent polymorphisms. For example, although the mitochondrial lineage of modern humans is only 100,000–200,000 years old, natural selection has maintained extensive polymorphisms among human MHC molecules (involved in the immune response against invading foreign substances), some of which predate the split between humans and chimpanzees (Ayala, 1995). The *P. falciparum* antigenic genes are under strong diversifying selection for evasion of human immune response (McCutchan and Waters, 1990; Miller *et al.*,

1993; Escalante *et al.*, 1998b), and so they too could be maintained despite a demographic bottleneck. The lack of silent-site differentiation among dimorphic forms of several of these antigenic determinants further supports this hypothesis.

In order to understand the evolutionary history of the antigenic alleles in *P. falciparum*, it is imperative to utilize a model that incorporates all biologically relevant information. A high level of amino acid polymorphism is evident in several antigenic genes that have been examined. Most of these amino acid changes have been mapped directly to B and T cell epitopes (Anders *et al.*, 1993). At the nucleotide level, the disproportionate number of nonsynonymous substitutions relative to synonymous substitutions indicates that these regions are under positive diversifying selection (Escalante *et al.*, 1998b). The requisite assumption of this model is that point mutations are equally likely to occur at any site but only those that favorably alter phenotype (amino acid) will be selected and hence be maintained, whereas the neutral sites (nonselected) will be lost or fixed at random. When deleterious mutations occur, negative selection will remove them, and the amino acid sequence will be conserved. This is the basic model of molecular evolution and, with various corrections for multiple nucleotide substitutions at individual sites, its validity has been confirmed in innumerable protein-coding genes in the great diversity of living species (Li, 1997).

However, not all DNA sequences adhere to this model. Consider, for example, the DNA repeat regions that make up micro- and minisatellite loci in various plant and animal species, including humans. Most of the variation within these repeats originates by a slipped-strand process that yields duplication and/or deletion of the repeated units. This process leads to rapid differentiation of alleles, wherein an individual mutational event can change several nucleotides at once, with greater impact on sequence divergence than the typical single-nucleotide mutation process. Moreover, these mutations occur at rates that are orders of magnitude greater than that of single nucleotide substitutions (for a review, see Hancock, 1999). For this reason, even closely related individuals in a population, which may be identical in their coding DNA at most loci, may show marked differentiation at microsatellite loci. That these loci are so highly polymorphic reflects the fact that repetitive elements are susceptible to frequent slipped-strand mutations. The diversification of DNA satellite sequences typically takes place in the absence of selection, because the loci themselves do not encode a protein product. For example, among the 27,336 bp of intronic sequence examined by Volkman *et al.* (2001), there was only one nucleotide substitution, but several mutations were detected in microsatellite repeats. The authors concluded that microsatellite

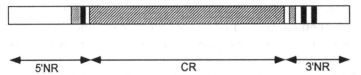

Figure 3.4. Structure of the *P. falciparum Csp* gene. Two nonrepeat regions (5′ NR and 3′ NR) brace a central region (CR, hatched) made up of a variable number of tandem repeats encoding 4-amino-acid-long motifs. The light gray boxes represent B-cell epitopes; the dark boxes represent T-cell epitopes.

mutations occur at rates that are much higher than that of single nucleotide substitutions.

A notable feature of *P. falciparum* surface proteins is the presence of certain repeating nucleotide sequences, which encode short iterative amino acid sequences (Dame *et al.*, 1984; Anders *et al.*, 1988). These antigenic repeat regions are highly polymorphic, yet the repeat regions are known to be in many instances under immune selection. This presents a novel situation to the molecular evolutionist, in that these loci behave as one would expect satellite DNA to behave with respect to the rapid mutation process and the generation of variable-length sequences, whereas the repeat portions encode part of the functional protein and so are subject to selection pressure.

The *Csp* gene has been extensively investigated in *P. falciparum* because it encodes the antigenic circumsporozoite protein, considered a likely target for vaccine development (Gramzinski *et al.*, 1997; Zevering *et al.*, 1998). The gene consists of two end-regions that are not repetitive (5′ NR and 3′ NR) but embrace a central region made up of a variable number (typically, between 40 and 50) of tandem repeats, each encoding one of two 4-amino-acid-long motifs (Figure 3.4). We have shown in Table 3.3 that there are not silent polymorphisms in the 5′ NR and 3′ NR regions and have used this evidence to infer the origin of *P. falciparum* populations from a single individual strain within the past several thousand years. The polymorphisms found in the B-cell and T-cell epitope regions can be attributed to antigenic natural selection (Rich *et al.*, 1998).

The variable number of repeats and the nucleotide variation among the repeats are difficult challenges when seeking the alignment of the repetitive central region (CR). We have sought to accomplish this and, more importantly, to understand the organization and evolution of the CR region, by proceeding in two steps: we align, first, the amino acid motifs and, second, the nucleotide variation within each motif. The two amino acid motifs in the *Csp* CR of *P. falciparum* are NANP and NVDP, represented by 1 and 2, respectively, in Table 3.5, which shows the organization of the

Table 3.5. Composition of the CR region of the *Csp* gene

Sequence	Repeat Motifs	Number of Repeats		
		1	2	3
M15505	1212111111111111112111111111111111111111111111111111	43	3	0
M83173	12121111111111111112111111111111111111111	43	3	0
M83149	12121211111111111111111111111111111111	41	3	0
M83150	12121111111111111111111211111111111111	44	3	0
M83156	1212111111111111111111111111111111111111111111111111	49	2	0
M83158	12121212111111111111111111111111111111	42	4	0
M83161	12121211111111111111121111111111111111	39	4	0
M83163	12121111111111111121111111111111111111111	43	3	0
M83164	121211111111111111121111111111111111111111	46	3	0
M83165	12121211111111111111111111111111111111	43	3	0
M83166	1212121111111111111111111111111111111111	42	4	0
M83167	12121212111111111111111111111111111111111	46	3	0
M83168	1212121111111111111111111111111111111111	42	4	0
M83169	12121211111111111111111111111111111111	41	3	0
M83170	12121212111111111111111111111111111111111	42	4	0
M83174	121212111111112111111111111111111111111	39	4	0
M19752	12121211111111111111111121111111111111111	41	3	0
M83172	12121211111111111111211111111111111111	38	4	0
K02194	121212111111111111112111111111111111	37	4	0
M57499	121212121111111111111111111111111111111	40	4	0
U20969	12121211111111111111121111111111111	36	4	0
M83886	1212121111111111111112111111111111111	38	4	0
M22982	121212111111111111111112111111111111111	40	4	0
X15363	12121211111111111111111112111111111111111	40	4	0
M57498	1212121111111111111111112111111111111111	37	4	0
P. reichenowi	12121212131212131313111111111111111111	26	5	4

Note: The repeat motifs NANP, NVDP, and NVNP are represented by 1, 2, and 3, respectively. A is alamine; D is aspartic acid; N is asparagine; P is proline; V is valine.

61

Table 3.6. Amino acid and nucleotide sequences of the repeat allotypes (RAT) and their incidences

	Motif		*falciparum*		*reichenowi*	
RAT	Amino Acid	Nucleotide	%	Number	%	Number
A	NANP	aatgcaaaccca	55.1	566	38.5	10
B	NANP	. . . . t . . t	16.1	165	30.8	8
C	NANP	. . . . t . . .	7.6	78	–	–
D	NANP	. . . . c . . t . . a	6.2	64	3.8	1
E	NANP	 c	6.2	64	–	–
F	NANP	. . c c	5.1	52	–	–
G	NANP	. . . . . c	3.1	32	7.7	2
H	NANP	. . . . t	0.3	3	–	–
I	NANP	. . c	0.2	2	3.8	1
J	NANP	. . . . c c	0.1	1	–	–
Z	NANP	. . . . t . . c	–	–	15.4	4
M	NVDP	. . . . t . g .t . . .	52.3	46	20.0	1
N	NVDP	. . . . t . g . t . . c	31.8	28	40.0	2
O	NVDP	. . c . t . g . t . . t	14.8	13	–	–
P	NVDP	. . . . t . g . t . . t	1.1	1	20.0	2
X	NVNP	. . . . t . . . t . . c	–	–	100.0	4

region in 25 *Csp* sequences of *P. falciparum* and one of *P. reichenowi*. A parsimonious interpretation of these arrangements is that length variation originates by duplication of the motif doublet 1–2 or simply of motif 1. We have introduced the concept of repeat allotype (RAT) to refer to particular 12-nucleotide-long sequences coding for a given amino acid motif. In *P. falciparum* there are 10 different RATs (A–J) coding for motif 1 and there are 4 RATs coding for motif 2 (Table 3.6). The distribution of RATs, even within a given motif, is very uneven. One particular RAT has an incidence greater than 50% over the whole set, whereas some RATs are present only once or a few times. Ayala *et al.* (1999) have shown a possible alignment of the *P. falciparum* RATs.

Which genetic mechanisms account for the variation in the number of RATs? As noted above, we have analyzed recombination in the *Csp* gene and inferred that variation in the CR region arises by mitotic intragenic recombination (Rich *et al.*, 1997). The model that we propose for RAT evolution in Figure 3.5 is simply an instance of the general slipped-strand model for generating length variation in repetitive DNA regions, such as the multiple repeats of short-length sequences characteristic of microsatellite

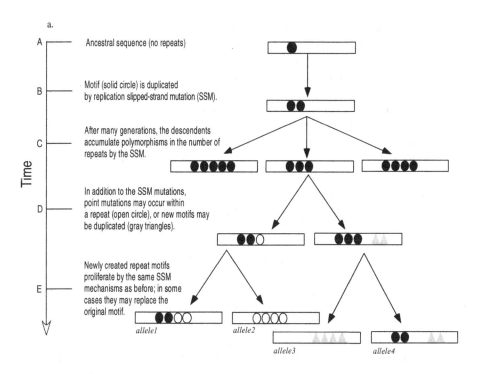

a.

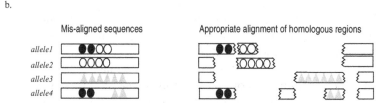

b.

Figure 3.5. (a) Model of nucleotide-repeat evolution. Rectangles represent the entire gene, which has a finite length such that proliferation of a repeating unit leads to loss of other repeats or loss of single-copy, nonrepeat regions. Three distinct repeat sequences, which differ from one another by at least one nucleotide, are shown as solid circles, open circles, and triangles. An arbitrary time scale is shown on the left. (b) An alignment of the four alleles from time-point E.

loci. New RATs can arise in this model by one of two processes:

(1) replacement or silent substitutions in a codon, and
(2) the slippage mechanism that leads to RAT proliferation. The two amino acid motifs and the different RAT types have arisen by the first process.

The variation in the number of RATs arises by the second process. As pointed out above and as observed in noncoding, satellite DNA (Schug *et al.*, 1998), process (2) occurs with a frequency several orders of magnitude greater than process (1).

How much of the variation now present in the *Csp* CR region of *falciparum* may have arisen by process (2)? Notice that only two amino acid motifs are present in the whole set of 25 *Csp* sequences and that both motifs are present in every one of the sequences (Table 3.5). Thus, there is no evidence that any replacement substitution has occurred in the recent evolution of *P. falciparum*, which we are proposing has evolved from one single strain. Notice also that the single *P. reichenowi* sequence available contains *three* amino acid motifs, including the same two motifs found in all *P. falciparum* strains.

The 25 *falciparum* CR sequences listed in Table 3.5 have a total of 14 different RATs distributed among two amino acid motifs (Table 3.6). The only *reichenowi* sequence available, although shorter than the *falciparum* sequences (35 vs. 40–51 repeats), consists of 10 different RATs and three amino acid motifs. Thus, even if we do not take into account new substitutions (process 1, which may actually have occurred in RATs H, I, J, and P), it is not unreasonable to assume that all CR variation observed in *falciparum* might have arisen by process (2) from a single ancestral strain, if this were about as heterogeneous as the extant *reichenowi* sequence.

As a second example, let us consider the alleles of a known *P. falciparum* antigen encoding gene: the merozoite surface protein-1 (*Msp-1*) locus. Tanabe and colleagues subdivided the protein into 17 *blocks*, which were labeled as "conserved," "semi-conserved," and "variable," based on the degree of polymorphism among various *Plasmodium* strains (Tanabe *et al.*, 1987). By examining the polymorphic blocks, i.e., the semi-conserved and variable blocks, Tanabe identified clear dimorphism that distinguished two groups, which we refer to herein as Group I and Group II. We examined numerous *Msp-1* sequences of Group I and Group II strains from the GenBank database to determine the distribution of polymorphisms within each block. Our findings are summarized in Table 3.7. We looked at the number of synonymous and nonsynonymous polymorphisms. It is clear from this analysis that the amount of nucleotide polymorphism is not uniform across the length of the molecule, and so we conclude that the different blocks may have quite distinct evolutionary histories.

For nearly every block, the degree of intragroup polymorphism is less than 0.05, and most are less than 0.01. The exceptional case is block 2,

Table 3.7. Nucleotide diversity within and between Group I and II alleles of the *P. falciparum Msp-1* genes

Block	Length (codons)*	N†	d_s			d_n		
			Group I	Group II	Group I + Group II	Group I	Group II	Group I + Group II
1	55	33	0.019	0.021	0.017	0.017	0.010	0.013
2	55	33	0.106	0.185	0.150	0.449	0.497	0.553
3	202	33	0.038	0.006	0.042	0.018	0.000	0.023
4	31	29	0.031	0.000	0.020	0.307	0.000	0.215
5	35	29	0.000	0.000	0.070	0.000	0.000	0.026
6	227	8	0.000	0.000	0.282	0.004	0.001	0.300
7	73	8	0.000	0.000	0.361	0.003	0.000	0.072
8	95	8	0.000	0.000	0.338	0.000	0.003	0.711
9	107	8	0.000	0.023	0.409	0.005	0.043	0.126
10	126	8	0.008	0.000	0.448	0.011	0.000	0.394
11	35	11	0.000	0.000	0.128	0.000	0.000	0.068
12	79	11	0.000	0.000	0.000	0.000	0.000	0.000
13	84	11	0.000	0.042	0.040	0.005	0.007	0.052
14	60	11	0.000	0.018	0.212	0.002	0.005	0.371
15	89	19	0.000	0.000	0.216	0.001	0.003	0.089
16	217	19	0.002	0.032	0.277	0.005	0.027	0.185
17	99	19	0.002	0.019	0.007	0.010	0.027	0.016

* Block length may vary between Group I and II alleles; the given value is the average length of Group I and II alleles.

† Only partial *Msp-1* sequences are available for some strains in GenBank; however, our analysis includes eight complete coding sequences, three of Group I and five of Group II.

Note: The values are the mean numbers of synonymous (d_s) or nonsynonymous (d_n) substitutions per site.

which shows markedly higher intra- and intergroup differences than any other block. This can be attributed to the tripeptide repeats in block 2, which create particular difficulty in determining appropriate alignment and therefore render these deceptively high values of d_s and d_n. Nucleotide repeats, and hence their corresponding peptide motifs, are susceptible to mutational mechanisms that occur at much greater frequency than singular point mutations. To discern the evolutionary history of regions containing these repeats, e.g., *Msp-1* block 2, we must consider the most likely model by which mutations accumulate. In Figure 3.5a, we have presented just such a model. Note that in the hypothetical ancestor, no repeats are present. Repeats arise first by duplication of a short sequence following mutation during replication of the DNA strand. In subsequent generations, additional copies of the repeat accumulate in some alleles by slipped-strand mutation (SSM), so that in a few generations various length polymorphisms may arise among the alleles (the same SSM process can lead to loss of a repeat copy; see Levinson and Gutman, 1987 for details of the SSM process). Novel repeats may also arise by either substitutions within the ancestral repeat (as with the open circles) or, alternatively, by precisely the means of the original repeat birth (shown here as gray triangles). This model is not particularly novel, because an analogous process is invoked to explain evolution of microsatellite loci. However, spurious conclusions arise when aligning the set of alleles that result from this process. This is demonstrated in Figure 3.5b, where we show the alignment of the four sequences from time-point E. Note that in the alignment on the left (Figure 3.5b), the various repeat motifs are aligned according to their position, but as we show in Figure 3.5a, these repeats are not homologous. In fact, the most likely homologue of a given repeat is an adjacent repeat from the same allele. The alignment on the right of Figure 3.5b shows how certain regions of individuals' alleles may lack areas of homology in other alleles due to the duplication and replacement of repeat units. It is exactly this kind of alignment artifact that explains the extraordinary intragroup polymorphism in block 2. We have shown similar patterns in nucleotide repeat regions of other antigenic genes as well, including *Csp* and *Msp-2* (Rich et al., 1997, 2000; Rich and Ayala, 2000).

With very few exceptions, the degree of difference between Group I and Group II far exceeds the amount of polymorphism within either group for both synonymous and nonsynonymous changes. As we have stated above, the remaining 16 blocks (i.e., other than block 2) of the *Msp-1* show very little within-group polymorphism. However, between groups, there is considerably more nucleotide polymorphism, both synonymous and nonsynonymous. Hughes (1992) has argued that polymorphisms observed between the two groups have been maintained within the species for millions of years by balancing selection.

For Hughes' model to be correct, one must invoke extraordinarily high rates of recombination and extreme selection coefficients. Given the biology of the parasite, and the likelihood of past and recurrent bottlenecks, we conjectured that this seemed rather implausible. This led us to propose that it is rate of evolution and not the age of these blocks that is so vastly different. We hypothesized that the SSM processes that caused the extreme polymorphism and inflated estimates of nonsynonymous nucleotide diversity (d_n) in block 2 may also have occurred elsewhere in the molecule. Our suspicions were confirmed when we identified repeats within several of the most polymorphic *Msa-1* blocks, in particular, blocks 4, 8, and 14, which were previously characterized as nonrepeat blocks.

Consider block 8, which is the block identified by Tanabe *et al.* (1987) as showing the lowest amino acid similarity between groups (10%), and which in our analysis is the most polymorphic in terms of nonsynonymous nucleotide diversity ($d_n = 0.711$). We have identified three group-specific repeats within this block (see Rich *et al.*, 2000). One 9-bp repeat (R2a) is present in all Group II alleles; and two repeats, of 6-bp (R1a) and 7-bp (R1b), are present in all Group I alleles. We have hypothesized that the occurrence of these repeats within this very short stretch of DNA is a highly significant departure from chance. To test this hypothesis, we searched the recently completed genomic sequences of *P. falciparum* chromosomes 2 and 3. The nucleotide sequences of repeats R1a, R1b, and R2a appear 25, 116, and 11 times, respectively, within the 947 kbp of chromosome 2. Within the 1,060 kbp of chromosome 3, the R1a, R1b, and R2a are present 39, 52, and 7 times, respectively. None of the three nucleotide repeats ever appears in tandem on either chromosome 2 or 3. Moreover, the average distance between each occurrence on these chromosomes is >20 kb, demonstrating that their repeated occurrence in the short 147-bp segment of *Msp-1* block 8 strongly departs from random expectation. It would therefore appear that, like block 2, this region has undergone rapid differentiation, driven by the SSM process outlined in Figure 3.5a. This rapid differentiation is the mechanism by which the parasite is able to generate such great antigenic diversity even in the face of recurrent demographic sweeps. The widespread occurrence of similar genomic regions of high mutability among potential vaccine targets will surely have some bearing on predicting the probability of success of protective vaccines.

REFERENCES

Anders, R. F., Coppel, R. L., Brown, G. V., and Kemp, D. J., 1988. Antigens with repeated amino acid sequences from the asexual blood stages of *Plasmodium falciparum*. *Prog. Allergy* 41:148–72.

Anders, R. F., McColl, D. J., and Coppel, R. L., 1993. Molecular variation in *Plasmodium falciparum*: polymorphic antigens of asexual erythrocytic stages. *Acta Trop.* 53:239–53.

Anderson, T. J., Su, X. Z., Bockarie, M., Lagog, M., and Day, K. P., 1999. Twelve microsatellite markers for characterization of *Plasmodium falciparum* from finger-prick blood samples. *Parasitology* 119:113–25.

Anderson, T. J., Su, X. Z., Roddam, A., and Day, K. P., 2000. Complex mutations in a high proportion of microsatellite loci from the protozoan parasite *Plasmodium falciparum*. *Mol. Ecol.* 9:1599–608.

Arnot, D. E., 1991. Possible mechanisms for the maintenance of polymorphisms in Plasmodium populations. *Acta Leiden* 60:29–35.

Ayala, F. J., 1995. Adam, Eve, and other ancestors: a story of human origins told by genes. *Pubbl. Stn. Zool. Napoli II* 17:303–13.

Ayala, F., Escalante, A., Lal, A., and Rich, S., 1998. Evolutionary relationships of human malarias. In *Malaria: Parasite Biology, Pathogenesis, and Protection*, I. W. Sherman, Ed., pp. 285–300. American Society of Microbiology Press, Washington, D.C.

Ayala, F. J., Escalante, A. A., and Rich, S. M., 1999. Evolution of Plasmodium and the recent origin of the world populations of *Plasmodium falciparum*. *Parassitologia* 41:55–68.

Babiker, H., and Walliker, D., 1997. Current views on the population structure of *Plasmodium falciparum*: implications for control. *Parasitol. Today* 13:262–7.

Barta, J. R., 1989. Phylogenetic analysis of the class Sporozoea (phylum Apicomplexa Levine, 1970): evidence for the independent evolution of heteroxenous life cycles. *J. Parasitol.* 75:195–206.

Barta, J. R., Jenkins, M. C., and Danforth, H. D., 1991. Evolutionary relationships of avian Eimeria species among other Apicomplexan protozoa: monophyly of the apicomplexa is supported. *Mol. Biol. Evol.* 8:345–55.

Bensch, S., Stjernman, M., Hasselquist, D., Ostman, O., Hansson, B., Westerdahl, H., and Pinheiro, R. T., 2000. Host specificity in avian blood parasites: a study of Plasmodium and Haemoproteus mitochondrial DNA amplified from birds. *Proc. R. Soc. B* 267:1583–9.

Bickle, Q., Anders, R. F., Day, K., and Coppel, R. L., 1993. The S-antigen of *Plasmodium falciparum*: repertoire and origin of diversity. *Mol. Biochem. Parasitol.* 61:189–96.

Coatney, G. R., 1976. Relapse in malaria – an enigma. *J. Parasitol.* 62:3–9.

Coluzzi, M., 1994. Malaria and the afro-tropical ecosystems impact of man-made environmental changes. *Parassitologia* 36:223–7.

Coluzzi, M., 1997. "Interazioni Evolutive Uomo-Plasmodío-Anophele." In *Evoluzione Biologica & i Grandi Problemi della Biologia*, pp. 263–85. Accademia dei Lincei, Rome.

Coluzzi, M., 1999. The clay feet of the malaria giant and its African roots: hypotheses and inferences about origin, spread and control of *Plasmodium falciparum*. *Parassitologia* **41**:277–83.

Coluzzi, M., Sabatini, A., della Torre, A., Di Deco, M. A., and Petrarca, V., 2002. A polytene chromosome analysis of the *Anopheles gambiae* species complex. *Science* **298**:1415–18.

Conway, D. J., and Baum, J., 2002. In the blood – the remarkable ancestry of *Plasmodium falciparum*. *Trends Parasitol.* **18**:351–5.

Conway, D. J., Fanello, C., Lloyd, J. M., Al-Joubori, B. M., Baloch, A. H., Somanath, S. D., Roper, C., Oduola, A. M. J., Mulder, B., Povoa, M. M., Singh, B., and Thomas, A. W., 2000. Origin of *Plasmodium falciparum* malaria is traced by mitochondrial DNA. *Mol. Biochem. Parasitol.* **111**:163–71.

Corliss, J. O., 1994. An interim utilitarian ("user-friendly") hierarchical classification and characterization of the protists. *Acta Protozool.* **33**:1–51.

Cowman, A. F., and Lew, A. M., 1989. Antifolate drug selection results in duplication and rearrangement of chromosome 7 in *Plasmodium chabaudi*. *Mol. Cell Biol.* **9**:5182–8.

Creasey, A., Fenton, B., Walker, A., Thaithong, S., Oliveira, S., Mutambu, S., and Walliker, D., 1990. Genetic diversity of *Plasmodium falciparum* shows geographical variation. *Am. J. Trop. Med. Hyg.* **42**:403–13.

Dame, J. B., Williams, J. L., McCutchan, T. F., Weber, J. L., Wirtz, R. A., Hockmeyer, W. T., Maloy, W. L., Haynes, J. D., Schneider, I., Roberts, D., Sanders, G. S., Reddy, E. P., Diggs, C. L., and Miller, L. H., 1984. Structure of the gene encoding the immunodominant surface antigen on the sporozoite of the human malaria parasite *Plasmodium falciparum*. *Science* **225**:593–9.

de Zulueta, J., 1994. Malaria and ecosystems: from prehistory to posteradication. *Parassitologia* **36**:7–15.

de Zulueta, J., Blazquez, J., and Maruto, J. F., 1973. Entomological aspects of receptivity to malaria in the region of Navalmoral of Mata. *Revista de Sanidad e Higiene Pública* (Madrid) **47**:853–70.

della Torre, A., Costantini, C., Besansky, N. J., Caccone, A., Petrarca, V., Powell, J. R., and Coluzzi, M., 2002. Speciation within *Anopheles gambiae* – the glass is half full. *Science* **298**:115–7.

Escalante, A. A., and Ayala, F. J., 1994. Phylogeny of the malarial genus Plasmodium, derived from rRNA gene sequences. *Proc. Natl. Acad. Sci. U.S.A.* **91**:11,373–7.

Escalante, A. A., and Ayala, F. J., 1995. Evolutionary origin of Plasmodium and other Apicomplexa based on rRNA genes. *Proc. Natl. Acad. Sci. U.S.A.* **92**:5793–97.

Escalante, A. A., Barrio, E, and Ayala, F. J., 1995. Evolutionary origin of human and primate malarias: evidence from the circumsporozoite protein gene. *Mol. Biol. Evol.* **12**:616–26.

Escalante, A. A., Freeland, D. E., Collins, W. E., and Lal, A. A., 1998a. The evolution of primate malaria parasites based on the gene encoding cytochrome b from the linear mitochondrial genome. *Proc. Natl. Acad. Sci. U.S.A.* **95**:8124–9.

Escalante, A. A., Lal, A. A., and Ayala, F. J., 1998b. Genetic polymorphism and natural selection in the malaria parasite *Plasmodium falciparum*. *Genetics* **149**:189–202.

Fandeur, T., Volney, B., Peneau, C., and De Thoisy, B., 2000. Monkeys of the rainforest in French Guiana are natural reservoirs for *P-brasilianum/P-malariae* malaria. *Parasitology* **120**:11–21.

Forsdyke, D., 2002. Selective pressures that decrease synonymous mutations in *Plasmodium falciparum*. *Trends Parasitol.* **18**:411.

Gardner, M. J., Hall, N., Fung, E., White, O., Berriman, M., Hyman, R. W., Carlton, J. M., Pain, A., Nelson, K. E., Bowman, S., Paulsen, I. T., James, K., Eisen, J. A., Rutherford, K., Salzberg, S. L., Craig, A., Kyes, S., Chan, M. S., Nene, V., Shallom, S. J., Suh, B., Peterson, J., Angiuoli, S., Pertea, M., Allen, J., Selengut, J., Haft, D., Mather, M. W., Vaidya, A. B., Martin, D. M. A., Fairlamb, A. H., Fraunholz, M. J., Roos, D. S., Ralph, S. A., McFadden, G. I., Cummings, L. M., Subramanian, G. M., Mungall, C., Venter, J. C., Carucci, D. J., Hoffman, S. L., Newbold, C., Davis, R. W., Fraser, C. M., and Barrell, B., 2002. Genome sequence of the human malaria parasite *Plasmodium falciparum*. *Nature* **419**:498–511.

Garnham, P. C. C., 1966. *Malaria Parasites and Other Hemosporidia*, pp. 60–84. Blackwell Scientific, Oxford.

Gilles, H. M., and Warrell, D. A., 1993. *Bruce-Chwatt's Essential Malariology*. Edward Arnold, London.

Gramzinski, R. A., Maris, D. C., Doolan, D., Charoenvit, Y., Obaldia, N., Rossan, R., Sedegah, M., Wang, R., Hobart, P., Margalith, M., and Hoffman, S., 1997. Malaria DNA vaccines in Aotus monkeys. *Vaccine* **15**:913–5.

Gysin, J., 1998. Animal models: primates. In *Malaria: Parasite Biology, Pathogenesis, and Protection*, I. W. Sherman, Ed., pp. 419–441. American Society of Microbiology Press, Washington, D.C.

Hancock, J. M., 1999. Microsatellites and other simple sequences: genomic context and mutational mechanisms. In *Microsatellites, Evolution and Applications*, D. B. Goldstein and C. Schlötterer, Eds., pp. 1–9. Oxford University Press, Oxford.

Hartl, D. L., Volkman, S. K., Nielsen, K. M., Barry, A. E., Day, K. P., Wirth, D. F., and Winzeler, E. A., 2002. The paradoxical population genetics *of Plasmodium falciparum*. *Trends Parasitol.* **18**:266–72.

Holt, R. A., Subramanian, G. M., Halpern, A., Sutton, G. G., Charlab, R., Nusskern, D. R., Wincker, P., Clark, A. G., Ribeiro, J. M. C., Wides, R., Salzberg, S. L., Loftus, B., Yandell, M., Majoros, W. H., Rusch, D. B., Lai, Z. W., Kraft, C. L., Abril, J. F., Anthouard, V., Arensburger, P., Atkinson, P. W., Baden, H., de Berardinis, V., Baldwin, D., Benes, V., Biedler, J., Blass, C., Bolanos, R., Boscus, D., Barnstead, M., Cai, S., Center, A., Chatuverdi, K., Christophides, G. K., Chrystal, M. A., Clamp, M., Cravchik, A., Curwen, V., Dana, A., Delcher, A., Dew, I., Evans, C. A., Flanigan, M., Grundschober-Freimoser, A., Friedli, L., Gu, Z. P., Guan, P., Guigo, R., Hillenmeyer, M. E., Hladun, S. L., Hogan, J. R., Hong, Y. S., Hoover, J., Jaillon, O., Ke, Z. X., Kodira, C., Kokoza, E., Koutsos, A., Letunic, I., Levitsky, A., Liang, Y., Lin, J. J., Lobo, N. F., Lopez, J. R., Malek, J. A., McIntosh, T. C., Meister, S., Miller, J., Mobarry, C., Mongin, E., Murphy, S. D., O'Brochta, D. A., Pfannkoch, C., Qi, R., Regier, M. A., Remington, K., Shao, H. G., Sharakhova, M. V., Sitter, C. D., Shetty, J., Smith, T. J., Strong, R., Sun, J. T., Thomasova, D., Ton, L. Q., Topalis, P., Tu, Z. J., Unger, M. F., Walenz, B., Wang, A. H., Wang, J., Wang, M., Wang,

X. L., Woodford, K. J., Wortman, J. R., Wu, M., Yao, A., Zdobnov, E. M., Zhang, H. Y., Zhao, Q., Zhao, S. Y., Zhu, S. P. C., Zhimulev, I., Coluzzi, M., della Torre, A., Roth, C. W., Louis, C., Kalush, F., Mural, R. J., Myers, E. W., Adams, M. D., Smith, H. O., Broder, S., Gardner, M. J., Fraser, C. M., Birney, E., Bork, P., Brey, P. T., Venter, J. C., Weissenbach, J., Kafatos, F. C., Collins, F. H., and Hoffman, S. L., 2002. The genome sequence of the malaria mosquito *Anopheles gambiae*. *Science* **298**:129–49.

Hudson, R. R., Sáez, A. G., and Ayala, F. J., 1997. DNA variation at the *Sod* locus of *Drosophila melanogaster*: an unfolding story of natural selection. *Proc. Natl. Acad. Sci. U.S.A.* **94**:7725–9.

Huff, C. A., 1938. Studies on the evolution of some disease-producing organisms. *Q. Rev. Biol.* **13**:196–206.

Hughes, A. L., 1992. Positive selection and interallelic recombination at the merozoite surface antigen-1 (MSA-1) locus of *Plasmodium falciparum*. *Mol. Biol. Evol.* **9**:381–93.

Hughes, A. L., 1993. Coevolution of immunogenic proteins of *Plasmodium falciparum* and the host's immune system. In *Mechanisms of Molecular Evolution*, N. Takahata and A. G. Clark, Eds., pp. 109–27. Sinauer Assoc., Sunderland, MA.

Hughes, A. L., and Hughes, M. K., 1995. Natural selection on *Plasmodium* surface proteins. *Mol. Biochem. Parasitol.* **71**:99–113.

Hughes, A. L., and Verra, F., 2001. Very large long-term effective population size in the virulent human malaria parasite *Plasmodium falciparum*. *Proc. R. Soc. London Ser. B-Biol. Sci.* **268**:1855–60.

Hughes, K. T., Cookson, B. T., Ladika, D., Olivera, B. M., and Roth, J. R., 1983. 6-Aminonicotinamide-resistant mutants of *Salmonella typhimurium*. *J. Bacteriol.* **154**:1126–36.

Kawamoto, F., Win, T. T., Mizuno, S., Lin, K., Kyaw, O., Tantular, I. S., Mason, D. P., Kimura, M., and Wongsrichanalai, C., 2002. Unusual Plasmodium malariae-like parasites in Southeast Asia. *J. Parasitol.* **88**:350–7.

Kemp, D. J., and Cowman, A. F., 1990. Genetic diversity of *Plasmodium falciparum*. *Adv. Parasitol.* **29**:75–133.

Kemp, D. J., Coppel, R. L., and Anders, R. F., 1987. Repetitive proteins and genes of malaria. *Ann. Rev. Microbiol.* **41**:181–208.

Krzywinski, J., and Besansky, N. J., 2002. Molecular systematics of Anopheles: from subgenera to subpopulations. *Ann. Rev. Entomol.* **27**:27.

Levine, N. D., 1988. *The Protozoan Phylum Apicomplexa*, Vol. 1. CRC Press, Boca Raton, FL.

Levinson, G., and Gutman, G. A., 1987. Slipped-strand mispairing: a major mechanism for DNA sequence evolution. *Mol. Biol. Evol.* **4**:203–21.

Li, W.-H., 1997. *Molecular Evolution*. Sinauer Assoc., Sunderland, MA.

Livingston, F. B., 1958. Anthropological implications of sickle cell gene distribution in West Africa. *Am. Anthropol.* **60**:533–60.

López-Antuñano, F., and Schumunis, F., 1993. Plasmodia of humans. In *Parasitic Protozoa*, J. Kreier, Ed., pp. 135–265. Academic Press, New York.

Manwell, R. D., 1955. Some evolutionary possibilities in the history of the malaria parasites. *Indian J. Malariol.* **9**:247–53.

Margulis, L., McKhann, H. I., and Olendzenski, L., 1993. *Illustrated Guide of Protoctista*. Jones & Bartlett, Boston.

Mason, S. J., Miller, L. H., Shiroishi, T., Dvorak, J. A., and McGinniss, M. H., 1977. The Duffy blood group determinants: their role in the susceptibility of human and animal erythrocytes to *Plasmodium knowlesi* malaria. *Brit. J. Haematol.* **36**:327–35.

Mattingly, P. F., 1965. In *Evolution of Parasites*, A. E. R. Taylor, Ed., pp. 29–45. Blackwell Scientific, Oxford.

McConkey, G. A., Waters, A. P., and McCutchan, T. F., 1990. The generation of genetic diversity in malarial parasites. *Ann. Rev. Microbiol.* **44**:479–98.

McCutchan, T. F., and Waters, A. P., 1990. Mutations with multiple independent origins in surface antigens mark the targets of biological selective pressure. *Immunol. Lett.* **25**:23–6.

McCutchan, T. F., Kissinger, J. C., Touray, M. G., Rogers, M. J., Li, J., Sullivan, M., Braga, E. M., Krettli, A. U., and Miller, L., 1996. Comparison of circumsporozoite proteins from avian and mammalian malaria: biological and phylogenetic implications. *Proc. Natl. Acad. Sci. U.S.A.* **93**:11,889–94.

Miller, L. H., Mason, S. J., Clyde, D. F., and McGinniss, M. H., 1976. The resistance factor to *Plasmodium vivax* in blacks. The Duffy-blood-group genotype, FyFy. *N. Engl. J. Med.* **295**:302–4.

Miller, L. H., Roberts, T., Shahabuddin, M., and McCutchan, T. F., 1993. Analysis of sequence diversity in the *Plasmodium falciparum* merozoite surface protein-1 (MSP-1). *Mol. Biochem. Parasitol.* **59**:1–14.

Miller, R. L., Ikram, S., Armelagos, G. J., Walker, R., Harer, W. B., Shiff, C. J., Baggett, D., Carrigan, M., and Maret, S. M., 1994. Diagnosis of *Plasmodium falciparum* infections in mummies using the rapid manual ParaSight-F test. *Trans. R. Soc. Trop. Med. Hyg.* **88**:31–2.

Morrison, D. A., and Ellis, J. T., 1997. Effects of nucleotide sequence alignment on phylogeny estimation: a case study of 18S rDNAs of apicomplexa. *Mol. Biol. Evol.* **14**:428–41.

Mu, J. B., Duan, J. H., Makova, K. D., Joy, D. A., Huynh, C. Q., Branch, O. H., Li, W. H., and Su, X. Z., 2002. Chromosome-wide SNPs reveal an ancient origin for *Plasmodium falciparum*. *Nature* **418**:323–6.

Powell, J. R., Petrarca, V., della Torre, A., Caccone, A., and Coluzzi, M., 1999. Population structure, speciation, and introgression in the *Anopheles gambiae* complex. *Parassitologia* **41**:101–13.

Qari, S. H., Collins, W. E., Lobel, H. O., Taylor, F., and Lal, A. A., 1994. A study of polymorphism in the circumsporozoite protein of human malaria parasites. *Am. J. Trop. Med. Hyg.* **50**:45–51.

Rich, S. M., and Ayala, F. J., 1998. The recent origin of allelic variation in antigenic determinants of *Plasmodium falciparum*. *Genetics* **150**:515–7.

Rich, S. M., and Ayala, F. J., 1999. Circumsporozoite polymorphism, silent mutations and the evolution of *Plasmodium falciparum*. Reply. *Parasitol. Today* **15**:39–40.

Rich, S. M., and Ayala, F. J., 2000. Population structure and recent evolution of *Plasmodium falciparum*. *Proc. Natl. Acad. Sci. U.S.A.* **97**:6994–7001.

Rich, S. M., Hudson, R. R., and Ayala, F. J., 1997. *Plasmodium falciparum* antigenic diversity: evidence of clonal population structure. *Proc. Natl. Acad. Sci. U.S.A.* **94**:13,040–5.

Rich, S. M., Licht, M. C., Hudson, R. R., and Ayala, F. J., 1998. Malaria's Eve: evidence of a recent bottleneck in the global *Plasmodium falciparum* population. *Proc. Natl. Acad. Sci. U.S.A.* **95**:4425–30.

Rich, S. M., Ferreira, M. U., and Ayala, F. J., 2000. The origin of antigenic diversity in *Plasmodium falciparum*. *Parasitol. Today* **16**:390–6.

Ricklefs, R. E., and Fallon, S. M., 2002. Diversification and host switching in avian malaria parasites. *Proc. R. Soc. London Ser. B-Biol. Sci.* **269**:885–92.

Sáez, A. G., Tatarenkov, A., Barrio, E., Becerra, N. H., and Ayala, F. J., 2003. Patterns of DNA sequence polymorphism at *Sod* vicinities in *Drosophila melanogaster*: unraveling the footprint of a recent selective sweep. *Proc. Natl. Acad. Sci. U.S.A.* **100**:1793–8.

Saul, A., 1999. Circumsporozoite polymorphisms, silent mutations and the evolution of *Plasmodium falciparum*. *Parasitol. Today* **15**:38–9.

Saul, A., and Battistutta, D., 1988. Codon usage in *Plasmodium falciparum*. *Mol. Biochem. Parasitol.* **27**:35–42.

Schug, M. D., Hutter, C. M., Noor, M. A., and Aquadro, C. F., 1998. Mutation and evolution of microsatellites in *Drosophila melanogaster*. *Genetica* **102–103**:359–67.

Sharakhov, I. V., Serazin, A. C., Grushko, O. G., Dana, A., Lobo, N., Hillenmeyer, M. E., Westerman, R., Romero-Severson, J., Costantini, C., Sagnon, N., Collins, F. H., and Besansky, N. J., 2002. Inversions and gene order shuffling in *Anopheles gambiae* and *A. funestus*. *Science* **298**:182–5.

Sherman, I. W., 1998. A brief history of malaria and the discovery of the parasite's life cycle. In *Malaria: Parasite Biology, Pathogenesis, and Protection*, I. W. Sherman, Ed., pp. 3–10. American Society of Microbiology Press, Washington, D.C.

Sinnis, P., and Wellems, T. E., 1988. Long range restriction maps of *Plasmodium falciparum* chromosomes: crossing over and size variation in geographically distant isolates. *Genomics* **3**:287–95.

Slatkin, M., and Hudson, R. R., 1991. Pairwise comparisons of mitochondrial DNA sequences in stable and exponentially growing populations. *Genetics* **129**:555–62.

Su, X., and Wellems, T. E., 1996. Toward a high-resolution *Plasmodium falciparum* linkage map: polymorphic markers from hundreds of simple sequence repeats. *Genomics* **33**:430–44.

Tanabe, K., Mackay, M., Goman, M., and Scaife, J. G., 1987. Allelic dimorphism in a surface antigen gene of the malaria parasite *Plasmodium falciparum*. *J. Mol. Biol.* **195**:273–87.

Tishkoff, S. A., Varkonyi, R., Cahinhinan, N., Abbes, S., Argyropoulos, G., Destro-Bisol, G., Drousiotou, A., Dangerfield, B., Lefranc, G., Loiselet, J., Piro, A., Stoneking, M., Tagarelli, A., Tagarelli, G., Touma, E. H., Williams, S. M., and Clark, A. G., 2001. Haplotype diversity and linkage disequilibrium at human G6PD: recent origin of alleles that confer malarial resistance. *Science* **293**:455–62.

Trigg, P. I., and Kondrachine, A. V., 1998. The current global malaria situation. In *Malaria: Parasite Biology, Pathogenesis, and Protection*, I. W. Sherman, Ed., pp. 11–22. ASM Press, Washington, D.C.

Van de Peer, Y., Van der Auwera, G., and De Wachter, R., 1996. The evolution of stramenopiles and alveolates as derived by "substitution rate calibration" of small ribosomal subunit RNA. *J. Mol. Evol.* **42**:201–10.

Vivier, E., and Desportes, I., 1989. Apicomplexa. In *Handbook of Protoctista*, L. Margulis, J. O. Corliss, M. Melkonia, and D. J. Chapman, Eds., pp. 549–73. Jones & Bartlett, Boston.

Volkman, S. K., Barry, A. E., Lyons, E. J., Nielsen, K. M., Thomas, S. M., Choi, M., Thakore, S. S., Day, K. P., Wirth, D. F., and Hartl, D. L., 2001. Recent origin of *Plasmodium falciparum* from a single progenitor. *Science* 293:482–84.

Walton, C., Handley, J. M., Tun-Lin, W., Collins, F. H., Harbach, R. E., Baimai, V., and Butlin, R. K., 2000. Population structure and population history of *Anopheles dirus* mosquitoes in Southeast Asia. *Mol. Biol. Evol.* 17:962–74.

Weisenfeld, S. L., 1967. Sickle-cell trait in human biological and cultural evolution. Development of agriculture causing increased malaria is bound to gene-pool changes causing malaria reduction. *Science* 157:1134–40.

Zevering, Y., Khamboonruang, C., and Good, M. F., 1998. Human and murine T-cell responses to allelic forms of a malaria circumsporozoite protein epitope support a polyvalent vaccine strategy. *Immunology* 94:445–54.

FOUR

Evolutionary Biology of Malarial Parasites

Ananias A. Escalante and Altaf A. Lal

INTRODUCTION

Analysis of *Plasmodium* spp. genome sequences would allow for the discovery of thousands of new genes and proteins, providing a unique opportunity for understanding the complex biology of malarial parasites. However, the identification of these new genes will be followed by the same old questions that researchers have faced for nearly three decades: how variable are the newly identified genes? How is such variation generated and maintained? How can this diversity affect intervention efforts?

Understanding the origin and extent of malarial parasites' genetic diversity, and the implications of these on the development of new intervention strategies, requires a close collaboration between biomedical researchers and evolutionary biologists. In the case of malaria research, as is the case in the study of many other infectious diseases, biomedical researchers/public health professionals and evolutionary biologists have traditionally worked in isolation.

Biomedical researchers and public health managers often seek answers to the following questions: What are the specific clinical end points in malaria? What level of efficacy can be expected from a multivalent vaccine? How does drug-resistance emerge? How quickly do drug-resistant parasites disperse? What is the impact of transmission pressure on the dispersal of drug-resistant parasites? Is it better to use one drug at a time or to use a "cocktail" before any of the drugs become ineffective? What is the impact of bed-nets on the selection of parasite lines?

Evolutionary biologists, who are interested in similar issues, would interpret these questions as follows: How does natural selection favor the fixation of a given mutation? What is the gene flow among populations?

What is the population genetic structure of the parasite? What are the dynamics between the genetic diversities of the parasite and the host? Do the intra-host dynamics relate to virulence? Do the transmission dynamics affect the recombination rate in the parasite population? Is a gene under positive selection or is the variation observed neutral?

These two perspectives cannot be reduced to semantic discrepancies. On the other hand, a joint approach offers fertile ground for finding answers to longstanding problems.

In recent years only a few studies have integrated molecular population genetics and evolutionary biology of human malaria parasites with end points of public health importance. For example, there are relatively few studies on parasite genetic diversity incriminating the host immune response as positive selective pressure maintaining the observed polymorphism (Conway 1997; Conway et al. 2000; Hughes and Hughes 1995; Escalante et al. 1998a, 2001, 2002a,b; Polley and Conway 2001). Even more noticeable is the fact that none of the field molecular epidemiologic studies on the origin of drug resistance includes any consideration of parasite population structure or effective population size, two concepts that are essential to understanding the probability of fixation and dispersion of a gene.

In this chapter, we review the current status of molecular evolutionary and population genetics research of malarial parasites. First, we review information on molecular systematic studies and discuss their implications for understanding the origin and phenotypic plasticity of malarial parasites. Second, we review population genetics studies, especially those aimed at understanding the parasite population structure and its genetic diversity. Finally, we discuss what we believe should be the research agenda of molecular evolutionary studies in malaria from a public health perspective, specifically, the need for longitudinal studies that link population-based information with clinical end points.

MOLECULAR PHYLOGENETIC STUDIES ON *PLASMODIUM*

Plasmodial parasites infect a wide variety of vertebrate hosts, including reptiles, birds, and mammals. *Plasmodium* life cycles in mammals are characterized by an invasive haploid stage, the sporozoite, which is inoculated into the vertebrate host by the bite of an infected female anopheline mosquito during its blood meal. These sporozoites are carried to the liver, where they become hepnozoites and further develop into liver merozoites, which start the erythrocytic life stage. Some merozoites differentiate into gametocytes, which may be taken up with the mosquito's blood meal. The zygote is formed in the mosquito where it develops into the ookinete and further differentiates into the oocyst, the only diploid stage of the parasite life cycle. Meiosis takes place in the oocyst, resulting in the formation of haploid sporozoites (Coatney et al. 1971).

Four recognized *Plasmodium* species are parasitic to humans: *P. falciparum*, *P. malariae*, *P. ovale*, and *P. vivax*. Of the four species, *P. falciparum* and *P. vivax* cause almost all of the disease, with *P. falciparum* causing the most severe clinical manifestations (WHO 1997). A potential fifth species termed *P. vivax*–like has been identified on the basis of molecular evidence (Qari et al. 1993a,b). The circumsporozoite protein of *P. vivax*–like is identical to that of *P. simiovale*, a parasite from a nonhuman primate.

Understanding the origin of human malaria and the study of plasmodial parasites in animal models motivated early taxonomic investigations (Coatney et al. 1971). However, formal molecular phylogenetic studies did not appear until the early 1990s when a substantial amount of genetic data on housekeeping and antigen genes became available.

The question we need to ask is why do we need to infer the phylogeny of malarial parasites. Phylogenetic information provides the basis for modern comparative studies, as well as for the identification of potential new pathogens and the development of animal models. In the specific case of malarial parasites, molecular phylogenies have shown that host switching is possible on a time scale that may have public health implications (Qari et al. 1993a; Escalante et al. 1995).

Comparative genetic studies led to the discovery of a *P. vivax*–like parasite showing that the entity actually called *P. vivax* could be more than a single species (Coatney et al. 1971; Qari et al. 1993a,b; Escalante et al. 1995). In addition, comparative studies have allowed for a better understanding of the genetic diversity of human malarial parasites (Escalante et al. 1998a, 2001, 2002a).

Despite significant progress in phylogenetic studies of malarial parasites and other Apicomplexa, our knowledge is still incomplete. Information on several species has not progressed beyond their original description, lacking data on host range and other basic biological characteristics. Remarkable interest in biodiversity during the past two decades has generated several fauna surveys, but they seldom included parasitic protozoa or parasites of any kind. The lack of coordination between veterinarians working with wildlife and biomedical researchers has not allowed the sampling of valuable specimens that could be helpful in understanding the origin of human malaria, as well as the general evolutionary history of this important group of parasites (Wolfe et al. 1998). In addition, the expertise to properly identify and describe species is affected by the limited number of trained protozoologists. All these limitations have negative impacts on evolutionary studies of malaria.

The better known groups are the malarial parasites of mammals, even though there are complete groups for which we do not have basic information, such as *Plasmodium* from gibbons and lemurs. Molecular and biologic data from avian and lizard parasites are becoming available, providing valuable information for better understanding the evolution of malarial parasites (McCutchan et al. 1996; Bensch et al. 2000; Perkins 2000; Ricklefs and Fallon 2002).

Initial phylogenetic investigations focused on the origin of *P. falciparum* (Waters et al. 1991; Escalante and Ayala 1994; Escalante et al. 1995, 1997; Qari et al. 1996; McCutchan et al. 1996). The two genes originally used for such studies were the 18S small subunit ribosomal RNA (18S SSU rRNA), expressed in the asexual stage, and the circumsporozoite protein (CSP) gene. Phylogenetic trees obtained with these genes were congruent (Escalante and Ayala 1994, 1995; Escalante et al. 1995; McCutchan et al. 1996); however, no appropriate outgroup was included in initial analyses (Siddall and Barta 1992; Escalante and Ayala 1994). The basic lesson learned from these early phylogenetic studies is that the four *Plasmodium* species parasitic to humans arose independently as human pathogens. Independent origin is clear even for parasites such as *P. falciparum* and *P. vivax*, whereas early literature suggested that *P. ovale* could have been derived from *P. vivax*; this hypothesis was completely ruled out by the molecular data (Qari et al. 1996).

One concern in the use of the 18S SSU rRNA is the complex pattern of expression of this gene in *Plasmodium* (McCutchan et al. 1995). Report of partial gene conversion among the 18S SSU rRNA genes expressed during different parasite stages (Corredor and Enea 1993) raised doubts regarding phylogenetic trees derived from this gene. Nonhomologous copies of genes, which are subjected to low levels of conversion, can provide misleading phylogenetic results (Escalante et al. 1997).

The CSP, a polymorphic and immunogenic surface protein expressed at the sporozoite stage, is considered to be under selective pressure to accumulate polymorphism (Hughes 1991; Escalante et al. 2002b). Because natural selection could affect the observed phylogeny, this gene is not appropriate for phylogenetic studies. In addition, its tandem repeat motifs do not allow for the alignment of the full-length CSP gene (Escalante et al. 1995; McCutchan et al. 1996), and homologous genes for the CSP have not been identified in other Apicomplexa, making it difficult to investigate the relationships of *Plasmodium* with other genera.

There are several genes from which phylogenetic information can be derived, such as the genes encoding the caseinolutic protease ClpC from the plastid (Rathore et al. 2001) and the adenylosuccinate lyase (Kedzierski et al. 2002). However, they have essentially reproduced the results

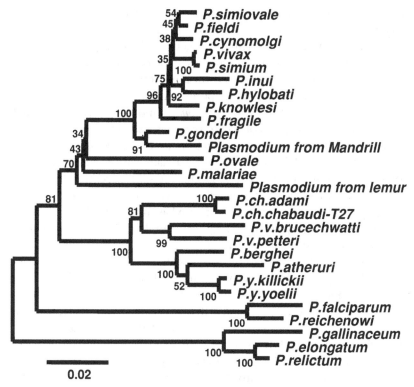

Figure 4.1. Phylogenetic tree inferred from the gene encoding cytochrome b. The tree was estimated using the neighbor-joining (NJ) method on a Tamura distance matrix using only v. The numbers on the nodes of the tree are percentages of bootstrap values based on 1,000 pseudo-replications.

obtained from the CSP and the SSU rRNA given the limited number of species from which data are currently available.

Cytochrome b has several advantages over antigens such as CSP or genes with a complex pattern of evolution such as the SSU rRNA. First, most of the substitutions observed are synonymous, suggesting that this gene is not under positive selection (Escalante et al. 1998b). Second, cytochrome b has approximately the same A + T content across several *Plasmodium* species, reducing the risk of potential artifacts due to convergence in base composition given differences in A + T content among the nuclear genomes of *Plasmodium* lineages. Finally, there are homologous genes in other Apicomplexa, making it a suitable target for phylogenetic studies.

The phylogeny of cytochrome b is depicted in Figure 4.1. The phylogeny was estimated using the neighbor-joining (NJ) method on a Tamura distance matrix using only transversions. The bootstrap values were calculated using 1,000 pseudo-replicates. Biologic characteristics of the species are described in Table 4.1. This phylogeny includes not only primate malarial parasites but also rodent and a sample of lemur parasites.

The Southeast Asian parasites form a monophyletic group, together with *P. vivax* and *P. simium*. These species form a sister group of two primate parasite, species from Africa, *P. gonderi* and a *Plasmodium* isolated from a mandrill. The group of rodent parasites also form a monophyletic group, including the species *P. atheruri* from porcupine. Finally, a *Plasmodium* species obtained from lemurs forms a clade with the other primate parasites, making a monophyletic group, the only exception being the *P. falciparum–P. reichenowi* lineage, which diverged early on in the evolution of the genus. Avian parasites were used as the root in the phylogeny. Even though this phylogeny is relatively simple, it is very informative regarding the evolutionary history of this genus.

Traditionally, the origin of primate malarial parasites, with the exception of *P. falciparum*, has been placed in Southeast Asia, given the high diversity of malarial parasites found in that region. In the case of *P. vivax* generally, it had been accepted that this parasite originated in Asia (Garnham 1966; Coatney et al. 1971). However, this conclusion was controversial given the high frequency in Africa of a Duffy blood group allele that does not allow for invasion of erythrocytes by *P. vivax* (Livingstone 1984). Assuming that the Duffy blood group was selected by *P. vivax*, the high frequency suggests a long association of the African human population with this parasite.

Our phylogenetic analysis solves this apparent controversy. First, our results suggest an African origin for primate malarial parasites. The finding of common ancestors for Southeast Asian primate parasites in African primate parasites may explain why the Duffy polymorphism has been maintained in African primates, including man (Palatnik and Rowe 1984). Parasites similar to *P. vivax* may have been associated with African primates before the lineage leading to *P. vivax* was introduced into Southeast Asia. That is, the observation of ancestors of the lineage leading to *P. vivax* in Africa does not reject the origin of *P. vivax* as a human-infecting parasite in Asia. The lineage of *P. vivax* could have been introduced into Asia by any of the primates that also had their origin in Africa (Shoshani et al. 1996) before the host switch from nonhuman primates to humans in Asia. Indeed, our most recent investigation includes two species of Orangutan parasites, *P. selvaticum* and *P. pitheci*, as part of this radiation event (Wolfe et al. submitted personal communication). Our data indicate that the lineage that originated as *P. vivax* was part of the parasite speciation process that

Table 4.1. Biologic characteristics and geographic distribution of malarial parasites

Plasmodium Species	Natural Hosts	Known Geographic Range	Periodicity	Relapse
falciparum	*H. sapiens*	Tropics worldwide	Tertian	No
vivax	*H. sapiens*	Tropical and subtropical regions	Tertian	Yes
malariae	*H. sapiens*	Tropical and subtropical regions	Quartan	No
ovale	*H. sapiens*	Tropics of Asia and Africa	Tertian	No
cynomolgi	Multiple macaque species, *Presbytis cristatus* *P. entellus*	India, Sri Lanka, Malaysia, Indonesia (Java and the Celebes), Assam	Tertian	Yes
fieldi	Multiple macaque species	Malaysia	Tertian	Yes
cragile	*Macaca radiata*	Sri Lanka, India	Tertian	No
hylobati	*H. moloch*	East Malaysia	Tertian	U
inui	Multiple macaque species, *P. obscura, P. cristatus, P. melalophos*	Sri Lanka, India, Malaysia, Philippines, Indonesia, Taiwan	Quartan	No
knowlesi	*M. nemestrina, P. melalophus*	Malaysia	Quotidian	No
simiovale	*M. sinica*	Sri Lanka	Tertian	Yes
gonderi	*Cercocebus sp* and *Mandrillus leucophaceus*	Central Africa	Tertian	No

(continued)

81

Table 4.1 (continued)

Plasmodium Species	Natural Hosts	Known Geographic Range	Periodicity	Relapse
Plasmodium sp1	M. leucophaceus	Gabon	*	*
simium	Alouatta fusca, Brachiteles arachinoides	Brazil	Tertian	U
Plasmodium sp2	Lemur macaco	Madagascar	*	*
berghei	Grammomys surdaster	Central Africa	24h	NA
chabaudi	Thamnomys rutilans	Central Africa	24h	NA
vinckei	T. rutilans	Central Africa	24h	NA
yoelii	T. rutilans	Central Africa	24h	NA
gallinaceum	Gallus gallus	Asia	36h	NA
elongatum	Passer domesticus	North America	24h	NA
relictum	Several avian species	Worldwide	36h	NA

Note: *, Insufficient information; NA, not applicable.

took place in the region over the past 2–3 million years (Escalante et al. 1998b).

Another important finding derived from the cytochrome b phylogeny is that disease characteristics, such as periodicity and virulence, have little value in drawing evolutionary inferences about the phylogenetic relationships of malarial parasites. For instance, the length of periodicity is a convergent characteristic. The quartan parasites *P. inui* and *P. malariae* do not form a monophyletic group; *P. knowlesi*, the only quotidian primate parasite (24-hour erythrocytic cycle) shares this characteristic with several species parasitic to birds (Hewitt 1940). Another example is the capacity for relapse. Parasites that share this characteristic do not form a monophyletic group.

The original evolutionary inferences regarding the origin of *P. falciparum* malaria considered its high virulence to be evidence of its recent origin as a human parasite (Boyd 1949). Our phylogenetic analysis shows no evidence in malaria that indicates an evolutionary trend toward reduced virulence with age of the host–parasite relationship. There is evidence of host switching between humans and monkeys, as is the case with *P. simium/P. vivax* and *P. malariae/P. brasilianum* (Escalante et al. 1995, 1997), but no differences in severity of disease have been observed (Coatney et al. 1971). Virulence is the result of genetic and ecological interactions between the host and the parasite, and in the case of malaria, an important factor is the intra-host dynamics of a genetically diverse parasite population (Ewald 1994; Ebert and Herre 1996; Gupta and Hill 1995; Read and Taylor 2001).

Regarding overall phylogenetic relationships among the different malarial parasites, our phylogenetic analysis allowed us to conclude that rodent parasites are a monophyletic group but they are not related to the lemur parasites as the classical taxonomy suggested. On the contrary, the location in the phylogeny of the lemur sequence suggests that all primate parasites, with the exception of *P. falciparum* and *P. reichenowi*, derived from a common ancestor. However, lack of information on parasites from lemurs does not allow us to be more conclusive about this statement.

Our analysis also suggests that the lineage of *P. falciparum–P. reichenowi* is ancient in the evolutionary history of this parasite. This result is consistent with the analysis of partial sequences with the inclusion of a broader sample of avian malarial parasites. Ricklefs and Fallon (2002) found mammalian parasites to be a monophyletic group. This group included *Plasmodium* spp and at least one species of the genus *Hepatocystis*. This finding suggests that the genus is paraphyletic, which is in agreement with

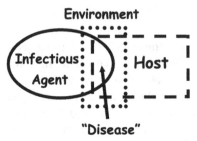

Figure 4.2. Diagram of the host–parasite interaction: a definition of disease based on the concepts of phenotype and genotype.

our findings, although we lacked an appropriate outgroup for rooting the tree in our analyses. The analysis made by Ricklefs and Fallon (2002) also showed an early divergence of the lineage leading to *P. falciparum* and *P. reichenowi* in the group of *Plasmodium* parasitic to mammals. This ancient origin of the *P. falciparum* lineage could partially explain some "avian-like" characteristics in this lineage as primitive characteristics in the group.

The fact that *P. falciparum* diverged early in the evolution of mammalian parasites makes it difficult to extrapolate and apply information derived from animal models to the understanding of human falciparum malaria. Malaria, as with any disease, can be seen as a complex phenotype that is the product of a genetically diverse population of parasites interacting with a genetically diverse population of hosts in a given environment (Figure 4.2). If we keep this schema in mind, we can correlate the plasticity in life histories observed in *Plasmodium*, as evidenced by our phylogenetic analysis (Escalante et al. 1998b), with the available biologic information (Coatney et al. 1971). This kind of comparison allows us to see that some clinical manifestations of disease are convergent phenotypes that may have very different genetic and immunologic bases. This convergence affects the notion of a "laboratory model" for malaria, especially in the case of *P. falciparum* which has no closely related species of *Plasmodium* parasitic to closely related hosts. In the case of *P. vivax*, because several closely related species of *Plasmodium* are natural parasites of hosts closely related to humans, such as primates, the issue is still a valid concern that needs to be evaluated. Comparative immunology and parasite biology will allow us to assess the extent to which information derived from animal models can be used to understand human malaria.

POPULATION GENETIC STUDIES OF *PLASMODIUM*

Studies of genetic diversity at the molecular level have focused mostly on *P. falciparum*. Two approaches have been followed:

(a) understanding the stability of strains, whether they are defined as multi-locus association of genetic markers or phenotypic entities with distinctive biological and immunological characteristics; and

(b) assessing the diversity of genes that encode vaccine antigens and those involved in drug resistance.

Studies focusing on genes provide information about how alleles are generated and maintained in the population. The "strain" approach usually relates to the issue of parasite genetic structure: it quantifies how overall genetic diversity is organized at the population level. These two approaches answer key questions related to the development and deployment of intervention strategies in malaria control programs. The term "strain," however, has been used in several contexts with very fuzzy boundaries, making it difficult to distinguish among isolates, strains, and alleles.

Specifically, in terms of the parasite population structure, the term "strain" refers to a population subdivided into units that have some temporal and spatial stability, usually identified as multi-locus genotypes (Tibayrenc et al. 1990; Tibayrenc and Ayala 2002). The term "strain" is also used in the context of a phenotype whose transmissibility can be measured (Gupta et al. 1994). These two concepts of "strain" are not exchangeable even when they can be related due to the action of natural selection (Gupta et al. 1996).

A population subdivided into genetically stable entities will be effectively clonal from a population genetic point of view. This clonal population structure does not imply lack of genetic sex, only that it is not enough to break the stability of these genetic lineages. The original notion of a clonal population structure for malarial parasites was highly controversial (Tibayrenc and Ayala 2002) and, as we discuss below, has turned into a sophisticated conceptual model in which transmission dynamics and inbreeding are taken into account (Awadalla et al. 2001). Nevertheless, the idea of detecting multi-locus associations motivated several molecular epidemiologic studies, changing our perspective of malarial parasites, genetic diversity, and population dynamics.

The currently held opinion is that, given the fact that meiosis takes place in the vector as part of the *Plasmodium* life cycle, the rate of genetic recombination and the effective population size are linked to transmission intensity. Studies carried out on *P. falciparum* have shown high genetic diversity and have failed to detect multi-locus associations in Africa, which has a high transmission rate. On the other hand, researchers have found low genetic diversity and strong linkage disequilibria in areas with low transmission, probably due to high inbreeding (Paul et al. 1995; Babiker and Walliker 1997; Arez et al. 1999; Anderson et al. 2000; Urdaneta et al. 2001; Tami et al. 2002). These geographic differences in heterozygocity can be observed in gene-encoding antigens (Escalante et al. 2001, 2002a,b).

An alternative explanation for differences in heterozygocity found in natural populations is that *P. falciparum* most likely originated in Africa (Escalante et al. 1998b), so the derived populations in Southeast Asia and the Americas may exhibit less genetic diversity due to bottlenecks during the colonization process. Geographic distribution of transmission intensities correlates with the historical process of malaria dispersion throughout the world; thus both scenarios make the same predictions. Nevertheless, the issue from a public health perspective, no matter which factor is more important, is that African populations exhibit overall more genetic diversity and this variability needs to be taken into account in vaccine development and testing. However, there are some exceptions, which we discuss later in this chapter (Escalante et al. 2002b).

Our perspective on what a malaria infection [is] has also changed in the past decade. Malaria infections are most often a mixture of different genetic types of *Plasmodium* parasites. It is widely accepted that complex intra-host dynamics cannot be reduced to a measure of parasitemia, and understanding these dynamics is considered essential in investigating the origin of drug-resistant parasites (Hastings and Mackinnon 1998; Hastings and D'Alessandro 2000; Hastings et al. 2002), in unveiling the interaction of the host immune system with parasite genetic variants (Ekala et al. 2002a,b; Eisen et al. 1998; Branch et al. 2001; Read and Taylor 2001), and in understanding the role played by the parasite genetic diversity in malarial parasites' virulence and disease manifestations (Read and Taylor 2001).

In terms of assessing nucleotide diversity at specific genes, information is limited for *P. falciparum*. Early studies on genetic diversity lacked formal population genetic analyses and included samples with a strong geographic bias, with a poor representation of isolates from Africa where most of the *P. falciparum*–associated mortality takes place. Even with limited sample size, it was obvious from these early investigations that *P. falciparum* exhibits an extensive amount of genetic diversity (McCutchan et al. 1988). Our own investigation of 10 loci from *P. falciparum* suggests that loci encoding proteins expressed at the surface of the sporozoite and merozoite are more diverse than those expressed during sexual stages or inside the parasite (Escalante et al. 1998a). These results agree with the general observation that stage-specific surface proteins will show high polymorphism when compared with internal antigens (McCutchan et al. 1988).

However, in terms of vaccine effectiveness, the most important issue is whether the vaccine-elicited immune response will be equally effective against all genetic variants found in nature. The first step in answering this question is to assess the nature and extent of the genetic variation in order to explore the factors involved in its maintenance. These kinds of studies demand the use of concepts and tools developed in population genetics.

High levels of genetic diversity observed in vaccine candidate antigens have been attributed to the action of natural selection imposed by the host immune system (Anders and Saul 1994). The logic behind this premise is as follows: Given that these genes encode antigenic proteins that elicit an immune response, the observed high levels of heterozygosity and rates of evolution can be accounted for by the accumulation of suitable mutations that allows the parasite to escape host immune recognition. Paradoxically, this widely accepted premise did not translate into an extensive literature of formal analyses using the neutral theory as the null hypothesis (Hughes 1991, 1992; Hughes and Hughes 1995; Conway 1997; Escalante et al. 1998a).

The neutral model states that genetic polymorphism is the result of random accumulation of mutations with minimal or no effect on the fitness of organisms (Kimura 1983). The neutral model makes several predictions, among them a faster random accumulation of synonymous substitutions if they are not under any constraint (Kimura 1983). Synonymous substitutions are likely to be neutral, or nearly so, whereas nonsynonymous substitutions may be functionally constrained and thus subject to negative natural selection and elimination from the population (Kimura 1977). An excess of nonsynonymous substitutions can be taken as evidence that these mutations may confer an adaptive advantage; therefore they are maintained by natural selection.

The bottom line is that selection acts on phenotypes so, in the absence of any other constraint, only mutations that change the protein can be the target of positive or negative selection. This link between selection and a phenotype is not a trivial matter when we try to explain how a given polymorphism is under positive natural selection in the context of vaccine development; appropriate immunologic information is available for a very limited number of antigens. Even though there is a paucity of immunologic evidence, natural selection is often considered the driving force behind the extensive polymorphism observed in some regions of the genome of malarial parasites. This premise has gained acceptance due to the limited number of polymorphic sites in other regions of the genome that do not encode antigens (Rich et al. 1998; Hartl et al. 2002; Mu et al.

2002), opening up the possibility of exploring patterns using genome-wide analysis of polymorphism that could detect the presence of proteins involved in the host–parasite immunologic interplay (Wootton et al. 2002). One way to validate the available tools used to incriminate the immune response as the driving selective force maintaining a given polymorphism is to investigate polymorphisms from which immunologic information is available so that we can establish the methodology necessary to identify variants of immunologic interest.

The ratio of synonymous to nonsynonymous substitutions is the most widely used criterion in malarial parasites for detecting positive natural selection. The nonsynonymous to synonymous substitution ratio has been used in *Drosophila* and other organisms for testing neutrality (Kreitman and Akashi 1995; Endo et al. 1996; Otha 1996). In the specific case of *P. falciparum*, a complication is that synonymous substitutions appear to be under selection by the strong codon bias and A + T richness of its genome (Escalante et al. 1998a). However, from comparative analysis with *P. reichenowi*, we can see how this constraint cannot account for the observed paucity of synonymous substitutions in *P. falciparum* antigens as we discuss below.

In addition to the synonymous to nonsynonymous substitution ratio, there are a variety of tests for detecting departures from neutrality that are less conservative (Kreitman and Akashi 1995). It is important to note that each test has its own assumptions and power limitations. The use of tests for detecting departure from neutrality needs to be done carefully because several alternative scenarios, that may not have any immunologic meaning, can reject the null hypothesis of neutrality. These tests could be sensitive to strong overdominance (heterozygous advantage), selective sweeps (favorable mutations sweep through the population and reach fixation), and the hitchhiking effect (during a selective sweep of a favorable allele any linked neutral allele goes along), as well as confounding factors such as population structure and population growth. Given these limitations, we believe that it is important to correlate the observed pattern of heterozygocity with immunologic information whenever possible. That is, we need to go back to the basics and use immunologic evidence to identify the potential phenotype on which selection is acting.

We discuss evidence for positive natural selection on *P. falciparum* antigens. Specifically, we present our studies on the *P. falciparum* genes encoding the apical membrane antigen 1 (AMA-1), the circumsporozoite protein (CSP), and the major surface protein 1 (MSP-1). The finding that immunologic and genetic evidence correlates across several genes validates the use of population genetics tools to study heterozygocity patterns that may be associated with eliciting a host immune response, a basic premise of genome-wide analysis directed at detecting new targets for vaccine development.

The Apical Membrane Antigen 1 (AMA-1)

The AMA-1, also known as PF83, is a protein of 622 residues and a molecular weight of 83 KDa, with three major domains defined by eight disulfide bonds (Deans et al. 1988; Hodder et al. 1996). This antigen first appears in the apical complex and then migrates to the merozoite's surface. Data from animal models suggest that this protein elicits protective immune responses (Deans et al. 1988; Kocken et al. 1999).

Table 4.2 presents estimates of genetic diversity in the predicted and naturally immunogenic T and B cell epitopes of AMA-1 from a sample of 44 sequences from around the globe (Escalante et al. 2001). All epitopes are polymorphic and have more nonsynonymous than synonymous substitutions, the only exception being the epitope located within the residues 571–588 in the cytoplasmic tail, which exhibits more synonymous than nonsynonymous replacements. The observation that there are more nonsynonymous than synonymous substitutions may be suggestive of positive natural selection acting on these epitopes; however, this observation alone is not a formal test of neutrality, at least until statistical analyses are performed using the neutral theory as the null hypothesis.

We compare the rate of synonymous (Ks) versus nonsynonymous substitutions (Kn). Under neutrality, it is expected that Ks/Kn will be equal to or higher than 1. This test is not affected by population structure or population growth. Table 4.3 presents estimates of synonymous and nonsynonymous substitutions per site across AMA-1 domains and compares them with *P. reichenowi*. Overall, the number of nonsynonymous substitutions is almost three times the number of synonymous substitutions within *P. falciparum*, a difference that is statistically significant ($p < 0.05$ for a Z test as suggested by Kumar and Nei 2000); that is, Ks/Kn < 1 (Table 4.3). However, the slow accumulation of synonymous substitutions could be the result of strong codon bias in the *P. falciparum* genome (Escalante et al. 1998a); that is, the rejection of the null hypothesis of Ks/Kn $\geq$ 1 expected under neutrality may be a measure of selective constraints imposed by the genome on silent substitutions rather than positive selection of polymorphisms leading to amino acid changes. In order to explore if the paucity of synonymous substitutions can be accounted for by codon bias, a comparison with *P. reichenowi* was performed.

Contrary to the pattern observed within *P. falciparum*, the divergence of the human parasite alleles from *P. reichenowi* exhibits, overall, more

Table 4.2. Polymorphism observed in the T and B cell epitopes of the *AMA-1* gene in *P. falciparum*

Residues	Sequence	Epitope	π(SD)	Kn	Nns	Ns
14–35	EFTYMINFGRGQNYWEHPYQNS	T	0.00899(0.00106)	0.01033	2	1
41–57	INEHREHPKEYEYPLHQ	B	0.01857(0.00265)	0.02274	6	0
259–271	GPRYCNKDESKRN	T	0.01694(0.00248)	0.02099	3	0
279–288	AKDISFQNYT	B & T	0.04031(0.00367)	0.05028	3	0
317–334	DGNCEDIPHVNEFPAIDL	B	0.00967(0.00318)	0.01101	6	1
321–338	EDIPHVNEFPAIDLFECN	B	0.00883(0.00295)	0.00981	5	1
348–366	DQPKQYEQHLTDYEKIKEG	T	0.00159(0.00105)	0.00196	2	0
444–461	SLYKDEIMKEIERESKRI	T	0.01218(0.00266)	0.01523	3	0
571–588	GNAEKYDKMDEPQDYGKS	B & T	0.01987(0.00216)	0.02041	2	4

Note: Sequence, sequence on the clone 3D7; underlined residues correspond with amino acid substitutions. π(SD), nucleotide diversity and its standard deviation within parentheses; Kn, nucleotide diversity of nonsynonymous substitutions per nonsynonymous site using the Nei & Gojobori method; Nns, number of nonsynonymous mutations; Ns, number of synonymous mutations; B, B cell epitope; T, T cell epitope.

Table 4.3. Polymorphism found on different regions of the *AMA-1* gene in *P. falciparum*

Region	Residues	Ks	Kn	Ks(Pre)	Kn(Pre)
Extracellular	1–546	0.00546	0.01926	0.07121	0.06250
Domain I	149–302	0.00505	0.03184	0.05469	0.05528
Domain II	320–420	0.00157	0.00916	0.01802	0.02238
Domain III	443–490	0.00000	0.01111	0.03912	0.03944
Transmembrane	547–567	0.00000	0.00000	0.00000	0.04800
Cytoplasmic Tail	568–622	0.01977	0.01596	0.05851	0.04897
Variable Region 1	160–210	0.00884	0.06026	0.05856	0.08225
All AMA-1	1–622	0.00463	0.01668	0.05774	0.04989

Note: Ks and Kn are the number of synonymous and nonsynonymous substitutions, respectively, estimated using the Nei & Gojobori method with the Jukes & Cantor correction. Ks(Pre) and Kn(Pre) are the synonymous and nonsynonymous divergence, respectively, from *P. reichenowi* using the Jukes & Cantor correction.

synonymous than nonsynonymous substitutions (Table 4.3). Because *P. reichenowi* has a G + C content and codon bias similar to *P. falciparum* (Escalante et al. 1998a), the fact that its divergence accumulates more synonymous than nonsynonymous substitutions can be taken as evidence that codon bias is not a significant constraint that can account for the slow accumulation of synonymous substitutions within *P. falciparum*, supporting the theory that positive natural selection is indeed acting on the observed polymorphism.

Second, we applied the D* and F* tests developed by Fu and Li (1993). They rejected the null hypothesis that the observed polymorphism can be explained by the neutral model. The D* test was significant, with a value of -2.42814 ($p < 0.05$), and the F* test was very close to significance with a value of -2.00450 with $0.10 > p > 0.05$. These tests are sensitive to a selective sweep in the population, that is, they could be detecting favorable mutations sweeping through the population and reaching fixation. However, they are also sensitive to population growth. To determine if there were specific regions of the gene that showed significant departure from neutrality, a sliding window analysis was performed using the D* test (Figure 4.3). The value of the statistic D* was calculated on a window of 120 basepairs, moving it in steps of 30 sites. D* values were plotted against the midpoint of the window. The polymorphism shows significant departures from neutrality in three major regions, between sites 61 and 180, 335 and 544, and between sites 875 and 1174. The regions between sites 61 and 180, and between sites 875 and 1174 include seven of the nine epitopes reported for AMA-1 (Lal et al. 1996), making evident

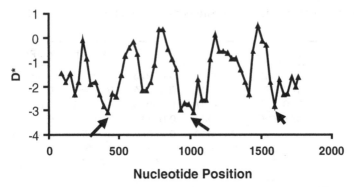

Nucleotide Position

Figure 4.3. Sliding window analysis of the test D* (Fu and Li 1993) gene using a window length of 120 basepairs and a step size of 30 basepairs. D* tests the hypothesis that all mutations are selectively neutral, using *P. reichenowi* as an outgroup.

that regions involved in the rejection of the null hypothesis are those that elicit a strong immune response (Escalante et al. 2001).

In summary, the analysis presented on the AMA-1 allowed for the identification of extensive polymorphism in its B and T cell epitopes. We also found that the polymorphisms observed in seven of the nine epitopes are maintained by positive natural selection, as indicated by the D* sliding window analysis (Figure 4.3), an observation corroborated by the comparison of the synonymous–nonsynonymous substitution rate. In the latter test, the possibility that synonymous substitutions could be under constraint was ruled out by comparison with *P. reichenowi*.

The Circumsporosoite Protein (CSP)

The CSP is the predominant protein found on the surface of the circumsporozoite; it has approximately 420 residues and a molecular weight of 58 KDa. The CSP can be subdivided into two nonrepetitive regions (5′ and 3′ ends) and a variable central region consisting of multiple repeats of four-residue-long motifs (McCutchan et al. 1988). There is substantial point mutation polymorphism in the 3′ region of the protein where T cell epitopes have been identified (Aidoo and Udhayakumar 2000); this polymorphism has been explained as a consequence of positive natural selection by the host immune system (Hughes 1991; Escalante et al. 2002b). Polymorphism in the CS protein is also observed in the number of tandem repeats in the central region.

We studied the genetic diversity of 75 complete and 174 partial sequences of the CS protein, excluding the repetitive region because it could not be unambiguously aligned. We did not detect synonymous single nucleotide polymorphisms (Escalante et al. 2002b); however, the rate of synonymous substitutions per synonymous site is not zero using the Nei & Gojobori method, as previously reported (Hughes 1991; Escalante et al. 1998a). Overall, the gene encoding the CSP showed a significantly higher rate of nonsynonymous substitutions per nonsynonymous site than the rate of synonymous substitutions per synonymous site when the complete sequences were studied (Kn = 0.01191 and Ks = 0.00431 for 75 complete sequences with $p < 0.05$ for a Z test, as suggested by Kumar and Nei 2000). This pattern is consistent with previous reports with smaller sample sizes (Hughes 1991; Escalante et al. 1998b).

The relationship between synonymous and nonsynonymous was not homogeneous across the gene. The 5' end has a higher rate of synonymous substitutions per synonymous sites when compared with the rate of nonsynonymous substitutions (Ks = 0.0074 and Kn = 0.0046, respectively); however, this difference is not significant. This trend shifted at the 3' end, where the rate of nonsynonymous substitutions was almost ten times higher than the rate of synonymous substitutions (Ks = 0.00192 and Kn = 0.01823, respectively; statistically significant with $p < 0.01$), and with the Th2R and Th3R epitopes are located. A similar result was found using all the partial 3' end sequences, with Ks = 0.00534 and Kn = 0.05024 (significant with $p < 0.01$). This first comparison suggests that there is acceleration in the rate of nonsynonymous over synonymous substitutions in the region of the gene where epitopes are located when compared with the 5' end. This comparison alone indicates that selection is operating differently in the two parts of the gene. The 5' region appears to be neutral or under purifying selection (Ks/Kn < 1), whereas the 3' region is under positive natural selection (Ks/Kn < 1).

In order to understand the polymorphism observed in the CSP, we studied the pattern of linkage and recombination events across the gene. First we estimated the Rm, which measures the minimum number of recombination events on the sample of 75 sequences. We found eight putative recombination events, seven of which took place within sites in the 3' end and one of which was between the 5' end and the 3' end. The linkage among sites, the correlation between pairs of polymorphic sites, was also estimated using the statistic R2. The results were similar to those found for the recombination events: most of the linked sites were found in the 3' end.

The observation that recombination events and linked sites appeared in a very narrow fragment of the gene is primarily due to the fact that most of the polymorphic sites are restricted to the Th2R and Th3R epitopes. In

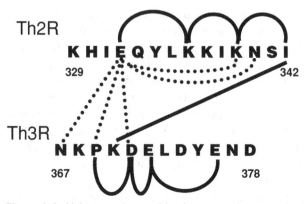

Figure 4.4. Linkage and recombination events between the Th2R and Th3R epitopes in the CS protein. The sequence of the 7G8 clone from Brazil is used as a reference for numbering the amino acid residues. Residues 329 to 342 and residues 367 to 378 correspond to epitopes Th2R and Th3R, respectively. Linkage was measured with the R2 statistic using the Fisher exact test for significance, and recombination is detected using the Rm statistic. The solid lines show the putative recombining residues and the dotted lines show the linked residues.

Figure 4.4, we plotted the recombinant and linked sites using the sequence of the clone 7G8 from Brazil as a reference. We found that recombination events interchange with linked sites, a pattern that is better explained by selection fixing convergent mutations than by a recombination event. Several putative recombination events are very close to each other in the sequence, as close as between two contiguous residues, as is the case with residue 371 (D in the sequence used in the figure) which "recombines" with position 372 of the Th3R epitope. The residue at position 332 in the epitope Th2R (E in Figure 4.4) appears linked with residues at positions 339 and 340 of the Th2R epitope and residues 367, 369, and 371 of the Th3R. This observation of convergent mutations due to selection that are taken as recombination events was made early using phylogenetic analysis (McCutchan et al. 1992). Direct evidence for intragenic recombination in a gene under selection depends on the observation of such events using synonymous substitutions (Escalante et al. 1998a); unfortunately, the lack of synonymous substitutions in the *P. falciparum* genome makes such observation difficult (Escalante et al. 2001, 2002a,b).

We have shown by comparing the rates of synonymous and nonsynonymous substitutions that the CS polymorphism at the 3′ end is maintained by positive natural selection. In addition, we have discussed how the complex pattern of linkage and putative recombination events in the 3′ end of the gene is better explained by positive natural selection even when this observation is not a formal test of the neutral model.

Table 4.4. Nucleotide diversity at the Th2R and Th3R epitopes in different geographic locations

Locality	n	Th2R Π	Th3R Π
Kenya	51	0.09772(0.00525)	0.07442(0.00687)
Gambia	17	0.08789(0.01110)	0.06454(0.01207)
Senegal	11	0.07446(0.01010)	0.03939(0.00544)
Cameroon	9	0.07143(0.01063)	0.04321(0.001926)
Brazil	9	0.02646(0.01979)	0.00(0.00)
Venezuela	10	0.08466(0.00898)	0.08025(0.01204)
India	12	0.01840(0.00686)	0.01305(0.00517)
Thailand	24	0.03990(0.00873)	0.04680(0.00742)
All	174*	0.07855(0.00326) 50	0.06110(0.00332)

Note: n, number of sequences sampled per locality; Π, nucleotide diversity. The standard deviation of Π is within parentheses.
* All sequences included.

As stated previously, overall, African malaria populations are more diverse than other populations around the globe. However, the CS protein appears to be an exception. Table 4.4 shows estimates of genetic diversity in the Th2R and Th3R epitopes. A total of 174 sequences of the 3′ end were included in this analysis. A first observation derived from Table 4.4 is that the uneven geographic distribution of the CSP alleles may jeopardize the deployment of vaccines directed to a specific locus because local variants may not be taken into account in the vaccine design.

Higher genetic diversity in the Th3R and Th2R epitopes is found in Kenyan isolates, $\pi = 0.09772$ and $\pi = 0.07442$, respectively, a level of diversity also evidenced by the haplotype diversity and the number of alleles. African populations appear to be more polymorphic when sequences from Kenya, Senegal, Gambia, and Cameroon are considered. Southeast Asian populations (India and Thailand) are less diverse, with just 50% of the diversity reported for African populations (Escalante et al. 2002b). Results from South America were contradictory. Contrary to information derived from other genes encoding antigens, such as the Pfs48/45 and the AMA-1 (Escalante et al. 2001, 2002a), a Venezuelan population showed as much diversity as an African population with a $\pi = 0.08466$, and a comparable sample from Brazil shows a very low genetic diversity, comparable to samples from Southeast Asia.

The Venezuelan population was more diverse than the sample from Cameroon on complete sequences and had polymorphisms comparable to other African populations when only the Th2R and Th3R epitopes were considered (Escalante et al. 2002b). The Venezuelan results contrast with previous studies from Brazil where low polymorphism had been system-atically observed (Yoshida et al. 1990; Shi et al. 1992); however, they are consistent with results previously reported using Random Amplified Poly-morphic DNA (RAPDs) markers (Urdaneta et al. 2001).

A possible explanation for this pattern is that the Venezuelan sample was collected from two different locations, so it is possible that the high genetic polymorphism is due to the combination of differentiated subpop-ulations. Our results suggest that low endemic areas need to be sampled extensively because low genetic diversity may be a local phenomenon due to an epidemic expansion of a few parental isolates. It is possible that several differentiated subpopulations coexist in a malaria-endemic area defined by a malaria control program.

The Merozoite Surface Protein 1 (MSP-1)

Among the antigens currently under consideration for incorporation in an antimalaria vaccine, the major merozoite surface protein 1 (MSP-1) is one of the most promising (Branch et al. 1998, 2000; Shi et al. 1999; Conway et al. 2000; Genton et al. 2002).

P. falciparum MSP-1 antigen is expressed as a large protein of 190–200 KDa on the parasite surface. This precursor undergoes two steps of proteolytic cleavage during merozoite maturation. First, it is cleaved into four major fragments of 83, 30, 38, and 42 KDa. Prior to erythrocyte inva-sion, the 42-KDa fragment undergoes a second cleavage, resulting in the generation of 33- and 19-KDa fragments. Tanabe and coworkers identified two major allele groups, MAD-20 and K1 (Tanabe et al. 1987). Further stud-ies revealed the presence of new allele families based on tandem repeats of the 83-KDa fragment. First, a new repeatless allele family was identified and named RO33 (Certa et al. 1987). We have identified a recombinant of MAD20 and RO33 that we have termed the MR allele (Takala et al. 2002).

We analyzed a total of 24 complete MSP-1 sequences and 108 partial sequences including only the 42-KDa fragment (Escalante et al. submit-ted). We studied the effect of natural selection using D^* statistics on the alignment of full-length sequences (Fu and Li 1993). The estimated value for D^* was 1.56234, which is statistically significant with $p < 0.02$. The re-sult of the D^* indicated that the gene is under positive selection, which is in agreement with previous investigations (Hughes 1992). We explored positive natural selection by comparing the rates of synonymous and nonsynonymous substitutions (Table 4.5). The numbers of synonymous and nonsynonymous substitutions per site were estimated by the Nei & Gojobori modified method with the Jukes & Cantor correction (Kumar and Nei 2000). This method uses the estimated transition/transversion

Table 4.5. Polymorphism found on different regions of the *MSP-1* gene in *P. falciparum*

Region	Residues	π(SD)	Ds	Dn	
MSP-1$_{83KDa}$	1–744	0.0468[0.0160]	0.0520[0.0052]	0.0517[0.0034]	ns (1)
MSP-1$_{30KDa}$	745–935	0.0622[0.0348]	0.0929[0.0152]	0.0821[0.0090]	ns (1)
MSP-1$_{38KDa}$	936–1423	0.0388[0.0219]	0.0393[0.0043]	0.0489[0.0038]	ns (2)
MSP-1$_{42KDa}$	1424–1635	0.0383[0.0195]	0.0421[0.0057]	0.0448[0.0043]	ns (2)
MSP-1$_{19KDa}$	1636–1749	0.0063[0.0010]	0.0029[0.0016]	0.0076[0.0030]	0.05 < p < 0.10 (2)
All MSP1	1–1749	0.0445[0.0204]	0.0504[0.0032]	0.0518[0.0019]	ns (2)

Note: The statistics were calculated on 24 complete sequences, 22 belonging to the MAD-20 and 2 belonging to the K1 allele family, respectively. π(SD) is nucleotide diversity with its standard deviation in parentheses; Ds and Dn are the number of synonymous and nonsynonymous substitutions, respectively, estimated using the Nei & Gojobori modified method with the Jukes & Cantor correction using the s/v ratio calculated for each region.

ratio, R, for the nucleotides under consideration. The 83-, 30-, and 38-KDa fragments showed more synonymous than nonsynonymous substitutions. The differences between the synonymous and nonsynonymous rates were not significant using a Z test (Kumar and Nei 2000).

The 42-KDa fragment had more nonsynonymous than synonymous substitutions, but this difference was not significant. The 33- and 19-KDa fragments of the 42-KDa secondary cleavage were further studied separately because they play different roles in parasite invasion of the red blood cell. The number of nonsynonymous substitutions was higher than the number of synonymous substitutions in the 19-KDa fragment, but this difference was not significant ($0.05 < p < 0.10$). A p value between 0.05 and 0.10 could indicate the limited statistical power of the tests used for detecting selection. The only way to circumvent this problem would be to increase the sample size, something that we could do on the 42-KDa fragment by using the sample of 108 sequences.

In the expanded sample of 42-KDa sequences, the number of synonymous substitutions per site exceeded the number of nonsynonymous, with $Ds = 0.0658 \pm 0.0087$ and $Dn = 0.0585 \pm 0.0057$, but the difference was still not significant. However, the Ds and Dn values for the 33-KDa fragment were 0.0887 ± 0.0088 and 0.0614 ± 0.0042, respectively, where Ds was significantly higher than Dn ($p < 0.05$). The opposite took place in the 19-KDa fragment, where the Ds and Dn values were 0.0009 ± 0.0005 and 0.0110 ± 0.0040, respectively, with Dn significantly higher than Ds ($p < 0.005$). These results suggest that while the 19-KDa fragment is under positive selection, the 33-KDa fragment is under purifying selection; that is, natural selection favors the maintenance of amino acid polymorphism in the 19-KDa fragment, whereas it maintains relatively low amino acid polymorphism in the 33-KDa fragment when compared with the overall accumulated diversity that can be accounted for by the random accumulation of neutral mutations. This finding supports the assertion that these two fragments play different roles in the immune response elicited by the MSP-1.

The comparison of the 33-KDa and the 19-KDa fragments is even more interesting in the context of vaccine development, where the issue of genetic diversity is of great concern. This study supports prior observations of limited genetic diversity in the 19-KDa fragment; as an example, it is clearly less polymorphic than the 33-KDa fragment ($\pi = 0.00860$ compared with $\pi = 0.06697$, respectively, on 108 sequences). However, our analysis points out that positive natural selection is operating on the more "conserved" fragment. This observation demonstrates that the extent of

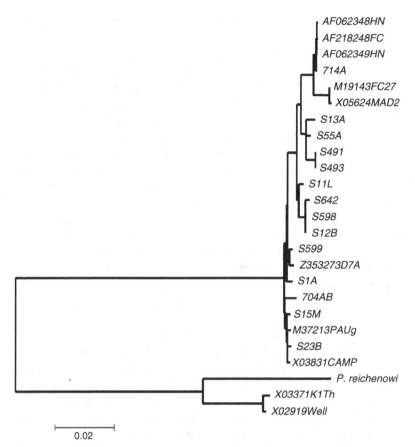

Figure 4.5. Neighbor-joining tree of the MSP-1 alleles using a Tamura three-parameter distance. The numbers on the nodes of the tree are percentages of bootstrap values based on 1,000 pseudo-replications. The sequences of *P. reichenowi* are used as an outgroup.

polymorphism alone is not an indicator of positive selection associated with the host immune response.

In order to further investigate the long-term evolution of the *P. falciparum* MSP-1, the repetitive region of the 83-KDa and the complete 42-KDa fragments were compared with a homologous sequence of *P. reichenowi*. Figure 4.5 shows a NJ tree using a Tamura three-parameter model. The *P. reichenowi* sequence was closer but outside the polymorphisms of the K1. The polymorphisms within MAD-20 and K1 for the aligned regions are 0.01435 and 0.08070, and the divergences with *P. reichenowi* were 0.21274 and 0.08646, respectively. This phylogenetic analysis suggested that the divergence of the two major allele groups took place before the divergence of the two species.

This result has important implications in the ongoing debate on the recent bottleneck proposed for *P. falciparum* populations. It has been argued that MSP-1 is one of the oldest polymorphisms found (Hughes 1992), a scenario that generates conflicts with the proposed bottleneck and/or recent population expansion of *P. falciparum* populations (Hughes and Verra 2001; Hartl et al. 2002). In order to explain the origin of the allele families, two hypotheses have been proposed. First, it has been argued that divergence can be explained by the fast evolution and rapid divergence of tandem repeats. Evidence for fast-evolving repeats can be found in other genes, such as the circumsporozoite protein in the primate malarias from Southeast Asia (Escalante et al. 1995); however, although this process of fast-evolving repeats can account for the divergence in block 2 and other repetitive sequences, it does not explain the extensive divergence in nonrepetitive regions (Hartl et al. 2002).

The second hypothesis implies that the two major groups evolved separately as paralogous genes after gene duplication from a homologous common ancestor, then a reciprocal loss of one of the copies took place, reverting the homology and generating two lineages with highly divergent alleles (Hartl et al. 2002). No evidence of gene duplication in the MSP-1 has been reported in several *Plasmodium* species where homologous genes have been identified. The data presented here suggest that if such an event took place, it happened before the *P. reichenowi–P. falciparum* divergence. The data from *P. reichenowi*, together with the evidence of positive natural selection, argue in favor of considering the MSP-1 to be an old polymorphism (Hughes 1992).

MOLECULAR EPIDEMIOLOGIC STUDIES IN MALARIAL PARASITES

The use of molecular tools to explore relationships between pathogens and clinical manifestations of illness is appealing, as evidenced by the extensive literature in this area of malaria research. Molecular markers allow testing of hypotheses and scenarios that otherwise could not be explored by descriptive epidemiologic investigations.

Molecular epidemiologic studies involve the investigation of phenotypes using appropriate molecular markers. The use of the term phenotype instead of "clinical manifestation of disease" allows us to underline the potential complex interactions that originate the wide range of characteristics observed in malaria. It is important to note that phenotypes are usually observed and defined in the host (ineffective drug, clinical end points, or immunity) but the genetic bases of such phenotypes are investigated in the parasite. This approach in malaria research makes the implicit assumption that the host genetic diversity, usually ignored due to logistic and technical limitations, is not playing a major role. The host population is considered the environment where the parasites' genotypes express their phenotypic differences. Following the argument, the parasite genotypes compete and exploit resources in a changing environment of hosts and vectors.

The conceptual framework discussed above provides valuable information in cases such as drug resistance where it is reasonable to assume that host diversity plays a secondary role if any. However, this framework faces major problems whenever the host immune response is involved, because this response is related to the host genetic background. There are several factors that need to be taken into account whenever a molecular epidemiologic study is designed.

First, a genetic marker needs a clearly defined phenotype, which in epidemiological terms means an appropriate case definition. For example, we could study the effect of parasite diversity on severe anemia due to malaria but the study might fail without a clear definition of severe anemia.

Second, we need to understand the genetic bases of the phenotype. In the specific case of drug resistance, for example, we could use as potential markers those mutations found in the gene encoding the protein that is the drug target (Kublin et al. 1998, 2002; Djimde et al. 2001). The phenotype "drug resistance" can be clearly defined in epidemiologic terms and the genetic basis of such resistance can potentially be identified. This is not the case for "strain-specific immunity" or the complexity of infection. In these cases, genetic diversity is measured in the parasite to identify the strain or level of diversity of the infection, and the phenotype is observed in the host population where the host's genetic diversity could be playing a determinant role in the immune response elicited by parasite genotypes (Farnert et al. 1999; Branch et al. 2000).

Third, the biologic characteristics and evolutionary processes involved in the diversity observed in the genetic marker need to be understood. Using microsatellites or SNPs is not equivalent to using the tandem repeat region of an antigen because the latter may be under selection by the host immune system. In addition, a microsatellite or SNP could be linked to a gene under selection, something that can be very useful in discovering genes linked to a specific gene under selection but can be misleading if we want to use the same marker for measuring gene flow.

Finally, the use of patterns of genetic diversity requires a good understanding of the analytical tools needed to answer a specific question. For example, an investigation aiming to separate a new infection from a recrudescent one needs markers as variable as possible because the molecular tools are needed to identify the parasite's origin. However, that is not the case for measuring gene flow or genetic structure because the use of highly variable markers presents a problem of inaccurate estimation of several low allelic frequencies, leading to the detection of population structure by chance, which is meaningless in the context of public health.

A major practical consequence of the parasite population and intrahost dynamics is that they hamper our capacity to use and develop genetic markers for molecular epidemiologic studies. In a population with a panmictic population structure and high genetic diversity, two isolates showing the same pattern at a given set of markers could be considered part of a common infection. These tools can be used to separate a recrudescent infection from a new infection, allowing the identification of drug resistance cases (Basco and Ringwald 2000a). A cautionary note, however, is that specific criteria about the degree of similarity between two samples need to be discussed because independent infections may share alleles by chance. The studies performed so far have not paid much attention to this issue because they focus mostly on high-transmission areas. This problem will be especially critical in low-transmission areas. We propose that the probability of matches in genetic profiles by chance increases in inverse proportion to the local parasite diversity and transmission dynamics. If this is the case, it may be difficult to establish whether two isolates with similar genetic profiles are part of two independent infections or of recrudescence of a previous one in a population with low diversity, where pseudo-clonal population expansions may take place, as has been the case in some areas of South America (Haddad et al. 1999; Urdaneta et al. 2001; Tami et al. 2002). The situation could be even more dramatic if some alleles are hard to reproduce due to technical limitations of the PCR.

Low-transmission areas or seasons may generate additional complications if we try to identify genetic markers linked with a phenotype, such as drug resistance, or a specific clinical end point, such as severe cases. Spurious linkage between markers and the phenotype can be obtained simply due to inbreeding. In such cases, linkage needs to be established between the genetic marker (alleles at a given locus) and the phenotype using different parasite populations, because populations rather than individuals should be used as the unit of observation.

Currently, only genes involved in drug resistance have been tested in different populations worldwide (Basco and Ringwald 2000b; Biswas et al. 2000; Cortese et al. 2002; Kublin et al. 2002) and they provide the strongest evidence available today of genes linked to phenotypes that are of public health relevance. An appropriate example is a study of alleles of the gene encoding the digestive vacuole transmembrane protein (pfcrt), which has been linked to choloroquine resistance. The first interesting result of these investigations was that the alleles involved in the Old and New World are different, consistent with the independent origin suggested by epidemiologic information (Fidock et al. 2000; Mehlotra et al. 2001). In addition, molecular population studies have also revealed the independent origin of chloroquine resistance, showing linkage disequilibrium consistent with independent selective sweeps in the pfcrt region (Wootton et al. 2002). This convergent phenotype from different genetic backgrounds provides additional support incriminating these genes with chloroquine resistance, providing a gold standard for the identification of genes linked with drug resistance/effectiveness, where samples from different geographic populations are used as the unit of observation.

A limitation of this approach is that the application of these criteria can be complicated if the phenotype involves several loci. The relative contribution of each locus may change locally, so a clear pattern demands enormous sample sizes and very likely such phenotypes will be almost impossible to track using simple molecular markers.

Another factor that has practical consequences in the use of molecular markers is intra-host diversity. Most studies including molecular markers rely on the extensive use of qualitative PCR. A PCR amplification provides evidence only that a given allele is present, not quantitative evidence of its proportion. A mutation can be present at low frequency so it may not have any relevant clinical effect. This problem may account in part for the lack of association between certain mutations and clinical drug failure.

An additional problem is that PCR amplification does not allow for a clear description of the alleles involved in an infection. For example, the presence of triple mutants in the DHFR loci and double mutants in the DHPS loci has been associated with antifolate resistance. However, the possibility that a mixed infection of double mutants may be erroneously taken as a triple mutant needs to be considered, especially in areas where infections by multiple genotypes are common.

An important effect of the intra-host and transmission dynamics on epidemiologic studies is the use of linkage disequilibrium to identify "strains" or genetic variants associated with specific clinical end points. Linkage in low-transmission areas or seasons may be the result of a small effective population size, generating spurious associations among loci and a clinical end point. The simple observation of linkage disequilibrium does not imply anything without a mechanistic explanation. For example, evidence of stable multi-locus associations has been observed linked to severe malaria cases in French Guiana (Ariey et al. 2001). The result appears to be robust but, the authors properly pointed out that the data only show the over representation of a parasite lineage and do not clearly implicate the genes used as markers in the study (a var allele and MSP-1) as virulent factors.

"Strains," in terms of multi-locus associations generated by genetic drift due to small population sizes, can be mistakenly considered the mechanistic explanation behind a given clinical end point rather than a natural consequence of the local transmission dynamics. The observation of stable genetic entities may increase if bed-nets are widely used to reduce transmission, so careful epidemiologic designs are needed to distinguish the effect of "strains" due to small population sizes from that of "strains" generated by selection due to their interaction with the host population generating a specific clinical end point.

The same logic can be applied to studies trying to observe strain-specific immunity because most of them lack appropriate controls to separately measure linkage due to an immune response (strain-specific immunity) and linkage due to genetic drift. Even when evidence for allele-specific immunity is available (Branch et al. 1998, 2000; Conway et al. 1999, 2000; Gupta et al. 1994, 1996; Aidoo and Udhayakumar 2000; Ekala et al. 2002a,b), a strain necessarily implies some genetic cohesion due to selection by host immune pressure and/or groups of alleles without cross immunity that are independently transmitted (Gupta et al. 1994; Gupta and Hill 1995). Studies based on linkage across several antigens (Eisen et al. 2002) can have explanations that do not claim that linkage is maintained by the host immune response. As explained above, it is likely that linkage across genes without any immunological implication will be found.

In this context, longitudinal studies are very valuable because they can allow the study of linkage due to chance and separate such linkages from groups of loci linked due to selection by the host immune response. Unfortunately, longitudinal data are scarce, and more likely, several populations have to be studied in order to unveil the effect of selection by the host immune response on the maintenance of linkage across several antigens. This is a task that can only be accomplished by coordinating the results obtained from different malaria settings and maintaining strong field programs for long periods of time. For the investigation of strain immunity, studies in animal models can provide valuable information on the role played by the host immune system in the maintenance of multi-locus associations and/or in the effect of parasite diversity (Read and Taylor 2001).

Changes in transmission will decrease exposure of the human population, and such a decrease can by itself modify clinical manifestations of disease (Gupta et al. 1996; Snow et al. 1997). Control measures that target specific age groups could have a similar effect because severity of clinical manifestations appears to be affected by age. Such control measures would affect the intra-host dynamics, changing the parasite composition by reducing the probability of infection by a given genotype, and they would also change the susceptibility of the risk group, modifying the environment in which human and parasite populations collide.

Evaluating the consequences of control measures in the parasite and human populations is a task that demands proficient knowledge of population biology and a clear understanding of the epidemiology and clinical and immunological characteristics of malaria. Predicting the consequences of a new intervention, as part of control programs, requires longitudinal studies planned by multidisciplinary groups in which clinical manifestations of disease are treated as the result of interactions between genetically diverse populations of hosts and parasites in an environment that will change with public health policies and interventions. Comparison of several epidemiologic settings may appear unrealistic given the amount of resources involved and the careful coordination needed. However, evolutionary biologists need to establish a common agenda with professionals involved in control programs, as well as with colleagues working on immunologic and epidemiologic investigations. Establishing this communication is our best chance for improving data and sample collections, allowing us to evaluate policies and strategies that may change the emergence of phenotypes as complex as severe malaria.

ACKNOWLEDGMENTS

This research is supported by a grant from The National Institutes of Health (R01 GM60740) to AAE. This work was supported in part by U.S. Agency for International Development grant HRN-60010-A-00-4010-00 to AAL. We thank Kimberly C. Brouwer for comments that helped to improve this manuscript.

REFERENCES

Aidoo M, Udhayakumar V (2000). Field studies of cytotoxic T lymphocytes in malaria infections: implications for malaria vaccine development. *Parasitol. Today* 16:50–6.

Anders RF, Saul AJ (1994). Candidate antigens for an asexual blood stage vaccine against falciparum malaria. In *Molecular Immunological Considerations in Malaria Vaccine Development*, edited by MF Good and AJ Saul, pp. 169–208. CRC Press, Boca Raton, FL.

Anderson TJ, Haubold B, Williams JT, Estrada-Franco JG, Richardson L, Mollinedo R, Bockarie M, Mokili J, Mharakurwa S, French N, Whitworth J, Velez ID, Brockman

AH, Nosten F, Ferreira MU, Day KP (2000). Microsatellite markers reveal a spectrum of population structures in the malaria parasite *Plasmodium falciparum*. *Mol. Biol. Evol.* 17:1467–82.

Arez AP, Snounou G, Pinto J, Sousa CA, Modiano D, Ribeiro H, Franco AS, Alves J, do Rosario VE (1999). A clonal *Plasmodium falciparum* population in an isolated outbreak of malaria in the Republic of Cabo Verde. *Parasitology* 118:347–55.

Ariey F, Hommel D, Le Scanf C, Duchemin JB, Peneau C, Hulin A, Sarthou JL, Reynes JM, Fandeur T, Mercereau-Puijalon O (2001). Association of severe malaria with a specific *Plasmodium falciparum* genotype in French Guiana. *J. Infect. Dis.* 184:237–41.

Awadalla P, Walliker D, Babiker H, Mackinnon M (2001). The question of *Plasmodium falciparum* population structure. *Trends Parasitol.* 17:351–3.

Babiker HA, Walliker D (1997). Current views on the population structure of *Plasmodium falciparum*: implications for control. *Parasitol. Today* 13:262–7.

Basco LK, Ringwald P (2000a). Molecular epidemiology of malaria in Yaounde, Cameroon. VII. Analysis of recrudescence and reinfection in patients with uncomplicated falciparum malaria. *Am. J. Trop. Med. Hyg.* 63:215–21.

Basco LK, Ringwald P (2000b). Molecular epidemiology of malaria in Yaounde, Cameroon. VI. Sequence variations in the *Plasmodium falciparum* dihydrofolate reductase-thymidylate synthase gene and in vitro resistance to pyrimethamine and cycloguanil. *Am. J. Trop. Med. Hyg.* 62:271–6.

Bensch S, Stjernman M, Hasselquist D, Ostman O, Hansson B, Westerdahl H, Pinheiro RT (2000). Host specificity in avian blood parasites: a study of *Plasmodium* and *Haemoproteus* mitochondrial DNA amplified from birds. *Proc. R. Soc. Lond. B, Biol. Sci.* 267:1583–9.

Biswas S, Escalante AA, Chaiyaroj S, Angkasekwinai P, Lal AA (2000). Prevalence of point mutations in the dihydrofolate reductase and dihydropteroate synthetase genes of *Plasmodium falciparum* isolates from India and Thailand: a molecular epidemiologic study. *Trop. Med. Int. Health* 5:737–43.

Boyd MF (1949). A comprehensive survey of all aspects of this group of diseases from a global standpoint. In *Malariology*, Vol. 1, edited by MF Boyd, pp. Saunders, PA.

Branch OH, Udhayakumar V, Hightower AW, Oloo AJ, Hawley WA, Nahlen BL, Bloland PB, Kaslow DC, Lal AA (1998). A longitudinal investigation of IgG and IgM antibody responses to the merozoite surface protein-1 19-kiloDalton domain of *Plasmodium falciparum* in pregnant women and infants: associations with febrile illness, parasitemia, and anemia. *Am. J. Trop. Med. Hyg.* 58:211–9.

Branch OH, Oloo AJ, Nahlen BL, Kaslow D, Lal AA (2000). Anti-merozoite surface protein-1 19-kDa IgG in mother-infant pairs naturally exposed to *Plasmodium falciparum*: subclass analysis with age, exposure to asexual parasitemia, and protection against malaria. V. The Asembo Bay Cohort Project. *J. Infect. Dis.* 181:1746–52.

Branch OH, Takala S, Kariuki K, Nahlen BL, Kolczak M, Hawley W, Lal AA (2001). *Plasmodium falciparum* genotypes, low complexity of infection, and resistance to subsequent malaria in participants in the Asembo Bay Cohort Project. *Infect. Immun.* 69:7783–92.

Certa U, Rotmann D, Matile H, Reber-Liske R (1987). A naturally occurring gene encoding the major surface antigen precursor p190 of *Plasmodium falciparum* lacks tripeptide repeats. *EMBO J.* 6:4137–42.

Coatney RG, Collins WE, Warren M, Contacos PG (1971). *The Primate Malaria*. U.S. Government Printing Office, Washington D.C.

Conway DJ (1997). Natural selection on polymorphic malaria antigens and the search for a vaccine. *Parasitol. Today* 13:26–9.

Conway DJ, Roper C, Oduola AMJ, Arnot DE, Kremsner PG, Grobusch MP, Curtis CF, Greenwood BM (1999). High recombination rate in natural populations of *Plasmodium falciparum*. *Proc. Natl. Acad. Sci. USA* 96:4506–11.

Conway DJ, Cavanagh DR, Tanabe K, Roper C, Mikes ZS, Sakihama N, Bojang KA, Oduola AM, Kremsner PG, Arnot DE, Greenwood BM, McBride JS (2000). A principal target of human immunity to malaria identified by molecular population genetic and immunological analyses. *Nat. Med.* 6:689–92.

Corredor V, Enea V (1993). Plasmodial ribosomal RNA as phylogenetic probe: a cautionary note. *Mol. Biol. Evol.* 10:924–26.

Cortese JF, Caraballo A, Contreras CE, Plowe CV (2002). Origin and dissemination of *Plasmodium falciparum* drug-resistance mutations in South America. *J. Infect. Dis.* 186:999–1006.

Deans JA, Knight AM, Jean WC, Waters AP, Cohen S, Mitchell GH (1988). Vaccination trials in rhesus monkeys with a minor, invariant, *Plasmodium knowlesi* 66 kD merozoite antigen. *Parasitol. Immunol.* 10:535–52.

Djimde A, Doumbo OK, Cortese JF, Kayentao K, Doumbo S, Diourte Y, Dicko A, Su XZ, Nomura T, Fidock DA, Wellems TE, Plowe CV, Coulibaly D (2001). A molecular marker for chloroquine-resistant falciparum malaria. *N. Engl. J. Med.* 344:257–63.

Ebert D, Herre EA (1996). The evolution of parasitic diseases. *Parasitol. Today* 12:96–100.

Eisen D, Billman-Jacobe H, Marshall VF, Fryauff D, Coppel RL (1998). Temporal variation of the merozoite surface protein-2 gene of *Plasmodium falciparum*. *Infect. Immun.* 66:239–46.

Eisen DP, Saul A, Fryauff DJ, Reeder JC, Coppel RL (2002). Alterations in *Plasmodium falciparum* genotypes during sequential infections suggest the presence of strain specific immunity. *Am. J. Trop. Med. Hyg.* 67:8–16.

Ekala MT, Jouin H, Lekoulou F, Mercereau-Puijalon O, Ntoumi F (2002a). Allelic family-specific humoral responses to merozoite surface protein 2 (MSP2) in Gabonese residents with *Plasmodium falciparum* infections. *Clin. Exp. Immunol.* 129:326–31.

Ekala MT, Jouin H, Lekoulou F, Issifou S, Mercereau-Puijalon O, Ntoumi F (2002b). *Plasmodium falciparum* merozoite surface protein 1 (MSP1): genotyping and humoral responses to allele-specific variants. *Acta Trop.* 81:33–46.

Endo T, Ikeo K, Gojorobi T (1996). Large-scale search for genes on which positive selection may operate. *Mol. Biol. Evol.* 13:685–90.

Escalante A, Ayala FJ (1994). Phylogeny of the malarial genus *Plasmodium*, derived from ribosomal gene sequences. *Proc. Natl. Acad. Sci. USA* 91:11,373–7.

Escalante A, Ayala FJ (1995). Origin of *Plasmodium* and other Apicomplexa based on rRNA genes. *Proc. Natl. Acad. Sci. USA* 92:5793–7.

Escalante A, Barrio E, Ayala FJ (1995). The phylogeny of the *Plasmodium* species based on the circumsporozoite protein gene. *Mol. Biol. Evol.* 12:616–26.

Escalante AA, Goldman IF, De Rijk P, De Wachter R, Collins WE, Qari SH, Lal AA (1997). Phylogenetic study of the genus *Plasmodium* based on the secondary structure-based alignment of the small subunit ribosomal RNA. *Mol. Biochem. Parasitol.* 90:317–21.

Escalante AA, Freeland DE, Collins WE, Lal AA (1998a). The evolution of primate malarial parasites based on the gene encoding cytochrome b from the linear mitochondrial genome. *Proc. Natl. Acad. Sci. USA* 95:8124–9.

Escalante AA, Lal AA, Ayala FJ (1998b). Genetic polymorphism and natural selection in the malaria parasite *Plasmodium falciparum*. *Genetics* 149:189–202.

Escalante AA, Grebert HM, Chaiyaroj SC, Riggione F, Biswas S, Nahlen BL, Lal AA (2002a). Polymorphism in the gene encoding the Pfs48/45 antigen of *Plasmodium falciparum*. XI. Asembo Bay Cohort Project. *Mol. Biochem. Parasitol.* 119:17–22.

Escalante AA, Grebert HM, Isea R, Goldman IF, Basco L, Magris M, Biswas S, Kariuki S, Lal AA (2002b). A study of genetic diversity in the gene encoding the circumsporozoite protein (CSP) of *Plasmodium falciparum* from different transmission areas. *Mol. Biochem. Parasitol.* 125:83–90.

Escalante AA, Grebert HM, Chaiyaroj SC, Magris M, Biswas S, Nahlen BL, Lal AA (2001). Polymorphism in the gene encoding the Apical Membrane Antigen-1 (AMA-1) of *Plasmodium falciparum*. VI. Asembo Bay Cohort Project. *Mol. Biochem. Parasitol.* 113:279–87.

Ewald PW (1994). *Evolution of Infectious Disease*. Oxford University Press, New York.

Farnert A, Rooth I, Svensson, Snounou G, Bjorkman A (1999). Complexity of *Plasmodium falciparum* infections is consistent over time and protects against clinical disease in Tanzanian children. *J. Infect. Dis.* 179:989–95.

Fidock DA, Nomura T, Talley AK, Cooper RA, Dzekunov SM, Ferdig MT, Ursos LM, Sidhu AB, Naude B, Deitsch KW, Su XZ, Wootton JC, Roepe PD, Wellems TE (2000). Mutations in the *P. falciparum* digestive vacuole transmembrane protein PfCRT and evidence for their role in chloroquine resistance. *Mol. Cell* 6:861–71.

Fu YX, Li WH (1993). Statistical tests of neutrality of mutations. *Genetics* 133:693–709.

Garnham PCC (1966). *Malaria Parasites and Other Haemosporidia*. Blackwell Scientific Publications, Oxford.

Genton B, Betuela I, Felger I, Al-Yaman F, Anders RF, Saul A, Rare L, Baisor M, Lorry K, Brown GV, Pye D, Irving DO, Smith TA, Beck HP, Alpers MP (2002). A recombinant blood-stage malaria vaccine reduces *Plasmodium falciparum* density and exerts selective pressure on parasite populations in a phase 1-2b trial in Papua New Guinea. *J. Infect. Dis.* 185:820–7.

Gupta S, Hill AVS (1995). Dynamic interactions in malaria: host heterogeneity meets parasite polymorphism. *Proc. R. Soc. Lond. B*, 261:271–7.

Gupta S, Trenholme K, Anderson RM, Day KP (1994). Antigenic diversity and the transmission dynamics of *Plasmodium falciparum*. *Science* 263:961–3.

Gupta S, Maiden MC, Feavers IM, Nee S, May RM, Anderson RM (1996). The maintenance of strain structure in populations of recombining infectious agents. *Nat. Med.* 2:437–42.

Haddad D, Snounou G, Mattei D, Enamorado IG, Figueroa J, Stahl S, Berzins K (1999). Limited genetic diversity of *Plasmodium falciparum* in field isolates from Honduras. *Am. J. Trop. Med. Hyg.* 60:30–4.

Hartl DL, Volkman SK, Nielsen KM, Barry AE, Day KP, Wirth DF, Winzeler EA (2002). The paradoxical population genetics of *Plasmodium falciparum*. *Trends Parasitol.* 18:266–72.

Hastings IM, Mackinnon MJ (1998). The emergence of drug-resistant malaria. *Parasitology* 117:411–7.

Hastings IM, D'Alessandro U (2000). Modelling a predictable disaster: the rise and spread of drug-resistant malaria. *Parasitol. Today* 16:340–7.

Hastings IM, Watkins WM, White NJ (2002). The evolution of drug-resistant malaria: the role of drug elimination half-life. *Phil. Trans. R. Soc. Lond. B, Biol. Sci.* 357:505–19.

Hewitt, R (1940). *Bird malaria*. American Journal of Hygiene monographic series, no. 15. Johns Hopkins University Press, Baltimore.

Hodder AN, Crewther PE, Matthew MLSM, Reid GE, Moritz RL, Simpson RJ, Anders RF (1996). The disulfide bond structure of *Plasmodium* apical membrane antigen-1. *J. Biol. Chem.* 271:29,446–52.

Hughes AL (1991). Circumsporozoite protein genes of malarial parasites (*Plasmodium* spp.): evidence for positive selection on immunogenic regions. *Genetics* 127:345–53.

Hughes AL (1992). Positive selection and intrallelic recombination at the merozoite surface antigen-1 (MSA-1) locus of *Plasmodium falciparum*. *Mol. Biol. Evol.* 9:381–93.

Hughes MK, Hughes AL (1995). Natural selection on *Plasmodium* surface proteins. *Mol. Biochem. Parasitol.* 71:99–113.

Hughes AL, Verra F (2001). Very large long-term effective population size in the virulent human malaria parasite *Plasmodium falciparum*. *Proc. R. Soc. Lond. B, Biol. Sci.* 268:1855–60.

Kedzierski L, Escalante AA, Isea R, Black CG, Barnwell JW, Coppel RL (2002). Phylogenetic analysis of the genus *Plasmodium* based on the gene encoding adenylosuccinate lyase. *Infect. Genet. Evol.* 4:297–301.

Kimura M (1977). Preponderance of synonymous changes as evidence for the neutral theory of molecular evolution. *Nature* 267:275–6.

Kimura M (1983). *The Neutral Theory of Molecular Evolution*. Cambridge University Press, Cambridge, UK.

Kocken CHM, Dubbeld MA, Van der wel A, Pronk JT, Waters AP, Langermans JAM, Thomas AW (1999). High-level expression of *Plasmodium vivax* apical membrane antigen 1 (AMA-1) in *Pichia pastoris*: strong immunogenicity in *Macaca mulatta* immunized with *P. vivax* AMA-1 and adjuvant SBAS2. *Infect. Immun.* 67:43–9.

Kreitman M, Akashi H (1995). Molecular evidence for natural selection. *Annu. Rev. Ecol. Syst.* 26:403–22.

Kublin JG, Witzig RS, Shankar AH, Zurita JQ, Gilman RH, Guarda JA, Cortese JF, Plowe CV (1998). Molecular assays for surveillance of antifolate-resistant malaria. *Lancet* 351:1629–30.

Kublin JG, Dzinjalamala FK, Kamwendo DD, Malkin EM, Cortese JF, Martino LM, Mukadam RA, Rogerson SJ, Lescano AG, Molyneux ME, Winstanley PA, Chimpeni P, Taylor TE, Plowe CV (2002). Molecular markers for failure of sulfadoxine-pyrimethamine and chlorproguanil-dapsone treatment of *Plasmodium falciparum* malaria. *J. Infect. Dis.* 185:380–8.

Kumar S, Nei M (2000). *Molecular Evolution and Phylogenetics*. Oxford University Press, New York.

Lal AA, Hughes MA, Oliveira DA, Nelson C, Bloland PB, Oloo AJ, Hawley WE, Hightower AW, Nahlen BL, Udhayakumar V (1996). Identification of T-cell determinants in natural immune responses to the *Plasmodium falciparum* apical membrane antigen (AMA-1) in an adult population exposed to malaria. *Infect. Immun.* 64:1054–9.

Livingstone FB (1984). The Duffy blood groups, vivax malaria, and malaria selection in human populations: a review. *Hum. Biol.* 56:413–25.

McCutchan TF, de la Cruz VF, Good MF, Wellems TE (1988). Antigenic diversity in *Plasmodium falciparum*. *Prog. Allergy* 41:173–92.

McCutchan TF, Lal AA, de Rosario V, Waters AP (1992). Two types of sequence polymorphism in the circumsporozoite gene of *Plasmodium falciparum*. *Mol. Biochem. Parasitol.* 50:37–46.

McCutchan TF, Li J, McConkey GA, Rogers MJ, Waters AP (1995). The cytoplasmic ribosomal RNAs of *Plasmodium* spp. *Parasitol. Today* 11:134–8.

McCutchan TF, Kissinger JC, Touray MG, Rogers MJ, Li J, Sullivan M, Braga EM, Krettli AU, Miller LH (1996). Comparison of circumsporozoite proteins from avian and mammalian malarias: biological and phylogenetic implications. *Proc. Natl. Acad. Sci. USA* 93:11,889–94.

Mehlotra RK, Fujioka H, Roepe PD, Janneh O, Ursos LM, Jacobs-Lorena V, McNamara DT, Bockarie MJ, Kazura JW, Kyle DE, Fidock DA, Zimmerman PA (2001). Evolution of a unique *Plasmodium falciparum* chloroquine-resistance phenotype in association with pfcrt polymorphism in Papua New Guinea and South America. *Proc. Natl. Acad. Sci. USA* 98:12,689–94.

Mu J, Duan J, Makova KD, Joy DA, Huynh CQ, Branch OH, Li WH, Su XZ (2002). Chromosome-wide SNPs reveal an ancient origin for *Plasmodium falciparum*. *Nature* 418:323–6.

Ohta T (1996). The neutralist-selectionist debate. *Bioessays* 18:673–7.

Palatnik M, Rowe AW (1984). Duffy and Duffy-related human antigens in primates. *J. Hum. Evol.* 13:173–9.

Paul RE, Packer MJ, Walmsley M, Lagog M, Ranford-Cartwright LC, Paru R, Day KP (1995). Mating patterns in malaria parasite populations of Papua New Guinea. *Science* 269:1709–11.

Perkins SL (2000). Species concepts and malarial parasites: detecting a cryptic species of *Plasmodium*. *Proc. R. Soc. Lond. B, Biol. Sci.* 267:2345–50.

Polley SD, Conway DJ (2001). Strong diversifying selection on domains of the *Plasmodium falciparum* apical membrane antigen 1 gene. *Genetics* 158:1505–12.

Qari SH, Shi YP, Povoa MM, Alpers MP, Deloron P, Murphy GS, Harjosuwarno S, Lal AA (1993a). Global occurrence of *Plasmodium vivax*-like human malaria parasite. *J. Infect. Dis.* 168:1485–9.

Qari SH, Shi YP, Goldman IF, Udhayakumar V, Alpers MP, Collins WE, Lal AA (1993b). Identification of *Plasmodium vivax*-like human malaria parasite. *Lancet* 341:780–3.

Qari SH, Shi YP, Pieniazek NJ, Collins WE, Lal AA (1996). Phylogenetic relationship among the malarial parasites based on small subunit rRNA gene sequences: monophyletic nature of the human malaria parasite, *Plasmodium falciparum*. *Mol. Phylogenet. Evol.* 6:157–65.

Qari SH, Shi YP, Goldman IF, Nahlen BL, Tibayrenc M, Lal AA (1998). Predicted and observed alleles of *Plasmodium falciparum* merozoite surface protein-1 (MSP-1), a potential malaria vaccine antigen. *Mol. Biochem. Parasitol.* 92:241–52.

Rathore D, Wahl AM, Sullivan M, McCutchan TF (2001). A phylogenetic comparison of gene trees constructed from plastid, mitochondrial and genomic DNA of *Plasmodium* species. *Mol. Biochem. Parasitol.* 114:89–94.

Read AF, Taylor LH (2001). The ecology of genetically diverse infections. *Science* 292:1099–102.

Rich SM, Licht MC, Hudson RR, Ayala FJ (1998). Malaria's Eve: evidence of a recent population bottleneck throughout the world populations of *Plasmodium falciparum*. *Proc. Natl. Acad. Sci. USA* 95:4425–30.

Ricklefs RE, Fallon SM (2002). Diversification and host switching in avian malarial parasites. *Proc. R. Soc. Lond. B, Biol. Sci.* 269:885–92.

Shi YP, Alpers MP, Povoa MM, Lal AA (1992). Diversity in the immunodominant determinants of the circumsporozoite protein of *Plasmodium falciparum* parasites from malaria endemic regions of Papua New Guinea and Brazil. *Am. J. Trop. Med. Hyg.* 47:844–55.

Shi YP, Sayed U, Qari SH, Roberts JM, Udhayakumar V, Oloo AJ, Hawley WA, Kaslow DC, Nahlen BL, Lal AA (1996). Natural immune response to the C-terminal 19-kilodalton domain of *Plasmodium falciparum* merozoite surface protein 1. *Infect. Immun.* 64:2716–23.

Shi YP, Hasnain SE, Sacci JB, Holloway BP, Fujioka H, Kumar N, Wohlhueter R, Hoffman SL, Collins WE, Lal AA (1999). Immunogenicity and in vitro protective efficacy of a recombinant multistage *Plasmodium falciparum* candidate vaccine. *Proc. Natl. Acad. Sci. USA* 96:1615–20.

Shoshani J, Groves CP, Simons EL, Gunnell GF (1996). Primate phylogeny: morphological vs molecular results. *Mol. Phylogenet. Evol.* 5:102–54.

Siddall ME, Barta JR (1992). Phylogeny of *Plasmodium* species: estimation and inference. *J. Parasitol.* 78:567–8.

Snow RW, Omumbo JA, Lowe B, Molyneux CS, Obiero JO, Palmer A, Weber MW, Pinder M, Nahlen B, Obonyo C, Newbold C, Gupta S, Marsh K (1997). Relation between severe malaria morbidity in children and level of *Plasmodium falciparum* transmission in Africa. *Lancet* 349:1650–4.

Takala S, Branch O, Escalante AA, Kariuki S, Wootton J, Lal AA (2002). Evidence for intragenic recombination in *P falciparum*: identification of a novel allele family in block 2 of merozoite surface protein-1. *Mol. Biochem. Parasitol.* 125:163–71.

Tami A, Grundmann H, Sutherland C, McBride JS, Cavanagh DR, Campos E, Snounou G, Barnabe C, Tibayrenc M, Warhurst DC (2002). Restricted genetic and antigenic diversity of *Plasmodium falciparum* under mesoendemic transmission in the Venezuelan Amazon. *Parasitology* 124:569–81.

Tanabe K, Mackay M, Goman M, Scaife JG (1987). Allelic dimorphism in a surface antigen gene of the malaria parasite *Plasmodium falciparum*. *J. Mol. Biol.* 195:273–87.

Tibayrenc M, Ayala F (2002). The clonal theory of parasitic protozoa: 12 years on. *Trends Parasitol.* 18:405–10.

Tibayrenc M, Kjellberg F, Ayala FJ (1990). A clonal theory of parasitic protozoa: the population structures of Entamoeba, Giardia, Leishmania, Naegleria, Plasmodium, Trichomonas, and Trypanosoma and their medical and taxonomical consequences. *Proc. Natl. Acad. Sci. USA* 87:2414–8.

Urdaneta L, Lal A, Barnabe C, Oury B, Goldman I, Ayala FJ, Tibayrenc M (2001). Evidence for clonal propagation in natural isolates of *Plasmodium falciparum* from Venezuela. *Proc. Natl. Acad. Sci. USA* 98:6725–9.

Waters AP, Higgins DG, McCutchan TF (1991). *Plasmodium falciparum* appears to have arisen as a result of lateral transfer between avian and human hosts. *Proc. Natl. Acad. Sci. USA* 88:3140–44.

Wolfe ND, Escalante AA, Karesh WB, Kilbourn A, Spielman A, Lal AA (1998). Wild primate populations in emerging infectious disease research: the missing link? *Emerging Infect. Dis.* 4:149–58.

Wootton JC, Feng X, Ferdig MT, Cooper RA, Mu J, Baruch DI, Magill AJ, Su XZ (2002). Genetic diversity and chloroquine selective sweeps in *Plasmodium falciparum*. *Nature* 418:320–3.

Yoshida N, Di Santi SM, Dutra AP, Nussenzweig RS, Nussenzweig V, Enea V (1990). *Plasmodium falciparum*: restricted polymorphism of T cell epitopes of the circumsporozoite protein in Brazil. *Exp. Parasitol.* 71:386–92.

FIVE

G6PD Deficiency and Malarial Resistance in Humans: Insights from Evolutionary Genetic Analyses

Sarah A. Tishkoff and Brian C. Verrelli

THE ROLE OF G6PD IN MALARIA RESISTANCE

Malaria, resulting from infection by the *Plasmodium falciparum* and *Plasmodium vivax* parasites, is the leading cause of death in the global human population. Each year 500 million people suffer from malaria, resulting in approximately 2 million deaths annually. During the course of human evolution in regions where malaria is prevalent, naturally occurring genetic defense mechanisms have evolved for resisting infection by the *Plasmodium* parasite. Most of the human genes that are thought to provide resistance against malarial infection are expressed in red blood cells or play a role in the immune system (Hill, 2001). These loci include: HLA, α- and β-globins, Duffy factor (FY), tumor necrosis factor (TNF), and glucose-6-phosphate dehydrogenase (G6PD).

G6PD is an important "housekeeping" enzyme in the glycolytic pathway for glucose metabolism. G6PD also plays a critical role in maintaining the balance of NADPH, a necessary cofactor for cell detoxification. Although several enzymes can recycle this essential cofactor, G6PD is the sole generator of NADPH in the red blood cells and alone may prevent oxidative damage and severe anemia. Because of its role in preventing oxidative stress within the cell, G6PD may also play a role in longevity (Ho et al., 2000).

G6PD deficiency is the most common enzymopathy of humans, affecting an estimated 400 million people worldwide (Vulliamy et al., 1992). G6PD enzyme deficiency has been associated with many clinical disorders, such as neonatal jaundice, hemolytic anemia, granulomatous disease, and several cardiovascular diseases, often triggered by certain drugs

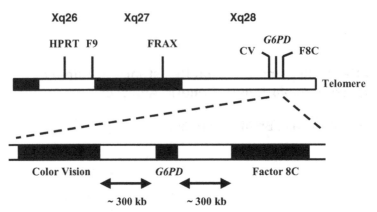

Figure 5.1. Diagram showing the proximity of *G6PD* to several genes of interest near the telomere of the X chromosome in region Xq28 (adapted from Vulliamy et al., 1992). HPRT, hypertension; F9, factor 9; FRAX, fragile X.

(particularly the antimalarial drug primaquine), foods, or infection, which cause oxidative stress to red blood cells (Vulliamy et al., 1992; Beutler, 1994; Ruwende and Hill, 1998). For example, digestion of fava beans by individuals with the common Mediterranean (Med) deficiency variant can trigger severe hemolytic anemia and death, causing the disease commonly referred to as "favism" (Vulliamy et al., 1992; Beutler, 1993). It is thought that Pythagoras was so distressed by this disorder that he was unwilling to cross a bean field while being attacked by an invading army, resulting in his death (Delwiche, 1978).

G6PD deficiency is a genetic disorder resulting from mutations within the *G6PD* gene, located on the telomeric region of the long arm of the X chromosome (Xq28) and flanked approximately 300 kb on either side by the factor VIII and red/green eye photopigment genes (Figure 5.1) that have been widely studied for their association with hemophilia (Toole et al., 1986; Jacquemin et al., 2000) and color-blindness, respectively (Nathans et al., 1986; Deeb et al., 1992). Nearly 400 G6PD variants have been identified to date based on electrophoretic and biochemical properties, and they have been grouped into classes based on the degree of enzyme deficiency and clinical symptoms (see reviews by Vulliamy et al., 1992; Beutler, 1993, 1994; Ruwende and Hill, 1998). The normal activity G6PD B variant is present worldwide, but other variants, particularly those resulting in enzyme deficiency, are restricted to specific geographic regions (Table 5.1), although they may occur at low frequency in regions where there has been recent gene flow (Vulliamy et al., 1992; Beutler, 1994; Verrelli et al., 2002). Many of the variants that were characterized as distinct at the biochemical level have turned out to be identical at the molecular level

Table 5.1. Several common G6PD deficiency mutations

Allele	Nucleotide Change[a]	Amino Acid Change[b]	Global Distribution
A	A / G (376)	Asn / Asp (126)	Africa
A-	A / G (376)	Asn / Asp (126)	Africa
	G / A (202)	Val / Met (68)	
Med	C / G (563)	Ser / Phe (188)	Middle East/Mediterranean
Seattle	G / C (844)	Asp / His (282)	Mediterranean
Mahidol	G / A (487)	Gly / Ser (160)	Southeast Asia
Union	C / T (360)	Arg / Cys (454)	Mediterranean/Southeast Asia

[a] Number in parentheses designates the nucleotide coding position.
[b] Number in parentheses designates the amino acid residue.
Source: adapted from Beutler (1994).

and vice versa (Beutler, 1993, 1994). At the molecular level, more than 130 different mutations have been identified in the *G6PD* gene that result in enzyme deficiency, nearly all of which are single-base substitutions in the coding sequence that cause a single amino acid substitution (Table 5.1; Beutler, 1994; Luzzatto et al., 2001). In addition, G6PD mutations that completely abolish the function of the protein are usually lethal (Vulliamy et al., 1993).

The distribution and frequency of G6PD deficiency are highly correlated with regions where malaria is currently (or previously has been) endemic (Figure 5.2). This observation has led to the widely accepted hypothesis that G6PD deficiency confers resistance to infection by the

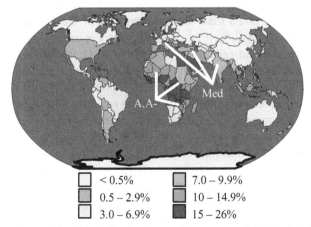

	< 0.5%		7.0 – 9.9%
	0.5 – 2.9%		10 – 14.9%
	3.0 – 6.9%		15 – 26%

Figure 5.2. Global distribution and frequency of G6PD deficiency (adapted from Vulliamy et al., 1992) with the location of common G6PD alleles.
For a colour version of this figure, see www.cambridge.org/9780521126557.

Plasmodium parasite (Allison, 1960; Allison and Clyde, 1961; Motulsky, 1961; Siniscalco et al., 1961; Luzatto et al., 1969). This correlation between allele frequency and incidence of malaria has been observed both on a global scale and on a micro-geographic scale. It has been noted that populations living at lower altitude, an environment more conducive for survival of the *Anopheles* mosquito vector that carries the malarial parasite, have a higher frequency of G6PD deficiency compared to high-altitude populations (Siniscalco et al., 1961; Ganczakowski et al., 1995). The hypothesis that mutations at *G6PD* confer resistance to malaria is supported by the observations that patients with G6PD deficiency have lower *P. falciparum* parasite loads than controls (e.g., Gilles et al., 1967) and that in heterozygotes for G6PD deficiency, more parasites are present in cells with normal enzyme activity than in cells with deficient enzyme activity (Luzzatto et al., 1969). *In vitro* culture studies demonstrate that parasite growth is inhibited in the first few cycles of infection in G6PD-deficient cells (Roth et al., 1983; Usanga and Luzzatto, 1985; Luzzatto et al., 1986). Additionally, a large case-control study of more than 2000 African children demonstrated that the most common form of G6PD deficiency in Africa (G6PD A-) is associated with a 46–58% reduction in risk of severe malaria for both female heterozygotes and male hemizygotes, although males may be more susceptible to hemopathologies (Ruwende et al., 1995). Thus, the selective advantage of the female heterozygote may be particularly strong because they are not as likely to suffer from disease caused by enzyme deficiency (Roth et al., 1983). The precise mechanism by which G6PD deficiency causes resistance to malaria remains unknown, although Friedman and Trager (1981) suggest that red cells deficient in G6PD may be placed under increased oxidative stress. Regardless of the mechanism, this oxidative stress may create a toxic environment for the *Plasmodium* parasites that cause malaria (Miller, 1994). Additionally, Cappadoro et al. (1998) observed that red blood cells that are deficient in G6PD enzyme and contain parasites at the ring stage of infection have a greatly impaired antioxidant defense and are more likely to undergo phagocytosis, thus decreasing the number of red cells containing mature parasites. Although G6PD enzyme deficiency may have detrimental effects, the benefit it provides in the presence of malaria suggests that G6PD deficiency may be maintained by balancing selection (Ruwende et al., 1995). Thus, the variability at the *G6PD* locus may be an example of a balanced polymorphism similar to the classic example of sickle cell anemia and thalassemia first described by Haldane (1949).

The *G6PD* gene spans approximately 18 kb and has 13 exons (Figure 5.3), coding for a protein consisting of 531 amino acids (Martini et al.,

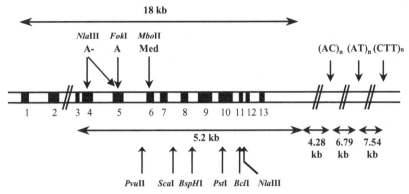

Figure 5.3. Diagram of *G6PD* gene structure showing the location of reported RFLPs and microsatellites (adapted from Tishkoff et al., 2001). Exons are shown as solid boxes. RFLPs that result in the three common deficiencies are shown, as well as six others (shown below gene) that result in silent SNPs. Exons 2 and 3 are separated by a 10-kb intron.

1986; Takizawa et al., 1986; Chen et al., 1991). The B variant, which possesses normal enzyme activity and dominates in frequency worldwide, is predicted to be the ancestral state by comparison with chimpanzee *G6PD* (Beutler et al., 1989; Kay et al., 1992; Verrelli et al., 2002). The two most common G6PD deficiency variants in Africa are G6PD A and G6PD A- (Table 5.1). The A variant has an 85% normal enzyme activity level and is not thought to confer resistance to malaria (Beutler et al., 1989; Ruwende et al., 1995). The A- deficiency was initially characterized by biochemical and electrophoretic analyses; however, recent studies have shown that this class of alleles is actually comprised of several different changes at the nucleotide sequence level (Beutler, 1994). The most common G6PD A- variant in Africa shares the mutation at nucleotide 376 that gives rise to the A allele, but it also has a mutation at nucleotide 202 (Table 5.1; Beutler et al., 1989; Beutler, 1994; Hirono et al., 2002). The two less common G6PD A- variants also share the mutation at nucleotide 376, but one has a T to C transition at nucleotide 968 and the other has a G to T transversion at nucleotide 680 (Beutler, 1994; Hirono et al., 2002). Throughout this review, we refer only to the most common G6PD A- variant as "G6PD A-." This G6PD A- variant has 12% of normal enzyme activity and is thought to provide resistance to malarial infection (Beutler et al., 1989; Ruwende et al., 1995). For G6PD B, frequencies range from 60–80%, for G6PD A from 15–40%, and for G6PD A- from 0–25% in sub-Saharan Africa, although frequencies of A and A- variants rarely reach more than 1–5%, respectively,

Table 5.2. RFLP identified at the *G6PD* locus

RFLP	Location	Nucleotide Change
*Nla*III	exon 4[a]	G / A (A-)
*Fok*I	exon 5[a]	A / G (A-, A)
*Pvu*II	intron 5	C / G
*Mbo*II	exon 6[a]	C / T (Med)
*Sca*I	intron 7	C / T
*Bsp*HI	intron 8	C / T
*Pst*I	exon 10/nt 1116[b]	G / A
*Bcl*I	exon 11/nt 1311[b]	C / T
*Nla*III	intron 11	C / T

[a] SNPs that change the amino acid sequence.
[b] SNPs at silent nucleotide sites in coding regions.

outside of Africa (Beutler, 1994; Ruwende et al., 1995; Ruwende and Hill, 1998). The G6PD Med variant has 3% normal enzyme activity and usually ranges in frequency from 2–20% but is as high as 70% among Kurdish Jews (Table 5.1; Oppenheim et al., 1993; Beutler, 1994). The A, A-, and Med variants can all be detected by RFLP analysis to distinguish these variants from the common G6PD B allele (Table 5.2).

MOLECULAR SIGNATURES OF NATURAL SELECTION

Although there are countless examples in the literature that document mutations in natural populations and their functional consequences, there is a great deal of information that this simple association does not convey. In many cases, we wish to know the age of this mutation or within what population it may have originated, and simply characterizing the presence or absence of a mutation in a collection of individuals is not sufficient. In fact, to understand something about the history or function of a gene mutation, it is also necessary to examine patterns of genetic variation in that gene in uninfected individuals to determine the impact of selection on the population as a whole. However, natural selection for an adaptive mutation can create many different patterns of nucleotide variability that differ from a simple neutral model of molecular evolution (Kimura, 1983; Hudson, 1990).

A general model of balancing selection predicts that no one single variant is functionally superior across all environments, and therefore we may discover more genetic variation at a locus than expected under neutrality. For example, if allelic diversity has been maintained in the population for an extended period of time, neutral variation can accumulate and "hitchhike" along with the selected variant (Kaplan et al., 1989; Hudson, 1990; Braverman et al., 1995). A classic example is the relatively ancient age of variants found at several MHC loci (e.g., Ayala et al., 1994). However, balancing selection need not imply that lineages are "old." If selection for functional diversity is relatively recent, then we may not find an excess of silent variation associated with the targeted balanced polymorphism simply because not enough time has passed. On the other hand, instead of preserving functional diversity at a locus, directional selection will effectively remove variation at a locus (Hudson, 1990; Akashi, 1999; Fay et al., 2001). This is often due to a mutation that rapidly reaches high frequency and replaces all other variants (i.e., becomes "fixed") because it is more adaptive in function. Only after there has been sufficient time will new mutations start to accumulate as "singletons" (i.e., found only once), and therefore the coalescence of all lineages in the gene genealogy will appear to be very recent. A definitive pattern associated with genes under recent directional selection is significant linkage disequilibrium (LD) due to the insufficient time for recombination to decay the association among variants (Braverman et al., 1995; Przeworski et al., 2000).

In a perfect scenario, selection events occur over very discrete time intervals in which we could easily detect the location and impact of adaptive mutations in the genome. However, a more realistic view considers that selection is operating on many genes and, in fact, different modes of selection may be operating at the same time; that is, even a single gene may have recently undergone both directional and balancing selection. In addition, we would like to be able to distinguish between patterns of variation that are the result of selection and those that are simply due to demographic processes. We expect that a demographic process may affect the genome as a whole, whereas selection is expected to impact only specific regions of the genome. Therefore, detecting the effects of a single selection event can be an arduous task and will rely on factors including the gene-specific mutation rate, long-term population size, and the strength of selection. Characterizing patterns of nucleotide sequence variation in human populations has just begun and it is unclear whether balancing or directional selection has played a larger role (Przeworski et al., 2000; Wall and Przeworski, 2000; Nachman, 2001), and several studies show no clear "signature" of selection (Harding et al., 1997; Aquadro et al., 2001). In the following discussion we present how different molecular approaches have been used to investigate the potential impact of selection at the *G6PD* locus.

RFLP HAPLOTYPE ANALYSIS AT *G6PD*

Although the *G6PD* locus has been the subject of active investigation among medical geneticists for the past 40 years and hundreds of variants causing enzyme deficiency have been identified in clinical cases, until very recently, relatively little was known about levels and patterns of molecular variation in non-coding regions of the gene. Analyses of patterns of genetic variation in both coding and non-coding regions of the *G6PD* gene are informative for reconstructing the evolutionary forces, such as mutation, migration, genetic drift, and selection resulting in the global distribution of G6PD deficiency mutations. Only six "silent" restriction fragment length polymorphisms (RFLPs) were identified prior to 2001, located either in introns or as synonymous substitutions in exons of the gene (Figure 5.3; Table 5.2).

These bi-allelic polymorphic sites have been studied as haplotypes to reconstruct the evolutionary history of mutation at the *G6PD* locus. Vulliamy et al. (1991) analyzed the *Nla*III and *Fok*I sites that distinguish the B, A, and A- variants, in addition to the *Pvu*II, *Sca*I, *Bsp*HI, *Pst*I, and *Bcl*I polymorphisms in 54 men of African origin. Because *G6PD* is located on the X chromosome, it is possible to unambiguously determine the order of alleles along the chromosome (e.g., haplotype phase) in men. Vulliamy et al. (1991) identified only 7 RFLP haplotypes: 3 associated with B chromosomes, 3 associated with A chromosomes, and only 1 associated with A- chromosomes. They conclude that the A- allele arose only once on an A chromosome and that all A- mutations can be traced back to a single ancestral chromosome in Africa. This conclusion has been supported by additional RFLP haplotype analyses (Coetzee et al., 1992; Kay et al., 1992; Tishkoff et al., 2001). Based on the number of estimated recombination events between RFLP sites, Kay et al. (1992) estimated that the A- mutation arose within the past 80,000 years but could have been considerably more recent. Beutler et al. (1989) proposed that the mutation giving rise to the A variant preceded the three A- mutations (at (nt) 202, nt 680, and nt 968) and that the A allele may have been more prevalent at the time that the A- mutations arose. Thus, they suggest that the A variant may have had an unknown selective advantage in an ancestral environment. In the majority of Mediterranean and Middle Eastern populations, a T is present at nt 1311 that is detected by a "*Bcl*I" restriction site (*Bcl*I+) using mismatched primers (Table 5.2), whereas in the Indian subcontinent, as well as in some Mediterranean populations, a C is present at nt 1311 (*Bcl*I−) (Beutler and Kuhl, 1990; Filosa et al., 1993; Cittadella et al., 1997). It was hypothesized that the presence of the Med mutation on two haplotype backgrounds could be due either to recurrent independent mutation in different populations or to a historic recombination event (Beutler and Kuhl, 1990; Filosa et al., 1993; Beutler, 1994). Interestingly, Filosa et al. (1993) observed a high frequency of the rare Med/*Bcl*I- haplotype in a population in southern Italy where it is always associated with color-blindness caused by mutation in a gene located over 300 kb away from *G6PD*. This strong association is likely the result of a founding event in this population but also indicates high levels of LD over an extended distance near *G6PD*. High levels of long-range LD in the region near *G6PD* have also been observed in recent studies by Saunders et al. (2002) and Sabeti et al. (2002).

MICROSATELLITE HAPLOTYPE ANALYSIS AT *G6PD*

Because of the low heterozygosity levels of the RFLPs identified at the *G6PD* locus and the strong LD between markers located within a 3-kb region of the gene (Vulliamy et al., 1991), the few RFLP haplotypes identified were not able to provide detailed resolution of the evolutionary history of *G6PD*. Additionally, because only one RFLP is polymorphic outside of Africa, RFLP haplotype analyses were not very informative for reconstructing the evolutionary history of the *G6PD* gene in non-African populations. The analysis of haplotypes consisting of both fast-evolving microsatellites as well as more slowly evolving markers, such as RFLPs, SNPs, and insertion/ deletion (In/Del) polymorphisms, can be particularly informative for reconstructing recent evolutionary history. Microsatellites have moderate to high mutation rates (on average .001 per meiosis) and are thought to mutate via the stepwise gain or loss of a repeat unit. The instability of microsatellites results in the formation of many alleles, and the stabile flanking markers allow greater certainty in tracing the lineage of each haplotype. For example, for two microsatellite markers, each with 10 alleles, there are $(10)^2$ possible haplotypes. Thus, if identical *G6PD* variants exist on a similar haplotype background, this would indicate that they descend from a common ancestral chromosome. By contrast, if identical *G6PD* variants exist on distinct haplotype backgrounds, this would indicate that they arose through recurrent mutation and had independent origins. The increased variability of haplotypes containing STRPs makes them particularly informative for reconstructing evolutionary events that have occurred on a recent time scale, such as the origin of mutations that provide resistance to malaria infection. After a mutation arises on a unique haplotype background, it will initially be in complete association (linkage disequilibrium) with flanking SNP and microsatellite markers. Linkage disequilibrium is expected to decay over time at a rate proportional to the rate of mutation at the microsatellite and recombination between the sites. Given an estimated mutation and recombination rate, LD can be used to estimate the time of origin of mutation events. Similar analyses (Slatkin and Rannala, 2000) have been used to estimate the time of origin of an *Alu* insertion polymorphism at the CD4 locus in non-African populations (Tishkoff et al., 1996), to estimate the age of the delta32 deletion at CCR5 in European populations that has been associated with resistance to HIV

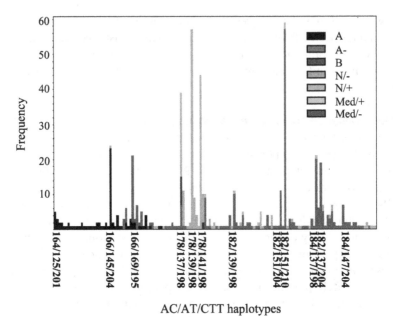

AC/AT/CTT haplotypes

Figure 5.4. Relative frequencies of AC/AT/CTT microsatellite haplotypes on B chromosomes ($n = 183$), A chromosomes ($n = 90$), A- chromosomes ($n = 42$), Norm/ *Bcl*I(−) chromosomes ($n = 188$), Norm/*Bcl*I(+) chromosomes ($n = 17$), Med/ *Bcl*I(+) chromosomes ($n = 63$), and Med/*Bcl*I(−) chromosomes ($n = 8$). Haplotypes are ordered by size of the AC, then the AT, then the CTT (data from Tishkoff et al., 2001). For a colour version of this figure, see www.cambridge.org/9780521126557.

(Stephens et al., 1998), and to estimate the age of the deltaF508 deletion that causes cystic fibrosis (Morral et al., 1994).

In order to obtain a more accurate estimate of the age of the major G6PD deficiency mutations in Africa (A-) and the Middle East and Mediterranean regions (Med), Tishkoff et al. (2001) identified three microsatellites within 19 kb downstream of the *G6PD* gene that could be used for microsatellite/RFLP haplotype analysis. The three microsatellite repeats (shown in Figure 5.3), referred to as "AC," "AT," and "CTT," were observed to have 10, 26, and 8 alleles, respectively, making them highly informative for reconstructing the evolutionary history of G6PD deficiency mutations. The "AC" repeat located 4.28 kb downstream of *G6PD* is a highly compound repeat consisting of the sequence $(TA)_5(AA)_1(TA)_6$ $(CA)_6(CT)_1(CA)_1(TA)_1(CA)_{10}$. PCR amplification using primers designed to flank the microsatellite repeat produces a product ranging from 164 to 188 bp in size. The $(AT)_n$ repeat, located 11.07 kb downstream from *G6PD*, consists of a perfect $(AT)_{14}$ repeat with alleles ranging from 125 to 179 bp in

Table 5.3. Microsatellite heterozygosity values for *G6PD* haplotypes

Haplotype[a]	Global Distribution	Heterozygosity[b]
B	African	0.96 ± 0.02
A	African	0.92 ± 0.04
Norm/*BclI*(+)	non-African	0.91 ± 0.10
Norm/*BclI*(−)	non-African	0.83 ± 0.03
A-	African	0.74 ± 0.08
Med/*BclI*(−)	Mediterranean	0.43 ± 0.17
Med/*BclI*(+)	Mediterranean	0.18 ± 0.04

[a] *Bcl*I designates the restriction site variant at the nucleotide coding position 1311.
[b] Mean ± standard error.
Source: adapted from Tishkoff et al. (2001).

size. The $(CTT)_n$ repeat, located 18.61 kb downstream from *G6PD*, is a compound repeat consisting of the sequence $(CTT)_{11}(ATT)_7$ with alleles ranging from 195 to 216 bp in size. Analysis of these microsatellites in a sample of 591 chromosomes originating from a worldwide global sample identified 149 AC/AT/CTT haplotypes (Tishkoff et al., 2001).

Microsatellite haplotype diversity was examined on chromosomes with normal enzyme activity (B chromosomes) as well as on chromosomes containing the A, A-, and Med deficiency mutations. B chromosomes outside of Africa (referred to as "norm" to distinguish them from B chromosomes in Africa) and Med chromosomes were further characterized by the presence (+) or absence (−) of the "*BclI*" site. The frequencies of AC/AT/CTT haplotypes on chromosomes with distinct *G6PD* alleles are shown in Figure 5.4. The pattern of haplotype variation and LD is strikingly different on chromosomes containing the deficiency mutations; there is reduced microsatellite haplotype diversity on chromosomes containing the Med and A- deficiency alleles (Table 5.3), and these alleles are associated with only a limited number of microsatellite haplotypes. Additionally, the microsatellite haplotype background is very distinct on deficiency versus normal chromosomes (Figure 5.4). The A- mutation appears to have derived from an A chromosome and is always associated with a 166-bp AC allele and a 195-bp CTT allele. A- mutations are associated with a broader range of large-sized AT alleles (which likely have a higher mutation rate), which are quite rare on other chromosomes. The majority of G6PD Med alleles are associated with the *BclI*(+) allele, and 57 out of 63 Med/*BclI*(+) chromosomes are associated with a 182/151/210 haplotype. Of the six chromosomes that do not have this association, five differ only by a single AC or AT repeat, and one (182/151/198) appears to be a recombinant at the CTT site. This haplotype background is very rare in the global sample and was observed on only two chromosomes from Cyprus and Italy.

EVOLUTIONARY HISTORY OF *G6PD* MICROSATELLITE HAPLOTYPES

All 90 A chromosomes reported by Tishkoff et al. (2001) were associated with a 166- or 164-bp allele, whereas all but one of the 183 B chromosomes were associated with a 176–186-bp allele. It is likely that the mutation resulting in a 166- or 164-bp allele occurred on an ancestral A chromosome. The highly compound AC repeat is expected to be more stable than the AT and CTT repeats, explaining the maintenance of LD between it and all *G6PD* variant alleles. However, there has been enough time to accumulate considerable variation at the AT and CTT microsatellites on both B and A chromosomes. The observation of only one recombinant chromosome between *G6PD* alleles and the AC repeat in Africa is consistent with previous reports indicating strong LD extending a distance of 350 kb at the *G6PD* locus (Filosa et al., 1993) as well as a study indicating regions of extensive LD on Xq28 (Taillon-Miller et al., 2000). Analysis of RFLP haplotypes by Tishkoff et al. (2001) identified nine RFLP haplotypes of moderate frequency that form two distinct clades consisting of A/A- and B chromosomes. Both the A and B clades have high levels of microsatellite haplotype diversity, supporting the hypothesis that the A variant may have an ancient origin (Beutler et al., 1989; Kay et al., 1992; Verrelli et al., 2002).

The A- mutation likely arose only once on an ancestral A chromosome containing a 164/169/195 haplotype (the most common A- haplotype) and then spread rapidly across a broad geographic range in Africa, with time for only a limited amount of variation at the AT repeat to accumulate. The few A and A- chromosomes observed outside of Africa have patterns of haplotype variation that are identical to that observed in Africa and likely originate from recent gene flow from Africa. The "Norm" chromosomes outside of Africa may descend from a subset of the B chromosomes that were carried by a small founding population during the migration of modern humans out of Africa within the past 100,000 years (Tishkoff and Williams, 2002). Genetic drift at the time of this founding event may have resulted in the distinct pattern of haplotype variability observed on normal chromosomes outside of Africa (Figure 5.4). The Med mutation likely arose on a normal chromosome with a 182/151/210 haplotype background (possibly on a Norm/*Bcl*I(+) chromosome) and spread rapidly throughout the

Middle East and Mediterranean region. The fact that Med and A- variants exist on very distinct microsatellite haplotypes supports the conclusion that they arose independently.

Coalescence modeling (Tishkoff et al., 2001) was used to determine whether the pattern of haplotype variability and LD on chromosomes containing A- and Med alleles is due to selection or neutral processes. This approach considers a sample of genes examined today and looks backwards in time at the pattern of common ancestry of those genes (Hudson, 1990). In the analysis by Tishkoff et al. (2001), 10,000 coalescence trees were simulated, conditional upon generating a gene genealogy that matched the observed number of B, A, A-, Norm, and Med alleles in the data set. Comparisons of the summary statistics from trees generated under a model of neutrality (both under an assumption of constant and expanding population size) with the observed data indicated that the high frequency and broad geographic range of G6PD deficiency mutations, in the face of low haplotype variability and high LD, are inconsistent with a model of neutrality. These initial results have been supported by similar analyses using SNP markers extending approximately 450-kb from G6PD (Sabeti et al., 2002). These data add support to the hypothesis that the A- and Med mutations have attained high frequency as a result of selection at this locus, most likely in response to malarial infection caused by the *Plasmodium* parasite.

Given that the ancestral A- mutation likely arose on an A chromosome containing a 164/169/195 haplotype background, and that the Med mutation likely arose on a normal chromosome with a 182/151/210 haplotype background, it is possible to estimate the time since the origin of these mutations by estimating the rate of decay of linkage disequilibrium due to mutations at the three microsatellites and recombination between the sites (Tishkoff et al., 2001). Tishkoff et al. (2001) simulated this process using a forward-in-time Poisson branching model, which considers the fate of a chromosome containing a mutation arising in an ancestral population of effective size 10,000 (the estimated effective population size of humans) that is rapidly growing. The ages of the A- and Med alleles (see Table 5.4) were estimated from these simulations that varied the mutation rate of the microsatellites, the rate of recombination between sites, and the strength of selection. The mean age of the A- allele in the simulated runs was 6,357 years, with a 95% credibility interval extending from 3,840 to 11,760 years. For the Med alleles, the mean age was 3,330 years with a 95% credibility interval of 1,600 to 6,640 years (Tishkoff et al., 2001). These age estimates imply a very recent origin for these two deficiency variants, and the narrow confidence intervals reflect the power that

Table 5.4. Various age estimates for *G6PD* alleles

Allele	Microsatellite LD Analysis (Tishkoff et al., 2001)	Sequence Analysis (Verrelli et al., 2002)	SNP LD Analysis (Saunders et al., in press)	SNP LD Analysis (Sabeti et al., 2002)
A-	6,357 (3,840–11,760)	45,000 (25,000–65,000)	10,000 (1,461–20,994)	2,325 (1,200–3,862.5)
Med	3,330 (1,600–6,640)	10,000 (0–35,000)	N/A	N/A
A	N/A	420,000 (300,000–540,000)	N/A	N/A

Note: estimates Age are given in years with 95% confidence intervals. N/A indicates that age of allele was not estimated.

haplotypes composed of quickly evolving markers can supply in estimating very recent events in evolutionary time.

NUCLEOTIDE DIVERSITY AT *G6PD*

By using rapidly evolving microsatellites in combination with *G6PD* single nucleotide polymorphisms (SNPs), the LD analysis of Tishkoff et al. (2001) showed that selection has had a large impact on the frequencies of the A- and Med deficiencies. However, several questions remained with respect to other aspects of variation at *G6PD*. First, most amino acid polymorphism at this locus had been discovered by biochemical analyses and it was unclear whether protein variation exists that would not have been picked up in functional studies. Second, although selection seems to have recently favored the A- and Med deficiencies, it is generally unknown what role, if any, other G6PD amino acid variants play. Therefore, by using a nucleotide sequencing approach, we can screen for all genetic variation at *G6PD* in a large random sample and determine if selection for malarial resistance is likely to explain patterns of variation associated with the gene genealogy.

To address some of these questions, Verrelli et al. (2002) surveyed *G6PD* nucleotide sequence variation from the region encompassing exons 3–13 for a total of 5.2 kb (Figure 5.3) from 216 male individuals from 13 populations, and Saunders et al. (2002) have surveyed nucleotide sequence variation in a 5.1-kb region of *G6PD* from 41 globally diverse male individuals (35 B, 2 A, and 4 A-). The sample used by Verrelli et al. (2002), which consists of 160 Africans (112 B, 32 A, and 16 A-) and 56 individuals from groups outside of sub-Saharan Africa (55 B and 1 Med), reflects the frequencies of the major G6PD alleles in natural populations, which is important for statistical tests that assume samples are unbiased (Hudson et al., 1994). Verrelli et al. (2002) and Saunders et al. (in press) found that *G6PD* segregates a typical level of silent site polymorphism when compared to even other X-linked loci (Przeworski et al., 2000; Nachman, 2001).

There are several statistical tests of neutrality that can be applied to data sets of nucleotide sequence variation (reviewed by Kreitman, 2000). For example, if a simple neutral explanation accounts for variation at *G6PD*, then we may expect that silent site polymorphism and divergence will be the same across loci in the human genome. The Hudson, Kreitman, and Aguade (HKA) test (Hudson et al., 1987) can be applied to *G6PD* in a comparison with other X-linked loci because they all share the relatively same effective population size (i.e., X chromosomes spend a third of their time hemizygous in males). Verrelli et al. (2002) and Saunders et al. (2002) also examined the SNP frequency distribution using the test of Tajima

(1989) to determine whether there is an excess of rare variants (due to directional selection) or high-frequency alleles (due to balancing selection) at *G6PD*. However, neither of these tests can significantly reject the neutral model for overall patterns of silent site variation at *G6PD*. A more practical approach may be to examine the patterns of variation associated with independent allele "lineages" (i.e., B, A, A-). A simulation test (Hudson, 1990) performed by Verrelli et al. (2002) found that given the overall level of variation at *G6PD*, there is proportionately less SNP variation associated with the A and A- alleles than one would expect by chance alone. This likely reflects the recent origin and rapid increase in frequency for the A-allele that was also found by Tishkoff et al. (2001). Because Verrelli et al. (2002) found low levels of SNP variation associated with the A allele as well, this allele may have also risen to high frequency recently. Although there has been little evidence for the adaptive value of this variant in regard to malarial resistance (Ruwende et al., 1995), the A variant may have had an adaptive function in response to an unknown selective force during the evolutionary history of humans in Africa.

Using all silent site variation at *G6PD*, a coalescent analysis can also be used to estimate the age of both the silent SNPs and the amino acid variants. Three approaches for dating the age of the A- variant gave consistent estimates for the age of this allele. Tishkoff et al. (2001) used LD analyses with flanking microsatellite alleles, and Saunders et al. (2002) as well as Sabeti et al. (2002), used LD analyses with SNPs identified in regions up to 556 kb from G6PD to estimate the ages of deficiency mutations (Table 5.4). In contrast, Verrelli et al. (2002) and Saunders et al. (in press) used the level of SNP variation associated with the different deficiency alleles in a maximum-likelihood approach that assumed complete neutrality (Figure 5.5; Griffiths and Tavare, 1997). Table 5.4 shows that the age of the A- variant estimated using all approaches is quite recent. However, the confidence intervals for the estimates based on SNP diversity are much larger, reflecting the high variance that is associated with recently derived SNP variants. Although the simulation "lineage" test (described above) found significantly less SNP variation associated with the A allele, the maximum-likelihood coalescent analysis of Verrelli et al. (2002) found that this variant may not be recent in origin.

Because G6PD enzyme deficiencies have both negative and positive (i.e., resistance to the parasite causing malaria) effects in the individuals who possess these alleles, allelic diversity at this locus is likely maintained in the population because no one phenotype is superior in all environments. The work of Ruwende et al. (1995) suggested that G6PD

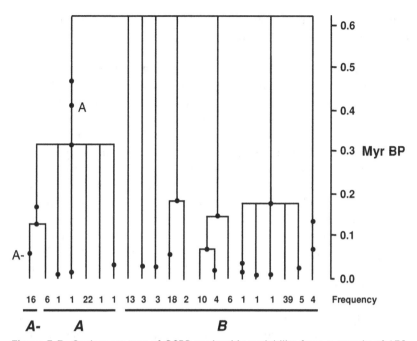

Figure 5.5. Coalescent tree of *G6PD* nucleotide variability from a sample of 158 Africans for a 5.2-kb region (data from Verrelli et al., 2002). Various silent mutation age estimates are labeled along the branches where they occur, and haplotype frequencies are shown.

amino acid variation was maintained by balancing selection because of the advantage it bestowed upon both males and females in the heterozygous state. To test this hypothesis, Verrelli et al. (2002) compared *G6PD* variation in human populations with chimpanzee (*Pan troglodytes*) *G6PD*. The use of a closely related outgroup for interspecific comparisons can tell us something about how variation has accumulated since species divergence. A McDonald–Kreitman (1991) test compares the ratio of silent and amino acid polymorphism within humans to the ratio of silent and amino acid fixed differences between humans and chimpanzees. Under neutrality we expect these ratios to be relatively equal. However, Verrelli et al. (2002) found a significant excess of amino acid polymorphism segregating within both African and non-African populations. There were no fixed amino acid differences between chimpanzees and humans, indicating functional constraint on the G6PD protein over the past 5–6 million years. Therefore, the excess of G6PD amino acid variation within humans is surprising and is likely a recent phenomenon. This result suggests a recent change in selection pressure favoring amino acid variation in humans and is consistent with balancing selection across many different geographic regions.

IMPLICATIONS FOR THE ORIGINS OF MALARIAL RESISTANCE

To summarize, studies of nucleotide diversity and microsatellite/RFLP haplotype diversity indicate low levels of genetic diversity on chromosomes containing A- and Med G6PD deficiency mutations, high levels of LD on these chromosomes, and an association of deficiency mutations with distinct haplotype backgrounds. Comparison of sequence diversity at *G6PD* in humans and chimpanzees indicates an excess of amino acid variation in humans relative to that expected based on levels of silent site variation and divergence. The pattern of genetic variation at *G6PD* is not consistent with a neutral model of evolution. Rather, it is most consistent with a model of balancing polymorphism whereby deficiency mutations are maintained at high frequency due to protection they provide against malarial infection but do not sweep to fixation because of their potentially serious negative effects, including severe anemia resulting in death.

Although the A- variant in Africa, which has a 12% enzyme activity levels, usually does not cause a severe phenotype, serious disease can be triggered by factors such as diet and infectious disease, the latter of which has historically been highly prevalent in African populations (Beutler, 1993, 1994). Additionally, because G6PD deficiency is an X-linked disease, there is a possibility that selection is acting differently in males and females. Females, who are usually heterozygous for the disorder, will have a mosaic of cells with normal and deficient enzyme activity due to the random nature of X-inactivation. The severity of disease in females will depend on the number of cells expressing deficiency alleles. Males, who are hemizygous, may be more likely to suffer the negative consequences of G6PD deficiency because all cells will have deficient enzyme activity levels. Although earlier reports indicated that only females heterozygous for G6PD deficiency were protected from malarial infection (Bienzle et al., 1975), later studies did not support these results and it seems likely that hemizygous males are also protected (Beutler, 1994; Ruwende et al., 1995). It is possible, however, that the more severe negative effects in males prevent G6PD deficiency alleles from reaching high frequency.

Coalescence analysis of nucleotide variability at the *G6PD* locus in Africans indicates that the A allele may not be recent in origin and may have risen in frequency after the divergence of modern human populations approximately 100,000 years ago (Verrelli et al., 2002; Saunders et al., in press). Interestingly, the age of the A allele predates the estimated age at which severe malaria likely had a major impact in humans. Although it may not be recent in age, it is likely that the A allele has recently attained a higher frequency in Africa and this may explain the absence of this variant in non-African populations. Functional studies suggested this variant has no association with malarial resistance; however, nucleotide analyses indicate that the A allele may be adaptive and is likely maintained by selection. Like the CCR5-Δ32 polymorphism that is not recent in origin, but is strongly associated with resistance to HIV infection (Martin et al., 1998), the A variant may have been historically maintained by an unknown selective pressure in African populations.

Studies of microsatellite/RFLP haplotypes (Tishkoff et al., 2001) and nucleotide diversity (Verrelli et al., 2002; Saunders et al., in press) indicate that G6PD deficiency mutations associated with resistance to malaria have originated and risen to high frequency very recently in human evolutionary history. Microsatellite haplotype analysis of A- and Med mutations indicates that they arose independently on chromosomes with distinct haplotype backgrounds (Tishkoff et al., 2001), likely because of a common selective pressure imposed by malarial infection. Distinct mutations have arisen in different geographic regions due to the stochastic nature of the mutation process. The diversity of deficiency variants identified across the globe suggests that the location of the amino acid substitutions, per se, may not be as important as their overall effect on enzyme activity level (Verrelli et al., 2002).

Based on decay of microsatellite haplotype diversity, Tishkoff et al. (2001) estimate that the A- allele originated in Africa within the past 3,840 to 11,760 years and that the Med mutation originated in the Mediterranean or Middle East within the past 1,600–6,640 years. After these mutations arose, they likely swept to high frequency across very broad geographic regions in a short period of time, indicating the strength of selection in response to the malarial parasite. Because of the recency of this event, the signature of selection cannot be detected by standard neutrality tests based on nucleotide diversity such as Tajima's D. However, microsatellites, which have a higher mutation rate, are more informative for reconstructing evolutionary events on such a recent time scale. In addition, neutrality tests like that of McDonald and Kreitman (1991) that do not rely solely on intraspecific variation but on interspecific divergence as well do possess the power to detect deviations from selection.

The estimate that G6PD deficiency mutations rose to high frequency within the past 12,000 years is consistent with estimates for a very recent origin of the sickle cell mutation in the β-globin gene within the past few thousand years, based on allele frequency data in Africa (Durham, 1992). These date estimates are also consistent with archeological evidence suggesting that malaria has only had a significant effect on humans in the past 10,000 years, after the advent of agriculture, animal domestication, and increased human population densities in the Middle East and Africa (Livingstone, 1971). Livingstone (1958, 1971) first proposed that there is a correlation between the prevalence of malaria and the practice of slash and burn agriculture in West African populations. According to Livingstone (1958, 1971), the introduction of slash and burn agriculture in West Africa approximately 2,000 to 4,000 years ago resulted in the clearing of tropical forests and an increase in sunlit pools of water, the preferred breeding place for *Anopheles gambiae*, the major vector for the *P. falciparum* parasite, the *Plasmodium* species associated with more severe, stable, hyperendemic malaria (Wiesenfeld, 1967). Additionally, agriculture enabled increased human population density, facilitating the spread of malaria and other infectious diseases. However, a number of factors may have caused malaria to become hyperendemic slightly earlier in Africa, as the date estimates of Tishkoff et al. (2001) suggest. Africa underwent an increase in both temperature and humidity between 12,000 and 7,000 years ago, along with a concurrent increase in the number of sunlit lakes and ponds, conditions that would support the spread and rapid adaptive speciation of the *A. gambiae* vector (Coluzzi, 1999). An increase in human population density likely occurred after the origination of plant and animal domestication in the Sahara and northeast Africa approximately 8,000–10,000 years ago (Livingstone, 1971; A. S. Brooks, personal communication, 2001), which could facilitate the spread of infectious disease. Secondly, archeological evidence indicates denser and more permanent populations around lakeshores due to the spread of fishing industries as well as to incipient cattle domestication in these regions (Tishkoff et al., 2001; A. S. Brooks, personal communication, 2001). These population settlements on or near lakeshores and water pools could have served as adequate preconditions for the spread of mosquito-borne pathogens (Tishkoff et al., 2001; A. S. Brooks, personal communication, 2001).

Estimates of genetic diversity in the mosquito vector *Anopheles gambiae* (Donnelly et al., 2001), as well as several genetic studies of the protozoan parasite *Plasmodium falciparum* that causes severe malaria (Anderson et al., 2000; Rich and Ayala, 2000; Volkman et al., 2001), are in accordance with the predicted recent expansion of human populations and may be due to the coevolution of these organisms. However, the evolutionary history of the *Plasmodium falciparum* genome remains controversial, and several studies (Verra and Hughes, 2000; Hughes and Verra, 2001; Mu et al., 2002) indicate that malaria caused by *P. falciparum* may be a more ancient disease.

It is possible that a form of *Plasmodium* causing less severe malaria has been infecting humans throughout much of their evolutionary history but that it did not become endemic until after 10,000 years ago, likely in response to climatic and/or cultural changes that facilitated population expansion and diversification of the *Anopheles* vector, the *P. falciparum* parasite, and the human host.

As discussed in Tishkoff et al. (2001), the more recent spread of the Med allele within the past 1,600–6,640 years is consistent with historical Greek and Egyptian documents indicating that, despite the earlier presence of more mild forms of malaria resulting from infection by *P. malariae* and *P. vivax*, the more severe *P. falciparum* malaria may not have been prevalent in the Mediterranean until after 500 B.C. (Sherman, 1998). It is possible that the recent and rapid spread of the Med allele corresponds with the spread of agriculture during a Neolithic expansion and migration across Europe from the Middle East 10,000–5,000 years ago (Cavalli-Sforza et al., 1994). However, the estimate that the mutation arose 1,600–6,640 years ago suggests that this mutation could have been spread by more recent migration events, perhaps as a result of the extensive trade routes and colonizations of the Greeks into these regions in the first several millennia B.C. (Tagarelli et al., 1991; Durando, 1997). It is even possible that the Med mutation was spread throughout this region by the army of Alexander the Great, which invaded and conquered territories ranging from the Mediterranean, to India, the Middle East, and even North Africa during the fourth century B.C. (Durando, 1997; Tishkoff et al., 2001).

FUTURE DIRECTIONS

Although these recent studies of molecular variation at *G6PD* have helped elucidate the evolutionary history of the deficiency-causing A, A-, and Med mutations, a number of unanswered questions remain. An extensive analysis of nucleotide variability in a 5.2-kb region of *G6PD* across geographically diverse African populations (Verrelli et al., 2002) identified only one severe deficiency mutation that was common (the A- variant). By contrast, in Mediterranean and Middle Eastern populations, approximately 85% of all deficiency alleles are caused by the Med mutation, but several other variants also occur at moderate frequencies, and in Asian populations, there are many deficiency alleles at moderate to high frequency (Calabro et al., 1998; Beutler, 1994). This observation raises the question of why more deficiency mutations have not become common in Africa in the past 10,000 years. It is possible that this reflects a bias due to the fact that non-African populations have been much more intensively studied, with biochemical and molecular screens of thousands of individuals with G6PD deficiency. Alternatively, does the A- variant have a particular selective advantage in an African environment? Or, is the high frequency of the A- allele across African regions simply a reflection of high levels of gene flow, genetic drift, and selection? Studies of haplotype variation in mtDNA, Y-chromosome, and autosomal loci indicate high levels of population subdivision in Africa (Tishkoff and Williams, 2002), which is not consistent with the shared pattern of genetic variation observed at *G6PD*. It is possible that the A- allele arose in a West African Bantu-speaking population (which first developed slash and burn agriculture) and then swept through other geographic regions due to the Bantu expansion across western and southern Africa that occurred within the past 3,000 years (Tishkoff and Williams, 2002). However, given that other studies do not find evidence for extensive levels of gene flow among African populations, these results suggest that low levels of gene flow, when combined with moderate to high levels of selection, can cause rapid changes in the genetic architecture of populations.

An additional question is whether mutations at *G6PD* interact epistatically with other malarial resistance genes at high frequency in African populations such as the sickle cell, HbC, and alpha and beta thalassemia mutations in the globin genes or genes involved in immune function such as HLA, TNF, and other cytokines. Several studies have observed a positive association between the frequency of G6PD deficiency and sickle cell anemia (Lewis and Hathorn, 1963; Piomelli et al., 1972; Beutler et al., 1974; Warsy, 1985; Samuel et al., 1986; Steinberg et al., 1988) whereas others show that the incidence of hemoglobin S and G6PD deficiency are independent (Steinberg and Dreiling, 1974; Bienzle et al., 1975; Bernstein et al., 1980; Nieuwenhuis et al., 1986; Saad and Costa, 1992). Additional studies of gene frequencies across loci in multiple ethnically diverse African populations will be required to adequately answer this question.

Haplotype analyses incorporating additional markers flanking *G6PD* will be useful for reconstructing the evolutionary history of many deficiency mutations in globally diverse populations. In particular, the question of whether the Med mutation arose independently on a distinct haplotype background or due to an historic recombination event between a Med and normal chromosome can be addressed by analyzing additional markers upstream of *G6PD*. A better understanding of genetic diversity at *G6PD* and the evolutionary forces (mutation, migration, drift, and selection) resulting in the global distribution of G6PD deficiency mutations will be informative for identifying naturally occurring mechanisms of resistance against malarial infection and for understanding which mutations may be most efficient at providing natural resistance. Functional studies of those variants could be informative for developing new treatments to prevent the spread of malaria.

ACKNOWLEDGMENTS
Funded by a Burroughs Wellcome Fund Career Award, a David and Lucile Packard Career Award, and NSF Grant BCS-9905396 to SAT. BCV was partially supported by NSF IGERT Training Grant BCS-9987590.

REFERENCES
Akashi, H. (1999). Inferring the fitness effects of DNA mutations from polymorphism and divergence data: statistical power to detect directional selection under stationary and free recombination. *Genetics*, **151**, 221–238.

Allison, A. C. (1960). Glucose-6-phosphate dehydrogenase deficiency in red blood cells of East Africans. *Nature*, **186**, 531–532.

Allison, A. C., and Clyde, D. F. (1961). Malaria in African children with deficient erythrocyte glucose-6-phosphate dehydrogenase. *Brit. Med. J.*, **1**, 1346–1356.

Anderson, T. J., Haubold, B., Williams, J. T., Estrada-Franco, J. G., Richardson, L., and 11 colleagues. (2000). Microsatellite markers reveal a spectrum of population structures in the malaria parasite *Plasmodium falciparum*. *Mol. Biol. Evol.*, **17**, 1467–1482.

Aquadro, C. F., Bauer DuMont, V., and Reed, F. A. (2001). Genome-wide variation in the human and fruitfly: a comparison. *Curr. Opin. Genet. Dev.*, **11**, 627–634.

Ayala, F. J., Escalante, A., O'Huigin, C., and Klein, J. (1994). Molecular genetics of speciation and human origins. *Proc. Natl. Acad. Sci. USA*, **91**, 6787–6794.

Bernstein, S. C., Bowman, J. E., and Noche, L. K. (1980). Population studies in Cameroon. *Hum. Hered.*, **30**, 7–11.

Beutler, E. (1993). Study of glucose-6-phosphate-dehydrogenase – history and molecular-biology. *Am. J. Hematol.*, **42**, 53–58.

Beutler, E. (1994). G6PD deficiency. *Blood*, **84**, 3613–3636.

Beutler, E., Johnson, C., Powars, D., and West, C. (1974). Prevalence of glucose-6-phosphate dehydrogenase deficiency in sickle cell disease. *N. Engl. J. Med.*, **290**, 826–828.

Beutler, E., Kuhl, W., Vives-Corrons, J. L., and Prchal, J. T. (1989). Molecular heterogeneity of glucose-6-phosphate dehydrogenase A-. *Blood*, **74**, 2550–2555.

Beutler, E., and Kuhl, W. (1990). The NT 1311 polymorphism of *G6PD*: *G6PD* Mediterranean mutation may have originated independently in Europe and Asia. *Am. J. Hum. Genet.*, **47**, 1008–1012.

Bienzle, U., Sodeinde, O., Effiong, C. E., and Luzzatto, L. (1975). Glucose-6-phosphate dehydrogenase deficiency and sickle cell anemia: frequency and features of the association in an African community. *Blood*, **46**, 591–597.

Braverman, J. M., Hudson, R. R., Kaplan, N. L., Langley, C. H., and Stephan, W. (1995). The hitchhiking effect on the site frequency spectrum of DNA polymorphisms. *Genetics*, **140**, 783–796.

Calabro, V., Mason, P. J., Filosa, S., Civitelli, D., Cittadella, R., and 4 colleagues. (1998). Genetic heterogeneity of glucose-6-phosphate dehydrogenase deficiency revealed by single-strand conformation and sequence analysis. *Am. J. Hum. Genet.*, **52**, 527–536.

Cappadoro, M., Giribaldi, G., O'Brien, E., Turrini, F., Mannu, F., and 4 colleagues. (1998). Early phagocytosis of glucose-6-phosphate dehydrogenase (G6PD)-deficient erythrocytes parasitized by *Plasmodium falciparum* may explain malaria protection in G6PD deficiency. *Blood*, **92**, 2527–2534.

Cavalli-Sforza, L. L., Piazza, A., and Menozzi, P. (1994). *History and Geography of Human Genes*. Princeton University Press, Princeton, NJ.

Chen, E. Y., Cheng, A., Lee, A., Kuang, W.-J., Hillier, L., and 4 colleagues. (1991). Sequence of human glucose-6-phosphate dehydrogenase cloned in plasmids and a yeast artificial chromosome (YAC). *Genomics*, **10**, 792–800.

Cittadella, B., Civitelli, D., Manna, I., Azzia, N., DiCataldo, A., Schiliroa, G. and Brancati, C. (1997). Genetic heterogeneity of glucose-6-phosphate dehydrogenase deficiency in south-east Sicily. *Ann. Hum. Genet.*, **61**, 229–234.

Coetzee, M. J., Bartleet, S. C., Ramsay, M., and Jenkins, T. (1992). Glucose-6-phosphate dehydrogenase (G6PD) electrophoretic variants and the *PvuII* polymorphism in Southern African populations. *Hum. Genet.*, **80**, 111–113.

Coluzzi, M. (1999). The clay feet of the malaria giant and its African roots: hypotheses and inferences about origin, spread and control of *Plasmodium falciparum*. *Parassitologia*, **41**, 277–283.

Deeb, S. S., Lindsey, D. T., Hibiya, Y., Sanocki, E., Winderickx, J., and 2 colleagues. (1992). Genotype-phenotype relationships in human red/green color-vision defects: molecular and psychophysical studies. *Am. J. Hum. Genet.*, **51**, 687–700.

Delwiche, C. C. (1978). Legumes – past, present, and future. *Bioscience*, **28**, 565–570.

Donnelly, M. J., Licht, M. C., and Lehmann, T. (2001). Evidence for recent population expansion in the evolutionary history of the malaria vectors *Anopheles arabiensis* and *Anopheles gambiae*. *Mol. Biol. Evol.*, **18**, 1353–1364.

Durando, F. (1997). *Ancient Greece; The Dawn of the Western World*. Syewart, Tabori, and Chang Publishers, New York.

Durham, W. H. (1992). *Coevolution: Genes, Culture, and Human Diversity*. Stanford University Press, Stanford, CA.

Fay, J. C., Wyckoff, G. J., and Wu, C.-I. (2001). Positive and negative selection on the human genome. *Genetics*, **158**, 1227–1234.

Filosa, S., Calabro, V., Lania, G., Vulliamy, T. J., Brancati, C., and 3 colleagues. (1993). *G6PD* haplotypes spanning Xq28 from *F8C* to red/green color vision. *Genomics*, **17**, 6–14.

Friedman, M. J., and Trager, W. (1981). The biochemistry of resistance to malaria. *Sci. Am.*, **244**, 154–164.

Ganczakowski, M., Town, M., Bowden, D. K., Vulliamy, T. J., Kaneko, A., and 3 colleagues. (1995). Multiple glucose-6-phosphate dehydrogenase-deficient variants correlate with malaria endemicity in the Vanuatu Archipelago (Southwestern Pacific). *Am. J. Hum. Genet.*, **56**, 294–301.

Gilles, H. M., Fletcher, K. A., Hendrickse, R. G., Lindner, R., Reddy, S., and colleagues. (1967). Glucose-6-phosphate dehydrogenase deficiency, sickling, and malaria in African children in South Western Nigeria. *Lancet*, **1**, 138–140.

Griffiths, R. C., and Tavare, S. (1997). Computational methods for the coalescent. In *Progress in Population Genetics and Human Evolution*, edited by P. Donnelly and S. Tavare, pp. 165–182. Springer-Verlag, New York.

Haldane, J. B. S. (1949). Disease and evolution. *La Ricerca Sci.*, **19**, 1–20.

Harding, R. M., Fullerton, S. M., Griffiths, R. C., Bond, J., Cox, M. J., and 3 colleagues. (1997). Archaic African and Asian lineages in the genetic ancestry of modern humans. *Am. J. Hum. Genet.*, **60**, 772–789.

Hill, A. V. (2001). The genomics and genetics of human infectious disease suscepti-bility. *Ann. Rev. Genomics Hum. Genet.*, **2**, 373–400.

Hirono, A., Kawate, K., Honda, A., Fujil, H., and Miwa, S. (2002). A single mutation 202G > A in the human glucose-6-phosphate dehydrogenase gene (*G6PD*) can cause acute hemolysis by itself. *Blood*, **99**, 1498–1499.

Ho, H. Y., Cheng, M. L., Lu, F. J., Chou, Y. H., Stern, A., and 2 colleagues. (2000). Enhanced oxidative stress and accelerated cellular senescence in glucose-6-phosphate dehydrogenase (G6PD)-deficient human fibroblasts. *Free Radic. Biol. Med.*, **29**, 156–159.

Hudson, R. R. (1990). Gene genealogies and the coalescent process. In *Oxford Series in Evolutionary Biology*, edited by D. J. Futuyma and J. Antonovics, pp. 1–44. Oxford University Press, New York.

Hudson, R. R., Kreitman, M., and Aguade, M. (1987). A test of neutral molecular evolution based on nucleotide data. *Genetics*, **116**, 153–159.

Hudson, R. R., Bailey, K., Skarecky, D., Kwiatowski, J., and Ayala, F. J. (1994). Evi-dence for positive selection in the superoxide dismutase (*Sod*) region of *Drosophila melanogaster*. *Genetics*, **136**, 1329–1340.

Hughes, A. L., and Verra, F. (2001). Very large long-term effective population size in the virulent human malaria parasite *Plasmodium falciparum*. *Proc. R. Soc. Lond. B.*, **268**, 1855–1860.

Jacquemin, M., Lavend'homme, R., Benhida, A., Vanzieleghem, B., d'Oiron, R., and 10 colleagues. (2000). A novel cause of mild/moderate hemophilia A: mutations scattered in the factor VIII C1 domain reduce factor VIII binding to von Willebrand factor. *Blood*, **96**, 958–965.

Kaplan, N. L., Hudson, R. R., and Langley, C. H. (1989). The "hitchhiking effect" revisited. *Genetics*, **123**, 887–899.

Kay, A. C., Kuhl, W., Prchal, J. T., and Beutler, E. (1992). The origin of glucose-6-phosphate dehydrogenase (*G6PD*) polymorphisms in Afro-Americans. *Am. J. Hum. Genet.*, **50**, 394–398.

Kimura, M. (1983). *The Neutral Theory of Molecular Evolution*. Cambridge University Press, London.

Kreitman, M. (2000). Methods to detect selection in populations with applications to the human. *Ann. Rev. Genomics Hum. Genet.*, **1**, 539–559.

Lewis, R. A., and Hathorn, M. (1963). Glucose-6-phosphate dehydrogenase defi-ciency correlated with S hemoglobin. *Ghana Med. J.*, **2**, 131–141.

Livingstone, F. B. (1958). Anthropological implications of sickle cell gene distribu-tion in West Africa. *Am. Anthropol.*, **60**, 533–562.

Livingstone, F. B. (1971). Malaria and human polymorphisms. *Ann. Rev. Genet.*, 5, 33–64.

Luzzatto, L., Usanga, E. A., and Reddy, S. (1969). Glucose 6-phosphate dehydrogenase deficient red cells: resistance to infection by malarial parasites. *Science*, 164, 839–842.

Luzzatto, L., O'Brien, S., Usanga, E., and Wanachiwanawin, W. (1986). Origin of G6PD polymorphism: malaria and G6PD deficiency. In *Glucose-6-Phosphate Dehydrogenase*, edited by A. Yoshida and E. Beutler, p. 181. Academic Press, Orlando, FL.

Luzzatto, L., Mehta, A., and Vulliamy, T. J. (2001). *The Metabolic and Molecular Bases of Inherited Disease*. McGraw-Hill, New York.

Martin, M. P., Dean, M., Smith, M. W., Winkler, C., Gerrard, B., and 11 colleagues. (1998). Genetic acceleration of AIDS progression by a promoter variant at CCR5. *Science*, 282, 1907–1911.

Martini, G., Toniolo, D., Vulliamy, T., Luzzatto, L., Dono, R., and colleagues. (1986). Structural analysis of the X-linked gene encoding human glucose-6-phosphate dehydrogenase. *EMBO J.*, 5, 1849–1855.

McDonald, J. H., and Kreitman, M. (1991). Adaptive protein evolution at the *Adh* locus in Drosophila. *Nature*, 351, 652–654.

Miller, L. H. (1994). Impact of malaria on genetic polymorphism and genetic diseases in Africans and African Americans. *Proc. Natl. Acad. Sci. USA*, 91, 2415–2419.

Morral, N., Bertranpetit, J., Estivill, X., Nunes, V., Casals, T., and 26 colleagues. (1994). The origin of the major cystic fibrosis mutation (delta F508) in European populations. *Nat. Genet.*, 7, 169–175.

Motulsky, A. G. (1961). Glucose-6-phosphate dehydrogenase deficiency haemolytic disease of newborn, and malaria. *Lancet*, 1, 1168–1169.

Mu, J., Duan, J., Makova, K. D., Joy, D. A., Huynh, C. Q., and 3 colleagues. (2002). Chromosome-wide SNPs reveal an ancient origin for *Plasmodium falciparum*. *Nature*, 18, 323–326.

Nachman, M. W. (2001). Single nucleotide polymorphisms and recombination rate in humans. *Trends Genet.*, 17, 481–485.

Nathans, J., Thomas, D., and Hogness, D. S. (1986). Molecular genetics of human color vision: the genes encoding blue, green, and red pigments. *Science*, 232, 193–202.

Nieuwenhuis, F., Wolf, B., Bomba, A., and De Graaf, P. (1986). Haematological study in Cabo Delgado province, Mozambique; sickle cell trait and G6PD deficiency. *Trop. Geogr. Med.*, 38, 183–187.

Oppenheim, A., Jury, C. L., Rund, D., Vulliamy, T. J., and Luzzatto, L. (1993). G6PD Mediterranean accounts for the high prevalence of G6PD deficiency in Kurdish Jews. *Hum. Genet.*, 91, 293–294.

Piomelli, S., Reindorf, C. A., Arzanian, M. T., and Corash, L. M. (1972). Clinical and biochemical interactions of glucose-6-phosphate dehydrogenase deficiency and sickle-cell anemia. *N. Engl. J. Med.*, 287, 213–217.

Przeworski, M., Hudson, R. R., and Di Rienzo, A. (2000). Adjusting the focus on human variation. *Trends Genet.*, 16, 296–302.

Rich, S. M., and Ayala, F. J. (2000). Population structure and recent evolution of *Plasmodium falciparum*. *Proc. Natl. Acad. Sci. USA*, **97**, 6994–7001.

Roth, E. F., Raventos Suarez, C., Rinaldi, A., and Nagel, R. L. (1983). The effect of X chromosome inactivation on the inhibition of *Plasmodium falciparum* malaria growth by glucose-6-phosphate-dehydrogenase- deficient red cells. *Blood*, **62**, 866–868.

Ruwende, C., and Hill, A. (1998). Glucose-6-phosphate dehydrogenase deficiency and malaria. *J. Mol. Med.*, **76**, 581–588.

Ruwende, C., Khoo, S. C., Snow, R. W., Yates, S. N., Kwiatkowski, D., and colleagues. (1995). Natural selection of hemi- and heterozygotes for G6PD deficiency in Africa by resistance to severe malaria. *Nature*, **376**, 246–249.

Saad, S. T. O., and Costa, F. F. (1992). Glucose-6-phosphate dehydrogenase deficiency and sickle cell disease in Brazil. *Hum. Hered.*, **42**, 125–128.

Sabeti, P. C., Reich, D. E., Higgins, J. M., Levine, H. Z., Richter, D. J., Schaffner, S. F., Gabriel, S. B., Platko, J. V., Patterson, N. J., McDonald, G. J., Ackerman, H. C., Campbell, S. J., Altshuler, D., Cooper, R., Kwiatkowski, D., Ward, R., Lander, E. S. (2002). Detecting recent positive selection in the human genome from haplotype structure. *Nature*, **419(6909)**, 832–837.

Samuel, A. P. W., Saha, N., Acquaye, J. K., Omer, A., Ganeshaguru, K., and 1 colleague. (1986). Association of red cell glucose-6-phosphate dehydrogenase with haemoglobinopathies. *Hum. Hered.*, **36**, 107–112.

Saunders, M. A., Hammer, M. F., and Nachman, M. W. (2002). Nucleotide variability at *G6PD* and the signature of malarial selection in humans. *Genetics*, **162**, 1849–1861.

Sherman, I. W. (1998). A brief history of malaria and discovery of the parasite's life cycle. In *Malaria: Parasite Biology, Pathogenesis, and Protection*, edited by I. W. Sherman, pp. 3–10. American Society for Microbiology, Washington, D.C.

Siniscalco, M., Bernini, L., Latte, B., and Motulsky, A. G. (1961). Favism and thalassaemia in Sardinia and their relationship to malaria. *Nature*, **190**, 1179–1180.

Slatkin, M., and Rannala, B. (2000). Estimating allele age. *Ann. Rev. Genomics Hum. Genet.*, **1**, 225–249.

Steinberg, M. H., and Dreiling, B. J. (1974). Glucose-6-phosphate dehydrogenase deficiency in sickle cell anemia. *Ann. Intern. Med.*, **80**, 217–220.

Steinberg, M. H., West, M. S., Gallagher, D., and Mentzer, W. C. J. (1988). The cooperative study of sickle cell diseases: effects of glucose-6-phosphate dehydrogenase deficiency upon sickle cell anemia. *Blood*, **71**, 748–752.

Stephens, J. C., Reich, D. E., Goldstein, D. B., Shin, H. D., Smith, M. W., and 34 colleagues. (1998). Dating the origin of the *CCR5-Delta32* AIDS-resistance allele by the coalescence of haplotypes. *Am. J. Hum. Genet.*, **62**, 1507–1515.

Tagarelli, A., Bastone, L., Cittadella, R., Calabro, V., Bria, M., and colleagues. (1991). Glucose-6-phosphate dehydrogenase (G6PD) deficiency in southern Italy: a study on the population of the Cosenza province. *Gene Geogr.*, **5**, 141–150.

Taillon-Miller, P., Bauer-Sardina, I., Saccone, N. L., Putzel, J., Laitinen, T., and colleagues. (2000). Juxtaposed regions of extensive and minimal linkage disequilibrium in human Xq25 and Xq28. *Nat. Genet.*, **25**, 324–328.

Tajima, F. (1989). Statistical method for testing the neutral mutation hypothesis by DNA polymorphism. *Genetics*, **123**, 585–595.

Takizawa, T., Huang, I. Y., Ikuta, T., and Yoshida, A. (1986). Human glucose-6-phosphate dehydrogenase: primary structure and cDNA cloning. *Proc. Natl. Acad. Sci. USA*, **83**, 4157–4161.

Tishkoff, S. A., Dietzsch, E., Speed, W., Pakstis, A. J., Kidd, J. R., and 5 colleagues. (1996). Global patterns of linkage disequilibrium at the *CD4* locus and modern human origins. *Science*, **271**, 1380–1387.

Tishkoff, S. A., Varkonyi, R., Cahinhinan, N., Abbes, S., Argyropoulos, G., and colleagues. (2001). Haplotype diversity and linkage disequilibrium at human *G6PD*: recent origin of alleles that confer malarial resistance. *Science*, **293**, 455–462.

Tishkoff, S. A., and Williams, S. M. (2002). Genetic analysis of African populations: human evolution and complex disease. *Nature Rev. Genet.*, **3**, 611–621.

Toole, J. J., Pittman, D. D., Orr, E. C., Murtha, P., Wasley, L. C., and Kantman, R. J. (1986). A large region (approximately equal to 95 kDa) of human factor VIII is dispensable for in vitro procoagulant activity. *Proc. Natl. Acad. Sci. USA*, **83**, 5939–5942.

Usanga, E. A., and Luzzatto, L. (1985). Adaptation of *Plasmodium falciparum* to glucose 6-phosphate dehydrogenase-deficient host red cells by production of parasite-encoded enzyme. *Nature*, **313**, 793–795.

Verra, F., and Hughes, A. L. (2000). Evidence for ancient balanced polymorphism at the Apical Membrane Antigen-1 (*AMA-1*) locus of *Plasmodium falciparum*. *Mol. Biochem. Parasit.*, **105**, 149–153.

Verrelli, B. C., McDonald, J. H., Argyropoulos, G., Destro-Bisol, G., Froment, A., and 5 colleagues. (2002). Evidence for balancing selection from nucleotide sequence analyses of human *G6PD*. *Am. J. Hum. Genet.*, **71**, 1112–1128.

Volkman, S. K., Barry, A. E., Lyons, E. J., Nielsen, K. M., Thomas, S. M., and colleagues. (2001). Recent origin of *Plasmodium falciparum* from a single progenitor. *Science*, **293**, 482–484.

Vulliamy, T. J., Othman, A., Town, M., Nathwani, A., Falusi, A. G., and 2 colleagues. (1991). Polymorphic sites in the African population detected by sequence analysis of the glucose-6-phosphate dehydrogenase gene outline the evolution of the variants A and A-. *Proc. Natl. Acad. Sci. USA*, **88**, 8568–8571.

Vulliamy, T., Mason, P., and Luzzatto, L. (1992). The molecular basis of glucose-6-phosphate dehydrogenase deficiency. *Trends Genet.*, **8**, 138–143.

Vulliamy, T., Beutler, E., and Luzzatto, L. (1993). Variants of glucose-6-phosphate-dehydrogenase are due to missense mutations spread throughout the coding region of the gene. *Hum. Mutat.*, **2**, 159–167.

Wall, J. D., and Przeworski, M. (2000). When did the human population size start increasing? *Genetics*, **155**, 1865–1874.

Warsy, A. S. (1985). Frequency of glucose-6-phosphate dehydrogenase deficiency in sickle-cell disease. *Hum. Hered.*, **35**, 143–147.

Wiesenfeld, S. L. (1967). Sickle-cell trait in human biological and cultural evolution. Development of agriculture causing increased malaria is bound to gene-pool changes causing malaria reduction. *Science*, **157**, 1134–1140.

SIX

The Enigma of *Plasmodium vivax* Malaria and Erythrocyte Duffy Negativity

Peter A. Zimmerman

INTRODUCTION

Humans have been plagued by malarial parasites for centuries and reference to maladies associated with malaria may be found in antiquities over the past 5,000 years (1). For much of this time the cause of the intermittent chills and fevers, splenomegaly, and mortality associated with malaria was unknown. However, with consistent identification of the brownish black pigment (hemozoin) found during autopsies of malaria victims from the early 1700s on, scientific discovery methodically began to dissect malarial parasites from the various secret hiding places of their complex life cycle. Alphonse Laveran first observed the tiny ring-stage parasites of the malaria blood-stage infection in 1880 (1), and Ronald Ross would reveal that the female anopheline mosquito was responsible for malaria transmission in 1897 (1). During the late 1800s and ending in 1922 individual discoveries illustrated that malaria in humans was caused by four distinct species of *Plasmodium* – *P. falciparum*, *P. vivax*, *P. malariae*, and *P. ovale* (2), and that fevers resulting from infection by these parasites would soon find their way into successful, albeit unorthodox, treatment of neurosyphilis.

The era of malaria therapy, launched in earnest by Julius Wagner von Jauregg in 1917 (3), paved the way for important advances in malaria research as observations resulting from thousands of treated patients provided the opportunity to study early stages of infection, development of immunity and characteristics of the immune response, and the efficacy of various antimalarial drugs (4). It was also through malaria therapy trials that many African American patients were observed to be highly resistant to infection by *P. vivax* but not to the other human malarial parasites (5–8). At the time, the mechanism underlying African American resistance to *P.*

vivax was unknown, and relationships between molecules and genes and between mutation and natural selection were not well understood.

Although important advances were accumulating in the identification of microbial pathogens through the methods and observations of Pasteur and Koch, application of the scientific method in the practice of medicine was still relatively new in the early 1900s. It is also useful to remember that the scientific community was largely unsettled with regard to Charles Darwin's theory that natural selection (9) was the driving force of evolution until the late 1940s. Ironically, from present-day perspectives, opposing factions of biologists argued that the mechanism of evolution was either mutation *or* selection. The population biologists Fisher, Haldane, and Wright were ultimately instrumental in promoting an "evolutionary synthesis" that came to recognize mutation(s) as the genetic raw material upon which natural selection was based, and they launched an important era that has come to define the gene and its molecular role in cell to organismal structure and function phenotypes (10,11). The evidence for, and practical implications of, mutation/selection and human population genetic diversity (polymorphism) arose with blood transfusion and transplantation medicine, which allowed many of the first human genetic polymorphisms to be identified through serological cross-reactivity recognizing variations in blood group antigens (12,13).

Of these human blood groups, Duffy blood group–negativity and its relationship with resistance to *P. vivax* blood-stage malaria has presented the most interesting puzzles to malariologists and population geneticists, and now to molecular biologists as well. Although we have come to understand details of very specific molecular interactions between *P. vivax* and the Duffy protein required to establish blood-stage infection, we have much to learn before we understand how, or if, this parasite has been involved with selection of a mutation responsible for Duffy negativity observed throughout much of malaria-endemic Africa. More intriguing are questions arising through identification of a new Duffy-negative allele in Papua New Guinea (14), where a complex constellation of human erythrocyte polymorphisms confront all four human malaria species, often in the same individual.

HISTORICAL PERSPECTIVE ON *PLASMODIUM VIVAX*

Many clinical, field, and laboratory observations were made well before modern biotechnology began to dissect the molecular interactions responsible for *P. vivax* infection of human erythrocytes. A review of the period between the 1920s and the 1960s when malaria therapy was performed for the treatment of neurosyphilis provides background that has shaped our present-day perspectives on the likelihood that *P. vivax* contributed to the evolution of human erythrocyte polymorphisms.

Malaria Fever Therapy Trials

The United States surgeon general Thomas Parran (Roosevelt administration, 1936–1948) estimated that syphilis accounted for 10% of the public drug bill, 1 in 14 mental hospital admissions, 20,000 deaths annually, and that the causative agent *Treponema pallidum* infected nearly 1.7 million Americans in 1937 (15). Treatment for syphilis at this time using combinations of arsenic, bismuth, and mercury was dangerous and questionably effective, and late-stage syphilis was not responsive to therapy (16,17). A scientific contemporary to Laveran and Ross, the Viennese psychiatrist Julius Wagner-Jauregg, believed that fever-inducing strategies would provide the missing cure for neurosyphilis (also referred to as demential paralytica, general paralysis of the insane, or GPI). Wagner-Jauregg experimented with a number of agents including tuberculin, typhoid bacterin, streptococcal vaccine, typhus vaccine, erysipelas, and malaria (3). As he reflected on the results of his first malaria therapy experiment during his Nobel Prize in Medicine Lecture (1927), he observed that "six of the nine cases of GPI treated with tertian malaria (*P. vivax*) showed extensive remission, and for three of the cases remission proved enduring," and these patients were remarkably able to take up their former occupations (18). No previous treatment for GPI had ever achieved this success.

Response in the medical community to this breakthrough was dramatic and widespread. Malaria therapy became a favorite treatment and was used by nearly all syphilologists around the world (19). In general, early malaria therapy treatment strategies for syphilis were performed by inoculating approximately 2–5 ml of blood from people infected with malarial parasites, allowing the patient to experience 10–12 cycles of fever and chills over a number of weeks, and then curing the patient of their malarial infection by treatment with quinine (4,20). Between the time when Wagner-Jauregg performed his first experiments in 1917 and the discovery of effective treatment of syphilis with penicillin in 1943, tens of thousands of patients were treated with malaria therapy (21). Overall summaries evaluating the efficacy of malaria treatment of GPI suggest that approximately one-third of treated patients experienced complete remission, one-third experienced incomplete remission but were able to return to work, and the remainder required permanent hospitalization; a very small percentage of patients died as a result of treatment (19,21). J. E. Moore, a professor at Johns Hopkins University and editor of the *American Journal of Syphilis*, evaluated malaria therapy from a different perspective by pointing out that when treatment of GPI was initiated within two months of the onset of symptoms, malaria therapy produced remission in 90% of cases, but when treatment was initiated after two years following the onset of symptoms, it resulted in remission for only 10–20% of cases (21).

Beyond the remarkable benefits of malaria therapy experienced by thousands of neurosyphilis patients, observations contributed from the many careful analyses performed on patients undergoing treatment (4) provided the foundation for the accelerating accomplishments in malaria research today. Advances made through, and in support of, malaria therapy programs have been enumerated by Chernin to include isolation and identification of the fourth human malarial parasite, *P. ovale*; isolation, use, and study of various strains of plasmodia; the extraordinary discovery of the sporozoite-induced exoerythrocytic cycle, which served to explain the source of relapses and the failure of quinine and certain other drugs to effect radical cures in *P. vivax* infections; establishment of the first laboratory colonies of mosquitoes used in transmission (*Anopheles quadrimaculatas* [New World] and *Ano. maculipennis* [Old World]). These studies also provide evidence that immunity to malaria develops with time and that it is mostly strain-specific. These studies also foreshadowed the frustrating obstacles encountered in subsequent eradication and control efforts as many strains of malaria were uncovered and observed to exhibit geographic differences in pathogenicity, drug resistance, and delayed primary attacks (latency) (4). It is difficult to assess the level to which current malaria research, which has recently led to completed sequencing of the *P. falciparum* (22), *Anopheles gambiae* (23), and soon-to-be-finished *P. vivax* (Jane Carlton, personal communication, 2002) genomes, would have been delayed without the insights provided through malaria therapy programs. It is therefore not surprising that these studies strongly influence what we know today about human malaria.

Exactly how relevant the observations from malaria therapy studies are to clinical malaria experienced in endemic regions is difficult to know. It is certain that malarial infections in children battling a variety of other infections and malnutrition do not receive the same medical attention experienced by adult patients receiving malaria therapy. As discussed more completely below, because observations pertaining to *P. vivax* malaria pathogenesis are based upon these very different study groups, numerous questions regarding the public health impact and selective burden on human populations imposed by *P. vivax* are difficult to answer.

Resistance to *Plasmodium vivax*

Plasmodium vivax was used most frequently in malaria therapy, although *P. falciparum* and *P. malariae* were available and used in the early 1920s.

Advantages of *P. vivax* included observations that the infection did not become life-threatening (in contrast to *P. falciparum*), repeated paroxysms occurred dependably, and the infection was easily treated with quinine. The only problem encountered with *P. vivax* was that the parasite was only weakly infectious to African Americans. Publications providing sketchy documentation of this phenomenon began to appear in the late 1920s (24,25). When more carefully controlled studies were performed, details showed that African Americans and Africans consistently displayed resistance to *P. vivax* strains regardless of the parasites' geographic origins and inoculation dose (8). As African Americans from nonmalarious regions of the United States were as refractory to *P. vivax* infection as those from malarious regions, this resistance was suggested to be natural rather than acquired (5–8,26). Of further interest, to determine if a *P. vivax* infection once established in an African American patient would acquire characteristics enabling more successful infection of resistant individuals, Young *et al.* used two blood samples from a *P. vivax*–infected African American containing 108 and 235 million parasites, respectively, to inoculate two individuals of the same race (8). Neither of the recipients developed blood-stage parasitemia following these inoculations despite receiving inocula 3 to 7 times higher than those that routinely causing blood-stage infection of Caucasian patients. From these results it was concluded that the *P. vivax* strain causing infection in the donor patient had not been transformed to acquire characteristics that would enable subsequent infection of resistant individuals (8).

Plasmodium vivax Malaria

The life cycle of *P. vivax* resembles that of the other human *Plasmodium* parasites. Following infection of liver cells, sporozoites undergo schizogony, and infectious merozoites go on to establish blood-stage infection. Unlike *P. falciparum*, a proportion of liver-stage *P. vivax* may not commence asexual reproduction but rather becomes dormant as hypnozoites (27,28). The latter may remain dormant in the liver for months, enabling the parasite to sustain infection through relapse. Upon reaching maturity in the liver (10–14 days), *P. vivax* merozoites specifically invade reticulocytes (29–31). Targeting of reticulocytes, which comprise ~1% of circulating erythrocytes in hematologically normal individuals, is likely to be a factor that limits the intensity of *P. vivax* asexual parasitemia. The average parasitemia of *P. vivax* is 2×10^4 merozoites/μl of blood, in contrast to parasitemia that may reach 5×10^5 merozoites/μl of blood for *P. falciparum*, which has merozoites able to invade erythrocytes of all ages (20).

From a clinical perspective, *P. vivax* malaria is generally regarded to be less severe than *P. falciparum* malaria, which is considered to be responsible for the majority of severe malaria morbidity and mortality throughout the world. As a consequence, *P. vivax* is characterized as a "benign," as opposed to "malignant," (*P. falciparum*) malarial parasite (32). With certainty, however, *P. vivax* infection by itself can lead to significant morbidity, including fever, respiratory distress, and anemia, particularly in nonimmune individuals. This is illustrated by the course of infection in adults who received blood-stage *P. vivax* for malaria therapy (33,34). In these trials, a prodromal period characterized by headache, malaise, and anorexia appeared approximately 12 days after parasite inoculation. This was followed by low-grade fever and intermittent "tertian" fevers (every second day) with parasitemia reaching $\sim 5 \times 10^4/\mu l$. Fevers of up to 105°F recurred over 35 to 40 days and were followed by gradual remission. In recent studies from Thailand, *P. vivax* infection was also reportedly associated with maternal anemia and an increase in low birth weight in babies born to *P. vivax*–infected mothers (35). As low birth weight is associated with increased infant mortality (36), it is important to acknowledge that *P. vivax* malaria may combine with other conditions that contribute significantly toward infant and childhood mortality in *P. vivax*–endemic communities.

Although *P. vivax* is estimated to cause 70–80 million cases of malaria each year (37), the manner in which disease caused by this parasite is manifested in epidemiological studies conducted in malaria-endemic regions is not well understood. Factors contributing to this poorly developed understanding of the community-based impact of *P. vivax* malaria include the following. Malaria epidemiology varies among endemic regions because of transmission seasonality, vector behaviors and competencies, exposure to other human pathogens, nutrition, development of immunity, and human genetic polymorphism. Because diagnosis of malarial infection is based primarily on blood smears, there may be significant discrepancies between the reported and actual incidence of *P. vivax* in particular. Furthermore, the overlapping distribution patterns of *Plasmodium* species in endemic regions make it difficult to study *P. vivax* in the absence of other human malarial parasite species.

In light of the well known technical limitations of malaria microscopy and other diagnostic strategies, it is important to note that there has been considerable interest in interactions between human malarial parasite species. Various epidemiological studies involving species co-occurrence have recorded fewer mixed infections (38–41) and attribute this to innate host factors that regulate parasite density (42), differences in parasite metabolism, or factors contributing to the acquisition of heterologous (or cross-species) immunity (43,44). Reports of *P. vivax* parasitemia following chemotherapeutic cure of *P. falciparum* (45), or the converse relationship

(46), have been used as evidence that one species may suppress infection of erythrocytes by other *Plasmodium* species. Of additional interest, other studies have reported that severe *P. falciparum* malaria was significantly reduced when co-infection occurred with *P. vivax* or *P. malariae* (47,48). Implications of these findings are important to the outcome of debates focused on developing strategies for malarial control in endemic regions. If interactions in mixed *Plasmodium* species infections influence clinical outcomes associated with malarial infections, it would be advisable for vaccine and drug treatment studies to include safeguards to ensure that the potential clinical benefits of mixed infections would not be disrupted thereby increasing the severity of malaria illness in treated communities. Beyond the practical considerations of malaria control, it may also be important to understand proposed mixed-species interactions in order to interpret the ongoing co-evolution of malarial parasites and human genetic polymorphism.

Distribution of *P. vivax*

P. vivax is endemic throughout much of the tropical world. As might be expected from the above-mentioned resistance to *P. vivax* infection in individuals of African ancestry, transmission of *P. vivax* is considered to be very rare or absent in many malaria-endemic regions of West Africa. Interestingly, however, it is not uncommon to observe reports of European travelers to West Africa returning with *P. vivax* infection (37,49). Outside of West Africa in regions of eastern and southern Africa and Madagascar, the prevalence of *P. vivax* has been observed to be responsible for as high as 20% of malaria infections; the annual number of *P. vivax* malaria cases in all of Africa is estimated to vary between 6 and 15 million (37). The majority of *P. vivax* cases are then distributed outside of Africa. Recent surveys of national statistics reported to the World Health Organization by Richard Carter estimated the annual number of *P. vivax* malaria cases to be 42 million in southeastern Asia and the western Pacific (49% of all malaria cases), 11 million in the eastern Mediterranean region (80% of all malaria cases), and 8–10 million in Central and South America (70–80% of all malaria cases) (37). Relative proportions of *P. falciparum* to *P. vivax* are suggested to vary from an excess prevalence of *P. falciparum* malaria in holoendemic and hyperendemic tropical regions to an excess prevalence of *P. vivax* malaria in subtropical and temperate regions (37).

Historically, *P. vivax* is believed to have been distributed over a wider range than that observed today. This range is known to have included much of eastern North America into Canada, northern Europe, and Asia. The noted British malariologist S. P. James concluded that local transmission of malaria in England was due entirely to *P. vivax* (50). Interestingly, distribution of *P. vivax* into isolated regions, or where transmission of the parasite would be interrupted by cold weather, may have been a factor responsible for geographic isolation and evolution of the *P. vivax* subspecies *P. vivax multinucleatum*, *P. vivax hibernans*, and *P. vivax collinsi* (51,52). In the broader context of host–parasite evolution, it is a necessary challenge to place *P. vivax* onto the African continent during the time frame that might overlap with evolution of resistance to the parasite. A survey of the evidence describing the evolution of *P. vivax* as a human parasite is provided below.

THE DUFFY BLOOD GROUP AND ITS MOLECULAR BIOLOGY

Beginning with the identification of the human ABO blood groups by Landsteiner in 1901 (53), a large number of erythrocyte membrane polymorphisms have been identified. By studying the prevalence and distribution of these blood group polymorphisms, human geneticists gained some of their first insights regarding genetic similarities and differences between races and ethnicities. These discoveries accumulated from 1900 to 1965 through observations that the serum of multiparous women and patients receiving blood transfusions carried antibodies to protein surfaces that were different from their own. Basic research studies have employed these cross-reacting antibodies to isolate and purify blood group proteins and identify their encoding genes. During this time period the Duffy blood group antigen was identified and found to play a significant role in the process whereby malarial parasites infect human erythrocytes *in vivo*. Through studies on the Duffy gene, mutations underlying resistance to blood-stage *P. vivax* malaria have been identified and insights regarding specific molecular interactions necessary for *Plasmodium* invasion of red blood cells have been gained.

Recognition of Duffy Blood Group Polymorphism – Serology

The Duffy blood group antigen (Fy^a) was first observed in 1950 on erythrocytes using allo-antisera found in a multiply transfused hemophiliac (from whom the blood group receives its name) in whom a hemolytic transfusion reaction was observed (54). The expected Fy^b antisera was discovered shortly thereafter (55) and surveys of British populations reported frequencies for the codominantly expressed Fy^a and Fy^b antigens of 0.41 and 0.59, respectively (56). Upon screening a series of blood samples from African American donors to the Knickerbocker Blood Bank of New York City, Sanger *et al.* observed that 68% of the samples did not react with either the Fy^a or the Fy^b antisera (57). Additional analyses performed on

Table 6.1. General guidelines for Duffy blood group nomenclature

Point of Reference

DNA/gene	*FY*	Allele = *FY*A, FY*B, FY*A^null, FY*B^null*	
Protein	Fy	Antigen = Fy^a, Fy^b or Fy$^{b \, weak}$	
Genotype		*FY*A/FY*A*	*FY*A/FY*B*	*FY*B/FY*B*
Phenotype		Fy(a+b−)	Fy(a+b+)	Fy(a−b+)

Note: This table follows the Guidelines for Human Gene Nomenclature (2002): http://www.gene.ucl.ac.uk/nomenclature/guidelines.html#2.%20Gene%20symbols.
Nomenclature recommendations from the International Society of Blood Transfusion (ISBT) prefer the use of *FY*1* for *FY*A*; *FY*2* for *FY*B*. The ISBT has not addressed further preferred nomenclature for the erythrocyte null expression alleles (Geoff Daniels, personal communication, 2002); http://www.iccbba.com/page25.htm.
For these alleles, the former *FY*0* designation is obsolete, as identification of the *FY*A^null* allele now mandates further information to specify the genetic basis of a Duffy-negative phenotype.

samples from Nigerian families showed that the null phenotype was inherited in Mendelian fashion and provided an opportunity to investigate Fy^a copy number. In an earlier study, Race *et al.* found that the antiserum, Pri, reliably distinguished between single and double donors for Fy^a (58). Results obtained from tests on three African Americans who were Fy(a+b+) and two who were Fy(a+b−) all suggested that Fy^a was observed to be present in single-dose quantity, as compared to double-dose quantities observed in Fy(a+b−) Europeans (57; Table 6.1 summarizes current Duffy blood group nomenclature). These observations were interpreted to suggest that individuals of African ancestry either possessed a different antigen, Fy^c, or they did not express the Duffy antigen and possessed a Duffy null allele, Fy. Blood group researchers hypothesized that identification of a Fy^c antigen would be relatively straightforward because of either anticipated frequent blood transfusion involving African and Caucasian donor–recipient pairs or attempts to stimulate anti-Fy^c reactivity through injection of Fy(a−b−) red cells into European volunteers (57); resolution of this hypothesis would ultimately have to await modern biotechnology and cloning of the. Duffy gene.

Population Surveys – Distribution of Duffy Blood Group Polymorphism

An extensive collection of human genetic polymorphism survey results from serology-based agglutination reactions (59) has come to reveal a very heterogeneous distribution of Duffy phenotypes among African, Caucasian, and Asian populations (Figure 6.1). Consistent with earlier studies, African populations have been characterized as Fy(a−b−); in fact, this

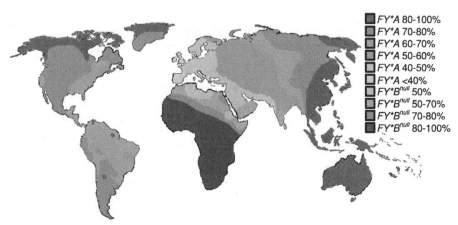

Figure 6.1. Distribution of *FY*A*, *FY*B*, and *FY*B^null^* alleles. In the regions of Africa where the *FY*B^null^* allele frequency drops below 80–100%, both *FY*A* and *FY*B* alleles are observed at low frequency. Outside of Africa, in regions where the *FY*A* allele frequency is below 80–100%, the *FY*B* allele is observed at frequencies of 1-*FY*A*. Adapted from Cavalli-Sforza *et al*. (59). For a colour version of this figure, see www.cambridge.org/9780521126557.

phenotype was observed at fixation in many West African populations (59). Caucasians expressed Fy(a+b−), Fy(a+b+), and Fy(a−b+) phenotypes. European and Asian populations were characterized by a west-to-east allele frequency gradient of decreasing Fy^b and increasing Fy^a. Surveys of Duffy blood group serology performed on Malaysian (60), Melanesian (61–66), and Oceanian populations (61,67,68) show that the Fy^a antigen, with an average frequency of 0.97, has reached fixation in this region of the world.

Surveys of South American populations now reveal an interesting admixture of Duffy allele frequencies. Consistent with the movement of African slaves to the New World, genetic polymorphisms, as well as the *P. vivax*–resistant phenotype observed in Africans displaced into North America, were also observed in South America (69,70). Also, as it is generally agreed that the original peopling of North and South America occurred as a result of northeastern Asian populations migrating overland through Beringia between 15,000 and 25,000 years ago, it is not surprising to observe genetic polymorphisms common to Asian populations in native North and South Americans (59). A representative survey of Duffy serological polymorphism in distinct populations of indigenous Chachi Amerindians and those of African ancestry living in Esmeraldas Province of northwest Ecuador reported the following results. The African population was predominantly Fy(a−b−) (phenotype frequency = 0.760), whereas the indigenous Chachi population was predominantly Fy(a+b−) (phenotype frequency = 0.710) (69). Admixture of the Fy(a−b−) phenotype into the Chachi population was observed in 6.4% of individuals surveyed (4/62) (69).

The Genetic Resistance Factor to *P. vivax* – The Duffy Blood Group

The impressive distribution of the Fy(a−b−) phenotype in diverse African populations has historically fascinated population geneticists and malariologists. Based upon the overlapping distribution of the Fy(a−b−) phenotype, the very low prevalence of *P. vivax* in African populations, and the desire to identify "receptors" used by malarial parasites to invade human erythrocytes, Louis Miller and colleagues at the National Institutes of Health performed a series of studies in the mid-1970s that suggested a remarkable *P. vivax* invasion pathway involving the Duffy blood group antigen. These studies showed first that the nonhuman primate malaria parasite *P. knowlesi* was not able to infect erythrocytes from Fy(a−b−) African Americans, *in vitro* (72); parallel studies showed that *P. knowlesi* easily infected erythrocytes from Fy(a+b−), Fy(a+b+), and Fy(a−b+) donors. A second study showed that *P. vivax* was not able to infect erythrocytes from Fy(a−b−) African Americans, *in vivo* (71). In this study, consenting subjects were first characterized for their Duffy blood group phenotype serologically. *P. vivax*–infected mosquitoes were then allowed to take blood meals, first from Fy(a−b−) African Americans, and then following interruption were allowed to continue feeding on Fy(a+b−), Fy(a+b+), and Fy(a−b+) Caucasian and African Americans. Results showed that none of the five Fy(a−b−) study subjects developed blood-stage parasitemia despite evaluation of daily blood smears for 90–180 days, whereas all of the Duffy-positive individuals developed blood-stage infection within 15 days (71).

Although this study showed that *P. vivax* requires the Duffy blood group antigen to be present on the erythrocyte surface to invade the cell successfully and continue its life cycle, little information beyond basic susceptibility to blood-stage infection was produced on individuals who were heterozygous for the Duffy-negative allele, expressing a single gene dose of the Duffy blood group protein.[1] Knowing the *P. vivax* susceptibility of these heterozygous individuals could have provided important insight toward understanding the selective advantage of the Duffy-negative allele.

---

[1] The authors did not indicate what antisera were used to characterize the Duffy blood group phenotypes so it may not have been possible to categorize individual study subjects beyond general Duffy positivity and Duffy negativity. If this study had enrolled Fy(a+b−) subjects preferentially, and the Pri, Fy^a-specific antiserum described by R. Sanger, R. R. Race, and J. Jack, in The "Duffy blood groups of the New York Negroes: the phenotype Fy(a−b−)," *British Journal of Haematology* 1 (1955): 370–4 (57), had been used to assess Duffy copy number, it may have been possible to evaluate further the influence of single gene vs. double gene dosage effects on susceptibility to *P. vivax* infection.

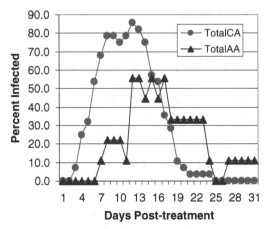

Figure 6.2. Summary of *P. knowlesi* infections in a cohort of malaria therapy patients (73). Caucasian (CA) and African American (AA) patients were compared based on the time to first blood-stage parasitemia, duration of blood-stage positivity, infectiousness to susceptible rhesus monkeys, and development of immunity to reinfection. For a colour version of this figure, see www.cambridge.org/9780521126557.

Of course assumptions can be made that most of the Duffy-positive African Americans in the study were heterozygous for the Duffy-negative allele, and that individuals carrying only one Duffy-positive allele were susceptible to *P. vivax* blood-stage infection; however, it is not possible to determine if these individuals were equally susceptible to the *P. vivax* merozoites compared to individuals carrying two Duffy-positive alleles. Data comparing Duffy-positive Caucasian (assuming two positive alleles) and African American (assuming one positive allele) study subjects could provide an estimate of potential Duffy genotype-based differences. The data from Miller *et al.* show that the average time to *P. vivax* blood smear positivity is 11 days, for the Duffy-positive Caucasians and 13 days (71) for the Duffy-positive African Americans (Mann–Whitney P-value $= 0.118$). Beyond this it is not possible to compare the average parasitemia or the length of the blood-stage infection. These additional comparisons may have provided further evidence for decreased susceptibility to *P. vivax* associated with Duffy gene dosage.

Interestingly, one study from malaria therapy trials may provide a glimpse at how data of this nature might look. In a study of susceptibility to *P. knowlesi* infection, Milam and Coggeshall found that duration of blood-stage infection was longer in Caucasian compared to African American patients (9.5 vs. 6.7 days; Mann–Whitney P-value $= 0.02$) (73). Also, results shown in Figure 6.2 illustrate that Caucasian patients displayed blood-stage parasitemia sooner than African American patients (average time to first blood-stage parasitemia: Caucasian $= 7.2$ days; African American $= 13.6$ days; Mann–Whitney P-value < 0.001) and that a higher percentage of patients were blood-smear positive over the time period in which parasitemia was observed (73). Consistent with *in vitro* findings from Miller *et al.* (73), three of the African American patients never became blood-smear positive. Limitations of this single study are easily recognized. These limitations include a lack of information on which *P. knowlesi* strain(s) were used, the method of inoculation, and the dose of exposure to infectious parasites; as the Duffy blood group had not been discovered, no Duffy genetic data were available. Despite these deficiencies, results from this study suggest that there were differences between Caucasian and African American patients regarding their susceptiblity to *P. knowlesi* blood-stage infection.

In an attempt to explain how the frequency of Duffy blood group negativity had risen to 100% corresponding with the absence of *P. vivax* from vast regions of malaria-endemic West Africa, Miller and colleagues offered two possibilities. First, "Although *P. vivax* infection rarely causes death, it may decrease survival in African children with malnutrition and other endemic diseases" (71). Second, despite observations showing that Fy(a–b–) individuals were susceptible to all other species of *Plasmodium*, the authors suggested that "the Duffy-negative phenotype may reduce the reproductive potential of *P. falciparum* and as a result may lower the mortality from this malignant infection" (71). Both of these hypotheses suggest that the Duffy-negative phenotype increased the fitness of human populations against malaria and would have led to the evolution of a human host population in which *P. vivax* was not able to reproduce with enough success to maintain its life cycle. These suggestions have fueled the debate surrounding the relationship between evolution of the Duffy blood group negative phenotype, its genetic fixation in African populations, the selective pressure that promoted the dramatic increase in phenotype frequency, and the molecular and cellular factors responsible for this form of resistance to malaria.

Identification of the Duffy Gene: Molecular Biology

As progress in attempts to identify an Fy^c antigen was not forthcoming, ultimate identification of the complete Duffy blood group antigen system would await development of tools and approaches to clone and study genes. Methodical progress toward cloning the Duffy gene can be observed in attempts to purify the Duffy protein (74–76) and in the identification of a series of Duffy antigens, Fy^3 (77), Fy^4 (78), Fy^5 (79), and Fy^6 (80), and their respective antisera. These antisera have been used to illustrate that the Duffy protein is characterized by a number of antigenic surfaces, all of which are absent from Fy(a−b−) African individuals. Beginning in 1988, Asok Chaudhuri and colleagues at the New York Blood Center described a series of experiments initiated by using the murine anti-Fy^6 monoclonal antibody to affinity purify Duffy antigens from solubilized erythrocytes (76). Chaudhuri *et al.* further studied these Duffy peptides by amino acid sequencing following cyanogen bromide/pepsin treatments and synthesis of a DNA probe to be used in screening a cDNA library (81). With this probe, a clone was identified representing a 338-codon open reading frame (ORF) sequence exhibiting significant sequence homology to the human interleukin 8 receptor, predicting seven transmembrane segments, an amino terminal head and three extracellular loop domains, and three intracellular loop domains and a carboxy-terminal cytoplasmic tail (Figure 6.3; 81). Further studies on the genomic organization of the Duffy gene sequence confirmed early predictions that the gene resided in a pericentromeric region of human chromosome 1 (1q22–23) (82–84); additional studies produced evidence that it was comprised of two exons (85). Exon one encodes seven amino acids, MGNCLHR; exon two encodes 338 amino acids. It has been shown that the primary transcript of the Duffy gene is comprised of codons 1–7 from exon one joined to codons 10–338 from exon two encoding a protein of 336 amino acids. These first studies characterizing the Duffy gene were also successful in identifying a single nucleotide polymorphism (SNP) in codon 42 associated with Fy^a (T<u>G</u>G; encodes Gly) and Fy^b (T<u>A</u>G; encodes Asp) antigens (86–88). After acknowledging no additional polymorphism compared to the *FY*B* allele, which suggests that Africans carried no important disruption in the Duffy ORF (86–89), Tournamille *et al.* identified a T to C SNP 33 nucleotides upstream from the primary transcription starting position (−33) in the Duffy gene promoter (originally positioned at nucleotide −46) (89), resulting in a *FY*B^{null}* allele. Duffy-negative Africans were homozygous for this polymorphism which was shown to occur in a tissue-specific GATA1 transcription factor binding motif. *In vitro* assays showed that this polymorphism blocked gene expression in erythroid lineage cells but did not block expression in nonerythroid cells. Results of this study therefore provided the genetic mechanism explaining Duffy negativity and therefore resistance to blood-stage *P. vivax* infection.

Following the results of Tournamille *et al.* (89), our own study sought to determine if the Duffy gene in Papua New Guineans living in

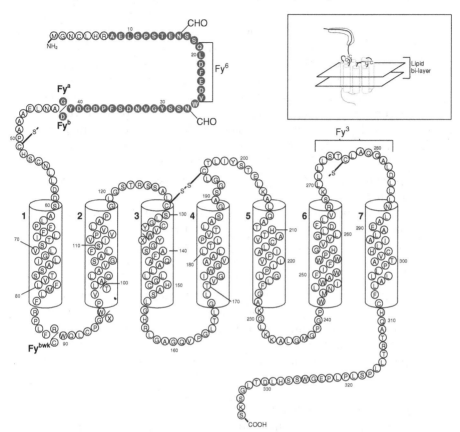

Figure 6.3. The Duffy antigen. The diagram illustrates the primary structure of the 236–amino acid, 36–46 kDa Duffy antigen, with seven predicted transmembrane domains, and extracellular and intracellular domains. Amino acids comprising the Fy⁶ and Fy³ antibody binding domains are marked by brackets. Amino acid sequence polymorphisms are identified at residues 42 (G vs. D; Fyᵃ vs. Fyᵇ), 89 (R vs. C; Fyᵇ vs. Fyᵇ ᵂᵉᵃᵏ), 100 (A vs. T), and two premature termination codons (W vs. X) at residue positions 96 and 134. Glycosylation sites are identified at amino acid residues N16 and N27. Disulfide bonds occurring between C129 (extracellular loop 2) and C195 (extracellular loop 3), and between C51 (amino terminal head) and C276 (extracellular loop 3) are predicted to contribute to further tertiary structure within the cell membrane as depicted in the inset. Amino acids predicted to comprise the *P. vivax* binding region are identified in red (108). For a colour version of this figure, see www.cambridge.org/9780521126557.

P. vivax–endemic regions would be characterized by the accumulation of any functional polymorphism. A survey of the known promoter and ORF polymorphisms identified previously revealed that the same promoter SNP identified in African populations on the *FY*B* allele was observed on the

FY^*A allele $(FY^*A^{null})^2$ at a frequency of 0.022 (23/1062 chromosomes) (14). By flow cytometry, comparing phycoerythrin-labeled Fy^6 (80) antibody binding to erythrocytes from 6 PNG homozygous wild type and 6 PNG heterozygous individuals, we showed that individuals with two erythroid-functional alleles expressed approximately twice the amount of the erythrocyte Fy^a antigen compared to individuals with one erythroid-functional allele. Our preliminary field studies suggested that the prevalence of *P. vivax* infection may be reduced in FY^*A/FY^*A^{null} compared to FY^*A/FY^*A individuals.[3] Prospective studies will further investigate the epidemiological significance of the emergence of the FY^*A^{null} in this unique study population.

Additional polymorphism in codon 89 is associated with distinguishing between the Fy^b (<u>C</u>GC encodes Arg) and $Fy^{b\ weak}$ (Fy^x; <u>T</u>GC encodes Cys) antigens (93–95). This polymorphism occurs within the first intracellular loop of the Duffy protein and appears to be associated with reduced cell surface expression of Duffy (93,95). The frequency of the FY^*B^{weak} allele is approximately 2% in Caucasians (95,96). Although the Fy(a–b^{weak}) and Fy(a–b–) phenotypes appear to be similar serologically, relative susceptibility to *P. vivax* infection in Fy(a–b^{weak}) individuals has not been adequately tested. The significance of additional *FY* DNA sequence polymorphisms (97) and their relationship to malaria exposure and susceptibility is not clear.

Duffy Expression and Cell Biology

In addition to erythroid cells, Duffy is expressed on endothelial cells of postcapillary venules (98,99), endothelial and epithelial cells in some non-erythroid organs (100), and Purkinje cells in the cerebellum (101). Because sequence analysis suggested significant homology between Duffy and the chemokine receptor family (alternative gene name: Duffy antigen/

---

[2] Shimizu *et al.* have described findings in Southeast Asian and Melanesian populations that may be consistent with identification of the FY^*A^{null} allele. Alternatively, their findings may suggest the identification of new polymorphism underlying a FY^*A^{weak} allele 90. For further discussion, see (90) Y. Shimizu, M. Kimura, W. Settheetham-Ishida, P. Duangchang, and T. Ishida, "Serotyping of Duffy blood group in several Thai ethnic groups," *Southeast Asian Journal of Tropical Medicine and Public Health* 28 (1997): 32–5 and (91). Y. Shimizu et al., "Sero- and molecular typing of Duffy blood group in Southeast Asians and Oceanians," *Human Biology* 72 (2000): 511–8.

[3] In a study from the Brazilian Amazon basin, Eugenio-Cavasini *et al.* do not observe that heterozygosity for the FY^*B^{null} allele provided significant protection from *P. vivax* infection. See (92) C. Eugenio-Cavasini, F. J. Tarelho-Pereira, W. Luidi-Ribeiro, G. Wunderlich, and M. Urgano-Ferreira, "Duffy blood group genotypes among malaria patients in Rondonia, Western Brazilian Amazon," *Revista da Sociedade Brasileira de Medicina Tropical* 34 (2001): 591–5.

receptor for chemokines (*DARC*; 102)), additional analyses have been performed to evaluate the Duffy antigen's chemokine receptor function. Interactions between chemokines and their receptors are shown to induce
rearrangement of the cytoskeleton and expression of adhesion molecules
and to provide traction as cells exhibit tropism toward sites of inflammation and/or infection along chemokine concentration gradients. *In
vitro* studies have shown that Duffy binds both α and β chemokines
but does not appear to communicate intracellularly through G protein–
coupled signal transduction (104).[4] Ordinarily, chemokine receptors
display faithful segregation in their interactions with either α chemokines
or β chemokines (α/C–X–C chemokines – interleukin 8 [IL-8], melanoma
growth stimulating activity [MGSA]; β/C–C chemokines – monocyte
chemotactic protein-1 [MCP-1], regulated on activation normal T cell expressed and secreted [RANTES] (104,105)); Duffy is the only chemokine
receptor known to bind members of both chemokine families. Although
these findings make it difficult to understand entirely the biological function of Duffy, *in vitro* studies suggest that this "receptor" can internalize lig-
and and have lead to the proposal that Duffy is expressed on erythrocytes
to clear excess chemokines from the blood and peripheral tissues (99).

Regardless of the overall role of Duffy as a chemokine receptor or sink,
it has been shown that *P. knowlesi* invasion of Duffy-positive human erythrocytes *in vitro* can be blocked by physiological concentrations of both
IL-8 and MGSA (106,107). Further observations from these studies have
identified the binding region within Duffy for both *P. knowlesi* and *P. vivax*
(108). These studies have also contributed to identification of the critical region of the parasite Duffy binding protein needed for erythrocyte
invasion. All of these studies indicate that the interaction between the
parasite and human erythrocytes is susceptible to disruption. Does this
suggest that interactions between erythrocytes expressing reduced levels
of the Duffy antigen decrease the efficiency of parasite invasion?

P. vivax Duffy Binding Protein
Several studies have now described the parasite ligand interacting with
Duffy. The Duffy binding proteins of *P. knowlesi* (PkDBP; 109,110) and
P. vivax (PvDBP; 111,112) have molecular weights of ~140 kD and are
characterized by two cysteine-rich regions sharing amino acid sequence

---

[4] Because the Duffy protein is observed to lack known signaling function, it is no longer
included in the chemokine receptor nomenclature system. See (103) P. M. Murphy,
"International Union of Pharmacology. XXX. Update on chemokine receptor nomenclature," *Pharmacology Review* 54 (2002): 227–9.

homology with other erythrocyte binding proteins encoded by malarial parasites (113). Recently a portion of DBP region II of ~138 amino acids has been determined to contain the minimal sequence necessary to enable *in vitro* interaction between transfected COS-7 cells and Fy-positive human erythrocytes (114,115). The relevance of this molecular interaction to parasite invasion of human erythrocytes was corroborated by competitively blocking Duffy/DBP binding with MGSA (EC$_{50}$ ≈ 1 nM for *P. vivax* DBP; EC$_{50}$ ≈ 0.2 nM for *P. knowlesi* DBP). Additional studies employing this *in vitro* binding affinity assay system have shown that erythrocytes heterozygous for a *FY^{null}* allele exhibit consistently lower affinity for cells transfected with the PvDBP (116). Additionally, elevated levels of amino acid sequence polymorphism in the DBP binding region, as well as antibody responses from people living in *P. vivax*–endemic regions recognizing DBP, suggest that this molecule may be under selective pressure (117–121) by the human immune system.

THE ASCENT OF *P. VIVAX* MALARIA: ORIGINS AND EVOLUTION OF DUFFY NEGATIVITY

General predictions of population genetics suggest that if an allele is present at low frequency and contributes nothing to the fitness of the genome, the allele will disappear as the result of genetic drift. It is assumed that the frequency of Duffy negativity was at low frequency at some time during human evolution because the phenotype has not been observed in primate ancestors of *Homo sapiens* (86). With this in mind, along with observations describing the inheritance of codominant *FY*A*, *FY*B*, and recessive *FY*A^{null}* and *FY*B^{null}* alleles, it would appear reasonable to hypothesize that at some time point during African human history the *FY*B^{null}* allele was present at a very low frequency and that all individuals carrying this allele were heterozygous. The fact that the *FY*B^{null}* allele has reached genetic fixation in many African populations is evidence that reproductive fitness was improved for heterozygous carriers.

Origin of the Human Malarial Parasite, *P. vivax*

As *P. vivax* is generally considered to have originated in Asia, implicating *P. vivax* as the selective agent responsible for the evolution of Duffy negativity requires that the parasite be placed among ancestral African populations. Evidence that coparasitism by *P. vivax*–related parasites of human and simian species has been possible for at least 20 million years is observed through homology between host receptors and parasite invasion ligands. The Fy^b and Fy^3 antigens, detected in Old and New World monkey species, date the origin of the Duffy antigen to the middle Eocene period (40 million years ago [Ma]) (122,123). Parallel observations show that homologues of the Duffy binding protein have been characterized for a number of the *P. vivax*–related parasite species (109,110,124). Moreover, observations showing that both *P. knowlesi* and *P. cynomolgi* are capable of causing infection of human erythrocytes (125–127) suggest that a Duffy antigen erythrocyte invasion pathway has been used by these parasites throughout the course of their evolution.[5]

Recent studies evaluating phylogenetic relationships among *Plasmodium* species by comparing DNA sequences of small subunit rRNA (129) genes, circumsporozoite (130), and cytochome b (131) genes illustrate that *P. vivax* and the simian parasites *P. knowlesi*, *P. simiovale*, and *P. cynomolgi* are very closely related, if not indistinguishable. Phylogenetic analysis of rRNA gene sequence polymorphism shows that *P. cynomolgi* clustered with *P. vivax* as opposed to other *Plasmodium* parasite species of macaques (*P. knowlesi* and *P. fragile*) (129). Overall, these studies suggest that divergence among these and other simian malarial parasites may have occurred up to as recently as 2–3 million years ago (Ma). These or closely related simian *Plasmodium* parasite species may have become human parasites at a time when ancestral humans and a simian host shared a common geographic area. The earliest this could have been possible was the early Miocene era (20 Ma) when the Tethys Sea, which separated terrestrial animals on the African and Eurasian land masses, disappeared (Richard Carter, personal communication, 2002; 123). Because these observations suggest that a biological mechanism for simian *Plasmodium* parasite species to make a lateral jump to become human parasites has been available, it is important to determine what time constraints would be imposed on the origin and later fixation of the FY^*B^{null} allele.

Origins of Duffy Negativity
Recent studies have begun to assemble observations on the diversity of Duffy gene sequence enabling tests to estimate when the FY^*B^{null} mutation event occurred and to determine if evidence consistent with directional selection can be observed. As shown in the allele frequency data summarized in Figure 6.1, the FY^*B^{null} allele is almost exclusively restricted to the African continent; FY^*B and FY^*A alleles are distributed in human

---

[5] It is of further interest to note that African Americans displayed resistance to *P. cynomolgi*, as observed with *P. knowlesi* and *P. vivax* in (128) J. K. Beye, M. E. Getz, G. R. Coatney, H. A. Elder, and D. E. Eyles, "Simian malaria in man," *American Journal of Tropical Medicine and Hygiene* 10 (1961): 311–16. This suggests that Duffy is the primary receptor for *P. cynomolgi* erythrocyte invasion and emphasizes homology with the *P. vivax* invasion pathway of human erythrocyte infection.

populations throughout the rest of the world. This pattern suggests that fixation of $FY*B^{null}$ in African populations occurred after early humans began migrating out of Africa, approximately 60,000–100,000 years ago. Microsatellite and DNA sequence polymorphism associated with $FY*B^{null}$ have been used to estimate that fixation of this allele in African populations occurred within the past 15,000–30,000 years (with 95% confidence intervals, within approximately the past 5,000–100,000 years) (132–134). This time frame is consistent with the distribution of $FY*A$, $FY*B$, and $FY*B^{null}$ alleles in human populations and would have occurred after P. vivax–related parasites had begun parasitizing humans.

Hamblin et al. have used the DNA sequence data to evaluate further the evidence that the $FY*B^{null}$ allele arose under the influence of natural selection to confer increased fitness (directional or positive selection) to ancestral African populations (132,133). Generally, a genetic locus under directional selection will exhibit lower diversity than a neutral locus because sequence conservation of the "more fit" allele would be favored. In their DNA sequence surveys, now inclusive of approximately 34 kb of the Duffy locus on chromosome 1, Hamblin et al. report a two-to-three–fold lower level of sequence variation within a 10-kb region of the $FY*B^{null}$ allele compared to the same region in a sample of Italian individuals (132,133). This observation is unusual given that genetic diversity in African populations is consistently higher than in non-African populations, as the former are older and have had greater time to diversify, and provides evidence for directional selection of $FY*B^{null}$.

Considering that fixation of the $FY*B^{null}$ allele may have occurred within the past 60,000–100,000 years, it is useful to consider a number of models that attempt to predict the rate and time frame over which the $FY*B^{null}$ allele frequency rose to fixation (Figure 6.4). By assuming differences in the relative fitness of $FY*B/B$, $FY*B/B^{null}$, and $FY*B^{null}/B^{null}$ genotypes, fixation could be predicted to occur within 100 generations (135). The speed with which the $FY*B^{null}$ allele becomes fixed in a population is influenced more by the fitness of the $FY*B/B^{null}$ than of the $FY*B^{null}/B^{null}$ genotype. Overall, this model is consistent with a time period of human history during which the size of populations increased and became more sedentary resulting from human development of agriculture. These conditions are suggested to have promoted increasing parasitism of humans by Plasmodium species.

Although it is not possible to determine what conditions led to the origin of the $FY*B^{null}$ allele or the absence of P. vivax from vast regions of Africa, familiar attempts to explain the pre-historic events underlying these evolutionarily related events include speculations that early strains of P. vivax were more lethal than modern strains and that some other

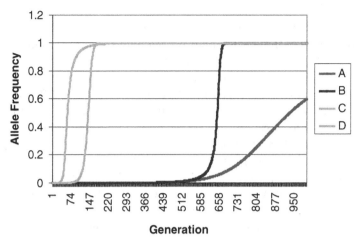

Figure 6.4. Models predicting the time frame over which the *FY*B^{null}* may have evolved to fixation. For each simulation the initial *FY*B^{null}* allele frequency was set to 0.00001. Assigned relative fitness scores for the *FY*B^{null}/B^{null}*, *FY*B/B^{null}*, and *FY*B/B* genotypes were 0.376, 0.375, and 0.370 (model A, red line); 0.470, 0.375, and 0.370 (model B, blue line); 0.470, 0.400, and 0.370 (model C, green line); and 0.470, 0.460, and 0.370 (model D, gold line). The number of generations required to reach a frequency of 0.500 was 934 (model A), 650 (model B), 141 (model C), and 58 (model D). Models were based upon standard Hardy–Weinberg equilibrium and relative fitness equations (135). For a colour version of this figure, see www.cambridge.org/9780521126557.

virulent erythroid pathogen, although not readily forthcoming, may have been the force of natural selection. If the former hypothesis prevails, Duffy negativity would have arisen to increase the fitness of ancestral African populations against *P. vivax* malaria. As the frequency of the *FY*B^{null}* allele increased, it became impossible for *P. vivax* to complete its life cycle and it was driven out of regions inhabited by Duffy-negative people. If the latter hypothesis is correct, attempts by *P. vivax* to establish infection in African populations have always been thwarted by the very high frequency of Duffy-negative people living in this region (136).

Emergence and Evolution of *FY*A^{null}* in *P. vivax*–Endemic Regions of Papua New Guinea

Given the preceding theoretical considerations, we are left to wonder what course the evolution of Duffy negativity will take in Papua New Guinea. Despite our observations that the *FY*A^{null}* allele reduces the level of erythrocyte Duffy antigen expression by 50%, and that it significantly reduces affinity of the *P. vivax* Duffy binding protein in heterozygous individuals, our ongoing studies have yet to identify an individual who is homozygous for the *FY*A^{null}* allele. To explain why our studies have not yet identified an

$FY*A^{null}/A^{null}$ individual we must consider at least two possibilities. First, the allele frequency in the population is not yet high enough to result in the $FY*A/A^{null}$ x $FY*A/A^{null}$ mating pairs necessary to produce $FY*A^{null}/A^{null}$ offspring. This is likely to be an important factor, as the allele frequency from our studies in the Wosera region of Papua New Guinea is 0.022. Second, Duffy negativity could reduce fitness in Papua New Guinea. With respect to this second possibility, we are reminded of a recent epidemiological study in Vanuatu suggesting that α^+-thalassemia may increase susceptibility to *P. vivax* in young children and in turn induce cross-species protection against subsequent severe *P. falciparum* malaria (137). Evidence gathered to support this hypothesis suggests involvement of cross-species immunity (42,138,139) where increased malarial parasite antigen may be expressed on infected thalassemic cells (140). However, epidemiological studies from other sites endemic for *P. vivax* and *P. falciparum* malaria have produced equivocal findings (42,48,138,141). In the Papua New Guinean populations we are studying, α^+-thalassemia is noted to approach fixation in some communities (14,142–144). These additional studies have observed an association between α^+-thalassemia and protection from severe malaria (142–144). Moreover, if α^+-thalassemia were involved in a complex array of factors that would predispose (137) children to *P. vivax* infection to later reduce the severity of *P. falciparum* malaria as proposed, in a region where both of these parasites were holoendemic, we would not expect the $FY*A^{null}$ allele to arise. Therefore, it is not clear that *P. vivax* plays a protective role against *P. falciparum* malaria or that it has played a role in shaping the human genome through natural selection. We expect that the $FY*A^{null}$ allele and other host genetic factors (145–151) are likely contributors to a complex network of factors evolving between malarial parasites and humans in Papua New Guinea.

CONCLUSIONS

P. vivax is an important human pathogen. Despite lacking the dramatic association with mortality observed for *P. falciparum* malaria, *P. vivax* imposes significantly on public health in the tropical world. Rather than classifying *P. vivax* malaria as a benign form of malaria in the endemic world, ongoing research must refine strategies to determine how *P. vivax* acts as an agent of natural selection on humans.

Although it is certain that Duffy negativity confers resistance to blood-stage *P. vivax* infection, much remains unclear in regard to the evolutionary path between the founding mutational event(s), fixation of the $FY*B^{null}$ allele in much of Africa, and the striking absence of this allele from human populations outside Africa. *In vitro* studies have shown that differences in Duffy expression levels on erythroid cells are associated with both promoter and ORF genetic polymorphisms. Although many laboratories have reported findings of this nature, very little data have been produced to translate these observations into consistent estimates of susceptibility to blood-stage *P. vivax* infection, let alone pathogenesis associated with *P. vivax* malaria. Future epidemiologic studies must look carefully to determine if heterozygosity for the Duffy-negative alleles is associated with decreased prevalence, parasitemia, duration, or morbidity of *P. vivax* infection or with increased intervals between *P. vivax* infection.

Whereas the evolutionary relationship between *P. vivax* and Duffy negativity will continue to promote fascinating academic debate, as development of *P. vivax*–specific vaccines advances, we will soon face malaria control issues that are closely integrated with the theoretical components of this discussion focused on human and parasite population biology. In areas coendemic for *P. vivax* and *P. falciparum* it will be important to determine if reduced *P. vivax* infection increases severity of *P. falciparum* malaria. If *P. vivax* infection attenuates *P. falciparum* malaria, the evolution of Duffy negativity would come at a cost to individual and population fitness and mandate that *P. vivax*–specific control measures be used with caution. Keeping in mind that *P. vivax* causes significant illness worldwide, if *P. vivax* infection does not reduce severity of *P. falciparum* malaria significantly, the evolution of Duffy negativity would not encounter the force of balancing selection and a gradual rise in Duffy-negative allele frequency would suggest that *P. vivax*–specific control measures should be developed without reservation.

ACKNOWLEDGMENTS

I thank L. H. Miller, W. E. Collins, J. W. Kazura, S. S. Patel, T. P. Zimmerman, and D. T. McNamara for helpful suggestions and criticisms leading to the completion of this manuscript. Additionally, I thank J. Carlton and R. Carter for allowing review and discussion of unpublished material and G. Daniels for advice on current Duffy blood group nomenclature. Most importantly, I thank the thousands of study volunteers who have contributed to our understanding of *P. vivax* malaria. Financial support was provided by the National Institute of Allergy and Infectious Diseases (Grant AI-46919).

REFERENCES

1. Bruce-Chwatt, L. J. (1988). History of malaria from prehistory to eradication. In *Malaria, Principles and Practices of Malariology*, Vol. 1 (Wensdorfer, W., and McGregor, I., Eds.), pp. 1–59. Churchill and Livingstone, Edinburgh.

2. Coatney, G. R., Collins, W. E., Warren, McW., and Contacos, P. G. (1971). *The Primate Malarias*. National Institute of Allergy and Infectious Diseases, Bethesda, MD.

3. Wagner-Jauregg, J. (1922). The treatment of general paresis by inoculation of malaria. *J. Nerv. Ment. Dis.* **55**, 369–75.

4. Chernin, E. (1984). The malaria therapy of neurosyphilis. *J. Parasitol.* **70**, 611–7.

5. Boyd, M. F., and Stratman-Thomas, W. K. (1993). Studies on benign tertian malaria. 4. On the refractoriness of negroes to inoculation with *Plasmodium vivax*. *Am. J. Hyg.* **18**, 485–9.

6. Young, M. D., Ellis, J. M., and Stubbs, T. H. (1946). Studies on imported malarias. 5. Transmission of foreign *Plasmodium vivax* by *Anopheles quadrimaculatus*. *Am. J. Trop. Med.* **26**, 477–82.

7. Becker, F. T., Read, H. S., and Boyd, M. F. (1946). Variations in susceptibility to malaria. *Am. J. Med. Sci.* **211**, 680–5.

8. Young, M. D., Eyles, D. E., Burgess, R. W., and Jeffery, G. M. (1955). Experimental testing of the immunity of negroes to *Plasmodium vivax*. *J. Parasitol.* **41**, 315–8.

9. Darwin, C. (1859). *The Origin of Species by Means of Natural Selection: or, The Preservation of Favored Races in the Struggle for Life*. John Murray, London.

10. Mayr, E., and Provine W. B. (Eds.) (1981). *The Evolutionary Synthesis: Perspectives on the Unification of Biology*. Harvard University Press, Cambridge, MA.

11. Futuyma, D. J. (1998). *Evolutionary Biology*. Sinauer Associates, Sunderland, MA.

12. Race, R. R., and Sanger, R. (1950). *Blood Groups in Man*. Blackwell Science, Oxford.

13. Mourant, A. E., Kopec, A. C., and Domaniewska-Sobczak, K. (1976). *The Distribution of the Human Blood Groups and Other Polymorphisms*. Oxford University Press, London.

14. Zimmerman, P. A., and 10 colleagues (1999). Emergence of $FY*A^{null}$ in a *Plasmodium vivax*-endemic region of Papua New Guinea. *Proc. Natl. Acad. Sci. USA* **96**, 13,973–7.

15. Parran, T. (1937). *A Shadow on the Land*, Reynal and Hitchcock, Inc., New York.

16. Arnold, H. L., Jr. (1984). Landmark perspective: penicillin and early syphilis. *Jama* **251**, 2011–2.

17. Sartin, J. S., and Perry, H. O. (1995). From mercury to malaria to penicillin: the history of the treatment of syphilis. *J. Am. Acad. Dermatol.* **32**, 255–61.

18. Wagner-Jauregg, J. (1927). The treatment of dementia paralytica by malaria inoculation. The Nobel Foundation (http://www.nobel.se/medicine/laureates/1927/wagner-jauregg-lecture.html).

19. Dennie, C. C. (1962). *A History of Syphilis*. Charles C Thomas, Springfield, Ill.

20. Kitchen, S. K. (1949). Symptomatology: general considerations. In *Malariology; A Comprehensive Survey of All Aspects of This Group of Diseases from a Global Standpoint* (Boyd, M. F., Ed.), pp. 966–94. W. B. Saunders, Philadelphia.

21. Moore, J. E. (1941). *The Modern Treatment of Syphilis*. Charles C Thomas, Springfield, Ill.

22. Gardner, M. J., and 44 colleagues (2002). Genome sequence of the human malaria parasite *Plasmodium falciparum*. *Nature* **419**, 498–511.

23. Holt, R. A., and 82 colleagues (2002). The genome sequence of the malaria mosquito *Anopheles gambiae*. *Science* **298**, 129–49.

24. O'Leary, P. A. (1927). Treatment of neurosyphilis by malaria: report on the three years' observation of the first one hundred patients treated. *J. Am. Med. Assoc.* **89**, 95–100.

25. Mayne, B. (1932). Note on experimental infection of *Anopheles punctipennis* with quartan malaria. *Public Health Rep.* **47**, 1771–3.

26. Bray, R. S. (1958). The susceptibility of Liberians to the Madagascar strain of *Plasmodium vivax*. *J. Parasitol.* **44**, 371–3.

27. Krotoski, W. A., and 10 colleagues (1982). Demonstration of hypnozoites in sporozoite-transmitted *Plasmodium vivax* infection. *Am. J. Trop. Med. Hyg.* **31**, 1291–3.

28. Krotoski, W. A. (1985). Discovery of the hypnozoite and a new theory of malarial relapse. *Trans. R. Soc. Trop. Med. Hyg.* **79**, 1–11.

29. Kitchen, S. K. (1938). The infection of reticulocytes by *Plasmodium vivax*. *Am. J. Trop. Med.* **18**, 347–53.

30. Mons, B., and 5 colleagues (1988). *Plasmodium vivax*: in vitro growth and reinvasion in red blood cells of Aotus nancymai. *Exp. Parasitol.* **66**, 183–8.

31. Galinski, M. R., Medina, C. C., Ingravallo, P., and Barnwell, J. W. (1992). A reticulocyte-binding protein complex of *Plasmodium vivax* merozoites. *Cell* **69**, 1213–26.

32. Boyd, M. F. (1949). Historical review. In *Malariology; A Comprehensive Survey of All Aspects of This Group of Diseases from a Global Standpoint* (Boyd, M. F., Ed.), pp. 3–25. W. B. Saunders, Philadelphia.

33. Boyd, M. F. (1942). Criteria of immunity and susceptibility in naturally induced *vivax* infections. *Am. J. Trop. Med.* **22**, 217–30.

34. Kitchen, S. K., and Putnam, P. (1946). Observations on the character of paroxysm in *vivax* malaria. *J. Natl. Malaria Soc.* **5**, 57–70.

35. Nosten, F., and 8 colleagues (1999). Effects of *Plasmodium vivax* malaria in pregnancy. *Lancet* **354**, 546–9.

36. Bloland, P., and 5 colleagues (1996). Rates and risk factors for mortality during the first two years of life in rural Malawi. *Am. J. Trop. Med. Hyg.* **55**, 82–6.

37. Mendis, K., Sina, B. J., Marchesini, P., and Carter, R. (2001). The neglected burden of *Plasmodium vivax* malaria. *Am. J. Trop. Med. Hyg.* **64**, 97–106.

38. Knowles, R., and White, R. S. (1930). Studies in the parasitology of malaria. *Indian Med. Res. Memoirs* **18**, 436.

39. Rosenberg, R., Andre, R. G., Ngampatom, S., Hatz, C., and Burge, R. (1990). A stable, oligosymptomatic malaria focus in Thailand. *Trans. R. Soc. Trop. Med. Hyg.* **84**, 14–21.

40. McKenzie, F. E., and Bossert, W. H. (1997). Mixed-species *Plasmodium* infections of humans. *J. Parasitol.* **83**, 593–600.

41. McKenzie, F. E., and Bossert, W. H. (1999). Multispecies *Plasmodium* infections of humans. *J. Parasitol.* **85**, 12–8.

42. Bruce, M. C., and 6 colleagues (2000). Cross-species interactions between malaria parasites in humans. *Science* **287**, 845–8.
43. Cohen, J. E. (1973). Heterologous immunity in human malaria. *Q. Rev. Biol.* **48**, 467–89.
44. Richie, T. L. (1988). Interactions between malaria parasites infecting the same vertebrate host. *Parasitology* **96 (Pt 3)**, 607–39.
45. Mayxay, M., and 5 colleagues (2001). Identification of cryptic coinfection with *Plasmodium falciparum* in patients presenting with *vivax* malaria. *Am. J. Trop. Med. Hyg.* **65**, 588–92.
46. Mason, D. P., and 9 colleagues (2001). Can treatment of *P. vivax* lead to an unexpected appearance of *falciparum* malaria? *Southeast Asian J. Trop. Med. Public Health* **32**, 57–63.
47. Luxemburger, C., and 5 colleagues (1997). The epidemiology of severe malaria in an area of low transmission in Thailand. *Trans. R. Soc. Trop. Med. Hyg.* **91**, 256–62.
48. Smith, T., and 5 colleagues (2001). Prospective risk of morbidity in relation to malaria infection in an area of high endemicity of multiple species of *Plasmodium*. *Am. J. Trop. Med. Hyg.* **64**, 262–7.
49. Gautret, P., Legros, F., Koulmann, P., Rodier, M. H., and Jacquemin, J. L. (2001). Imported *Plasmodium vivax* malaria in France: geographical origin and report of an atypical case acquired in Central or Western Africa. *Acta Trop.* **78**, 177–81.
50. James, S. P. (1929). The disappearance of malaria from England. *Proc. R. Soc. Med.* **23**, 71–87.
51. Garnham, P. C. C. (1988). Malaria parasites of man: life-cycles and morphology (excluding ultrastructure). In *Malaria, Principles and Practices of Malariology*, Vol. 1 (Wensdorfer, W., and McGregor, I., Eds.), pp. 61–96. Churchill and Livingstone, Edinburgh.
52. Li, J., and 5 colleagues (2001). Geographic subdivision of the range of the malaria parasite *Plasmodium vivax*. *Emerging Infect. Dis.* **7**, 35–42.
53. Landsteiner, K. (1901). Agglutination phenomena in normal human blood. *Wien. Klin. Wschr.* **14**, 1132–4.
54. Cutbush, M., Mollison, P. L., and Parkin, D. M. (1950). A new human blood group. *Nature* **165**, 188–9.
55. Ikin, E. W., Mourant, A. E., Pettenkofer, H. J., and Blumenthal, G. (1951). Discovery of the expected haemagglutinin, anti-Fyb. *Nature* **168**, 1077.
56. Ikin, E. W., Mourant, A. E., and Pettenkofer, H. J. (1955). Discovery of the expected haemagglutinin, anti-Fyb. *Nature* **168**, 1077.
57. Sanger, R., Race, R. R., and Jack, J. (1955). The Duffy blood groups of the New York Negroes: the phenotype Fy(a−b−). *Brit. J. Haematol.* **1**, 370–4.
58. Race, R. R., Sanger, R., and Lehane, D. (1953). Quantitative aspects of the blood-group antigen Fya. *Ann. Eugen.* **17**, 255.
59. Cavalli-Sforza, L. L., Menozzi, P., and Piazza, A. (1994). *The History and Geography of Human Genes*. Princeton University Press, Princeton.
60. Lewis, G. E., Jr., and 5 colleagues (1988). Duffy phenotypes in Malaysian populations: correction of previous unusual findings. *Trans. R. Soc. Trop. Med. Hyg.* **82**, 509–10.

61. Simmons, R. T., Gajdusek, D. C., Gorman, J. G., Kidson, C., and Hornabrook, R. W. (1967). Presence of the Duffy blood group gene Fyb demonstrated in Melanesians. *Nature* **213**, 1148–9.
62. Malcolm, L. A., Woodfield, D. G., Blake, N. M., Kirk, R. L., and McDermid, E. M. (1972). The distribution of blood, serum protein and enzyme groups on Manus Island (Admiralty Islands, New Guinea). *Hum. Hered.* **22**, 305–22.
63. Mourant, A. E., and 3 colleagues (1981). Red cell antigen, serum protein, and red cell enzyme polymorphisms in inhabitants of the Jimi Valley, Western Highlands, New Guinea. *Hum. Genet.* **59**, 77–80.
64. Mourant, A. E., and 3 colleagues (1982). Red cell antigen, serum protein and red cell enzyme polymorphisms in Eastern Highlanders of New Guinea. *Hum. Hered.* **32**, 374–84.
65. Booth, P. B., and 6 colleagues (1982). Red cell antigen, serum protein and red cell enzyme polymorphisms in Karkar Islanders and inhabitants of the adjacent North Coast of New Guinea. *Hum. Hered.* **32**, 385–403.
66. Long, J. C., and 6 colleagues (1986). Genetic characterization of Gainj- and Kalam-speaking peoples of Papua New Guinea. *Am. J .Phys. Anthropol.* **70**, 75–96.
67. Simmons, R. T., and Cooke, D. R. (1969). Population genetic studies in Australian aborigines of the Northern Territory. Blood group genetic studies in the Malag of Elcho Island. *Archaeol. Phys. Anthropol. Oceania* **4**, 252–9.
68. Booth, P. B., Faogali, J. L., Kirk, R. L., and Blake, N. M. (1977). HLA types, blood groups, serum protein, and red cell enzyme types among Samoans in New Zealand. *Hum. Hered.* **27**, 412–23.
69. Guderian, R., and Vargas, J. (1986). Duffy blood group distribution and the incidence of malaria in Ecuador. *Trans. R. Soc. Trop. Med. Hyg.* **80**, 162–3.
70. Spencer, H. C., and 3 colleagues (1978). The Duffy blood group and resistance to *Plasmodium vivax* in Honduras. *Am. J. Trop. Med. Hyg.* **27**, 664–70.
71. Miller, L. H., Mason, S. J., Clyde, D. F., and McGinniss, M. H. (1976). The resistance factor to *Plasmodium vivax* in blacks. The Duffy-blood-group genotype, FyFy. *N. Engl. J. Med.* **295**, 302–4.
72. Miller, L. H., Mason, S. J., Dvorak, J. A., McGinniss, M. H., and Rothman, I. K. (1975). Erythrocyte receptors for (*Plasmodium knowlesi*) malaria: Duffy blood group determinants. *Science* **189**, 561–3.
73. Milam, D. F., and Coggeshall, L. T. (1938). Duration of *Plasmodium knowlesi* infections in man. *Am. J. Trop. Med.* **18**, 331–8.
74. Moore, S., Woodrow, C. F., and McClelland, D. B. (1982). Isolation of membrane components associated with human red cell antigens Rh(D), (c), (E) and Fy. *Nature* **295**, 529–31.
75. Hadley, T. J., David, P. H., McGinniss, M. H., and Miller, L. H. (1984). Identification of an erythrocyte component carrying the Duffy blood group Fya antigen. *Science* **223**, 597–9.
76. Chaudhuri, A., and 6 colleagues (1989). Purification and characterization of an erythrocyte membrane protein complex carrying Duffy blood group antigenicity. Possible receptor for *Plasmodium vivax* and *Plasmodium knowlesi* malaria parasite. *J. Biol. Chem.* **264**, 13,770–4.

77. Albrey, J. A., and 6 colleagues (1971). A new antibody, anti-Fy3, in the Duffy blood group system. *Vox Sang* 20, 29–35.

78. Behzad, O., Lee, C. L., Gavin, J., and Marsh, W. L. (1973). A new anti-erythrocyte antibody in the Duffy system: anti-Fy4. *Vox Sang* 24, 337–42.

79. Colledge, K. I., Pezzulich, M., and Marsh, W. L. (1973). Anti-Fy5: an antibody disclosing a probable association between Rhesus and Duffy blood group genes. *Vox Sang* 24, 193–9.

80. Nichols, M. E., Rubinstein, P., Barnwell, J., Rodriguez de Cordoba, S., and Rosenfield, R. E. (1987). A new human Duffy blood group specificity defined by a murine monoclonal antibody. Immunogenetics and association with susceptibility to *Plasmodium vivax*. *J. Exp. Med.* 166, 776–85.

81. Chaudhuri, A., and 5 colleagues (1993). Cloning of glycoprotein D cDNA, which encodes the major subunit of the Duffy blood group system and the receptor for the *Plasmodium vivax* malaria parasite. *Proc. Natl. Acad. Sci. USA* 90, 10,793–7.

82. Donahue, R. P., Bias, W. B., Remwick, J. H., and McKusick, V. A. (1968). Probable assignment of the Duffy blood group locus to chromosome 1 in man. *Proc. Natl. Acad. Sci. USA* 61, 949–55.

83. Dracopoli, N. C., and 9 colleagues (1991). The CEPH consortium linkage map of human chromosome 1. *Genomics* 9, 686–700.

84. Mathew, S., Chaudhuri, A., Murty, V. V., and Pogo, A. O. (1994). Confirmation of Duffy blood group antigen locus (*FY*) at 1q22–>q23 by fluorescence in situ hybridization. *Cytogenet Cell Genet.* 67, 68.

85. Iwamoto, S., Li, J., Omi, T., Ikemoto, S., and Kajii, E. (1996). Identification of a novel exon and spliced form of Duffy mRNA that is the predominant transcript in both erythroid and postcapillary venule endothelium. *Blood* 87, 378–85.

86. Chaudhuri, A., Polyakova, J., Zbrzezna, V., and Pogo, A. O. (1995). The coding sequence of Duffy blood group gene in humans and simians: restriction fragment length polymorphism, antibody and malarial parasite specificities, and expression in nonerythroid tissues in Duffy-negative individuals. *Blood* 85, 615–21.

87. Iwamoto, S., Omi, T., Kajii, E., and Ikemoto, S. (1995). Genomic organization of the glycoprotein D gene: Duffy blood group Fya/Fyb alloantigen system is associated with a polymorphism at the 44-amino acid residue. *Blood* 85, 622–6.

88. Tournamille, C., Le Van Kim, C., Gane, P., Cartron, J. P., and Colin, Y. (1995). Molecular basis and PCR-DNA typing of the Fya/fyb blood group polymorphism. *Hum. Genet.* 95, 407–10.

89. Tournamille, C., Colin, Y., Cartron, J. P., and Le Van Kim, C. (1995). Disruption of a GATA motif in the Duffy gene promoter abolishes erythroid gene expression in Duffy-negative individuals. *Nat. Genet.* 10, 224–8.

90. Shimizu, Y., Kimura, M., Settheetham-Ishida, W., Duangchang, P., and Ishida, T. (1997). Serotyping of Duffy blood group in several Thai ethnic groups. *Southeast Asian J. Trop. Med. Public Health* 28, 32–5.

91. Shimizu, Y., and 7 colleagues (2000). Sero- and molecular typing of Duffy blood group in Southeast Asians and Oceanians. *Hum. Biol.* **72**, 511–8.

92. Eugenio-Cavasini, C., Tarelho-Pereira, F. J., Luidi-Ribeiro, W., Wunderlich, G., and Urgano-Ferreira, M. (2001). Duffy blood group genotypes among malaria patients in Rondonia, Western Brazilian Amazon. *Revista da Sociedade Brasileira de Medicina Tropical* **34**, 591–5.

93. Tournamille, C., and 7 colleagues (1998). Arg89Cys substitution results in very low membrane expression of the Duffy antigen/receptor for chemokines in Fy(x) individuals. *Blood* **92**, 2147–56.

94. Parasol, N., and 5 colleagues (1998). A novel mutation in the coding sequence of the *FY*B* allele of the Duffy chemokine receptor gene is associated with an altered erythrocyte phenotype. *Blood* **92**, 2237–43.

95. Olsson, M. L., and 7 colleagues (1998). The Fy(x) phenotype is associated with a missense mutation in the Fy(b) allele predicting Arg89Cys in the Duffy glycoprotein. *Br. J. Haematol.* **103**, 1184–91.

96. Chown, B., Lewis, M., and Kaita, H. (1965). The Duffy blood group system in Caucasians: evidence for a new allele. *Am. J. Hum. Genet.* **17**, 384–9.

97. Mallinson, G., Soo, K. S., Schall, T. J., Pisacka, M., and Anstee, D. J. (1995). Mutations in the erythrocyte chemokine receptor (Duffy) gene: the molecular basis of the Fya/Fyb antigens and identification of a deletion in the Duffy gene of an apparently healthy individual with the Fy(a–b–) phenotype. *Br. J. Haematol.* **90**, 823–9.

98. Hadley, T. J., and 6 colleagues (1994). Postcapillary venule endothelial cells in kidney express a multispecific chemokine receptor that is structurally and functionally identical to the erythroid isoform, which is the Duffy blood group antigen. *J. Clin. Invest.* **94**, 985–91.

99. Peiper, S. C., and 10 colleagues (1995). The Duffy antigen/receptor for chemokines (DARC) is expressed in endothelial cells of Duffy negative individuals who lack the erythrocyte receptor. *J. Exp. Med.* **181**, 1311–7.

100. Chaudhuri, A., and 5 colleagues (1997). Detection of Duffy antigen in the plasma membranes and caveolae of vascular endothelial and epithelial cells of nonerythroid organs. *Blood* **89**, 701–12.

101. Horuk, R., and 6 colleagues (1996). The Duffy antigen receptor for chemokines: structural analysis and expression in the brain. *J. Leukoc. Biol.* **59**, 29–38.

102. Murphy, P. M. (1996). Chemokine receptors: structure, function and role in microbial pathogenesis. *Cytokine Growth Factor Rev.* **7**, 47–64.

103. Murphy, P. M. (2002). International Union of Pharmacology. XXX. Update on chemokine receptor nomenclature. *Pharmacol. Rev.* **54**, 227–9.

104. Neote, K., Mak, J. Y., Kolakowski, L. F., Jr., and Schall, T. J. (1994). Functional and biochemical analysis of the cloned Duffy antigen: identity with the red blood cell chemokine receptor. *Blood* **84**, 44–52.

105. Chaudhuri, A., and 5 colleagues (1994). Expression of the Duffy antigen in K562 cells. Evidence that it is the human erythrocyte chemokine receptor. *J. Biol. Chem.* **269**, 7835–8.

106. Horuk, R., and 6 colleagues (1993). A receptor for the malarial parasite *Plasmodium vivax*: the erythrocyte chemokine receptor. *Science* **261**, 1182–4.

107. Horuk, R., Wang, Z. X., Peiper, S. C., and Hesselgesser, J. (1994). Identification and characterization of a promiscuous chemokine-binding protein in a human erythroleukemic cell line. *J. Biol. Chem.* **269**, 17,730–3.

108. Chitnis, C. E., Chaudhuri, A., Horuk, R., Pogo, A. O., and Miller, L. H. (1996). The domain on the Duffy blood group antigen for binding *Plasmodium vivax* and *P. knowlesi* malarial parasites to erythrocytes. *J. Exp. Med.* **184**, 1531–6.

109. Haynes, J. D., and 6 colleagues (1988). Receptor-like specificity of a *Plasmodium knowlesi* malarial protein that binds to Duffy antigen ligands on erythrocytes. *J. Exp. Med.* **167**, 1873–81.

110. Adams, J. H., and 6 colleagues (1990). The Duffy receptor family of *Plasmodium knowlesi* is located within the micronemes of invasive malaria merozoites. *Cell* **63**, 141–53.

111. Wertheimer, S. P., and Barnwell, J. W. (1989). *Plasmodium vivax* interaction with the human Duffy blood group glycoprotein: identification of a parasite receptor-like protein. *Exp. Parasitol.* **69**, 340–50.

112. Fang, X. D., Kaslow, D. C., Adams, J. H., and Miller, L. H. (1991). Cloning of the *Plasmodium vivax* Duffy receptor. *Mol. Biochem. Parasitol.* **44**, 125–32.

113. Adams, J. H., and 5 colleagues (1992). A family of erythrocyte binding proteins of malaria parasites. *Proc. Natl. Acad. Sci. USA* **89**, 7085–9.

114. Chitnis, C. E., and Miller, L. H. (1994). Identification of the erythrocyte binding domains of *Plasmodium vivax* and *Plasmodium knowlesi* proteins involved in erythrocyte invasion. *J. Exp. Med.* **180**, 497–506.

115. Ranjan, A., and Chitnis, C. E. (1999). Mapping regions containing binding residues within functional domains of *Plasmodium vivax* and *Plasmodium knowlesi* erythrocyte-binding proteins. *Proc. Natl. Acad. Sci. USA* **96**, 14,067–72.

116. Michon, P., and 5 colleagues (2001). Duffy-null promoter heterozygosity reduces *DARC* expression and abrogates adhesion of the *P. vivax* ligand required for blood-stage infection. *FEBS Lett.* **495**, 111–4.

117. Tsuboi, T., and 5 colleagues (1994). Natural variation within the principal adhesion domain of the *Plasmodium vivax* duffy binding protein. *Infect Immun.* **62**, 5581–6.

118. Ampudia, E., Patarroyo, M. A., Patarroyo, M. E., and Murillo, I. A. (1996). Genetic polymorphism of the Duffy receptor binding domain of *Plasmodium vivax* in Colombian wild isolates. *Mol. Biochem. Parasitol.* **78**, 269–72.

119. Fraser, T., and 6 colleagues (1997). Expression and serologic activity of a soluble recombinant *Plasmodium vivax* Duffy binding protein. *Infect. Immun.* **65**, 2772–7.

120. Michon, P., Arevalo-Herrera, M., Fraser, T., Herrera, S., and Adams, J. H. (1998). Serological responses to recombinant *Plasmodium vivax* Duffy binding protein in a Colombian village. *Am. J. Trop. Med. Hyg.* **59**, 597–9.

121. Cole-Tobian, J. L., and 7 colleagues (2002). Age-acquired immunity to a *Plasmodium vivax* invasion ligand, the Duffy binding protein. *J. Infect. Dis.* **186**, 531–9.

122. Palatnik, M., and Rowe, A. W. (1984). Duffy and Duffy-related human antiges in primates. *J. Hum. Evol.* **13**, 173–9.

123. Jones, S., Martin, R., and Pilbeam, D., Eds. (1992). *The Cambridge Encyclopedia of Human Evolution*. Cambridge University Press, Cambridge, UK.

124. Okenu, D. M., Malhotra, P., Lalitha, P. V., Chitnis, C. E., and Chauhan, V. S. (1997). Cloning and sequence analysis of a gene encoding an erythrocyte binding protein from *Plasmodium cynomolgi*. *Mol. Biochem. Parasitol.* **89**, 301–6.

125. Eyles, D. E., Coatney, G. R., and Getz, M. E. (1960). Vivax-type malaria parasite of macaques transmissible to man. *Science* **131**, 1812–13.

126. Butcher, G. A., Mitchell, G. H., and Cohen, S. (1973). Mechanism of host specificity in malarial infection. *Nature* **244**, 40–2.

127. Miller, L. H., Dvorak, J. A., Shiroishi, T., and Durocher, J. R. (1973). Influence of erythrocyte membrane components on malaria merozoite invasion. *J. Exp. Med.* **138**, 1597–1601.

128. Beye, J. K., Getz, M. E., Coatney, G. R., Elder, H. A., and Eyles, D. E. (1961). Simian malaria in man. *Am. J. Trop. Med. Hyg.* **10**, 311–16.

129. Escalante, A. A., and Ayala, F. J. (1994). Phylogeny of the malarial genus *Plasmodium*, derived from rRNA gene sequences. *Proc. Natl. Acad. Sci. USA* **91**, 11,373–7.

130. Escalante, A. A., Barrio, E., and Ayala, F. J. (1995). Evolutionary origin of human and primate malarias: evidence from the circumsporozoite protein gene. *Mol. Biol. Evol.* **12**, 616–26.

131. Escalante, A. A., Freeland, D. E., Collins, W. E., and Lal, A. A. (1998). The evolution of primate malaria parasites based on the gene encoding cytochrome b from the linear mitochondrial genome. *Proc. Natl. Acad. Sci. USA* **95**, 8124–9.

132. Hamblin, M. T., and Di Rienzo, A. (2000). Detection of the signature of natural selection in humans: evidence from the Duffy blood group locus. *Am. J. Hum. Genet.* **66**, 1669–79.

133. Hamblin, M. T., Thompson, E. E., and Di Rienzo, A. (2002). Complex signatures of natural selection at the Duffy blood group locus. *Am. J. Hum. Genet.* **70**, 369–83.

134. Seixas, S., Ferrand, N., and Rocha, J. (2002). Microsatellite variation and evolution of the human Duffy blood group polymorphism. *Mol. Biol. Evol.* **19**, 1802–6.

135. Hartl, D. L., and Clark, A. G. (1997). *Principles of Population Genetics*. Sinauer Associates, Inc., Sunderland, MA.

136. Livingstone, F. B. (1984). The Duffy blood groups, *vivax* malaria, and malaria selection in human populations: a review. *Hum. Biol.* **56**, 413–25.

137. Williams, T. N., and colleagues (1996). High incidence of malaria in alpha-thalassaemic children. *Nature* **383**, 522–5.

138. Jeffery, G. M. (1966). Epidemiological significance of repeated infections with homologous and heterologous strains and species of *Plasmodium*. *Bull. World Health Org.* **35**, 873–82.

139. Voller, A., and Rossan, R. N. (1969). Immunological studies with simian malarias. II. Heterologous immunity in the "cynomolgi" group. *Trans. R. Soc. Trop. Med. Hyg.* **63**, 57–63.

P. A. Zimmerman

P. A. Zimmerman

P. A. Zimmerman

P. A. Zimmerman

P. A. Zimmerman

. Luzzi, G. A., and 5 colleagues (1991). Surface antigen expression on *Plasmodium falciparum*-infected erythrocytes is modified in alpha- and beta-thalassemia. *J. Exp. Med.* **173**, 785–91.

. Mehlotra, R. K., and 7 colleagues (2000). Random distribution of mixed species malaria infections in Papua New Guinea. *Am. J. Trop. Med. Hyg.* **62**, 225–31.

. Oppenheimer, S. J., Higgs, D. R., Weatherall, D. J., Barker, J., and Spark, R. A. (1984). Alpha thalassaemia in Papua New Guinea. *Lancet* **1**, 424–6.

. Flint, J., and 9 colleagues (1986). High frequencies of alpha-thalassaemia are the result of natural selection by malaria. *Nature* **321**, 744–50.

. Allen, S. J., and 6 colleagues (1997). alpha+-Thalassemia protects children against disease caused by other infections as well as malaria. *Proc. Natl. Acad. Sci. USA* **94**, 14,736–41.

. Vulliamy, T., Mason, P., and Luzzatto, L. (1992). The molecular basis of glucose-6-phosphate dehydrogenase deficiency. *Trends Genet.* **8**, 138–43.

. Genton, B., and 6 colleagues (1995). Ovalocytosis and cerebral malaria. *Nature* **378**, 564–5.

. Mgone, C. S., and 7 colleagues (1996). Occurrence of the erythrocyte band 3 (AE1) gene deletion in relation to malaria endemicity in Papua New Guinea. *Trans. R. Soc. Trop. Med. Hyg.* **90**, 228–31.

. Patel, S. S., and colleagues (2001). The association of the glycophorin C exon 3 deletion with ovalocytosis and malaria susceptibility in the Wosera, Papua New Guinea. *Blood* **98**, 3489–91.

. Mayer, D. C., Kaneko, O., Hudson-Taylor, D. E., Reid, M. E., and Miller, L. H. (2001). Characterization of a *Plasmodium falciparum* erythrocyte–binding protein paralogous to EBA-175. *Proc. Natl. Acad. Sci. USA* **98**, 5222–7.

. Mayer, D. C., Mu, J. B., Feng, X., Su, X. Z., and Miller, L. H. (2002). Polymorphism in a *Plasmodium falciparum* erythrocyte-binding ligand changes its receptor specificity. *J. Exp. Med.* **196**, 1523–8.

. Maier, A. G., and 6 colleagues (2003). *Plasmodium falciparum* erythrocyte invasion through glycophorin C and selection for Gerbich negativity in human populations. *Nat. Med.* **9**, 87–92.

172 P. A. Zimmerman

PART THREE

OTHER PARASITES

SEVEN

Influenza Evolution

Robin M. Bush and Nancy J. Cox

THE VIRUS

The influenza viruses are classified in three genera of the family *Orthomyxoviridae*. The genera are referred to as "types" A, B, and C. The genome, about 14 KB in size, has eight single-stranded RNA segments of negative sense (seven segments in influenza C viruses). The influenza A genome encodes three polymerase proteins (PB1, PB2, and PA); two major surface glycoproteins, hemagglutinin (HA) and neuraminidase (NA); three structural proteins (NP, M1, and M2); and two non-structural proteins involved in nuclear export (NS1 and NS2) (Lamb, 1989). An eleventh open reading frame recently discovered within PB1 appears to code for a protein involved in host cell apoptosis (Chen et al., 2001).

Two surface glycoproteins have been the object of most evolutionary studies of influenza. Hemagglutinin (HA) is involved in binding to host cell surface receptors. Neuraminidase (NA) is necessary for release of daughter virions from host cells. These proteins protrude from the viral envelope and are exposed to host immune defenses. While the HA is the primary target for neutralizing antibodies, antibodies against NA also may reduce occurrence and severity of illness, and possibly prevent infection if present at high titer. The hemagglutinin esterase (HE) in influenza C assumes the functions of both HA and NA. Broad reviews of influenza biology can be found in Murphy and Webster (1996) and Glezen and Couch (1997).

SUBTYPES

Considerable genetic diversity exists among avian influenza A viruses (Webster et al., 1992). This variation has been categorized into "subtypes" based on antibody recognition of HA and NA. To date, influenza viruses bearing fifteen antigenically distinguishable HAs and nine antigenically distinguishable NAs have been isolated from birds. Viral subtypes are classified according to the particular HA and NA alleles they carry, e.g., subtype H3N2. A very limited number of HA and NA alleles circulate widely in non-avian hosts.

REASSORTMENT

Simultaneous infection of a host cell by two influenza viruses can produce viruses with mixtures of segments from both parents. This process is called genetic reassortment. Reassortment has been observed between influenza A subtypes and between antigenically distinct strains within both types B and C. There is no evidence of intra-typic reassortment between types A, B, and C (Xu et al., 1993; Peng et al., 1994; Scholtissek, 1998).

HOST RANGE

Influenza A viruses naturally infect a wide variety of birds, primarily waterfowl and gulls. These avian viruses are thought to be the ancestors of strains currently circulating in swine, horses, and humans. Influenza A and B viruses are responsible for winter epidemics in humans. Influenza C viruses are generally responsible for sporadic infections and local outbreaks of relatively mild respiratory disease in children. Influenza viruses have been isolated from a few other mammals: influenza A in seals, minks, and whales (Webster et al., 1992), influenza B in seals (Osterhaus et al., 2000), and influenza C in swine (Guo et al., 1983). The limited host range of influenza B and C viruses apparently prevents their involvement in antigenic shift because they normally infect only humans and thus limited opportunity exists for cross-species transfer.

We begin with a review of the natural history and evolution of influenza in birds, swine, and horses. We then discuss the biological basis of host specificity. We finish with a survey of human influenza in both its pandemic and epidemic forms.

AVIAN INFLUENZA

Influenza A viruses infect a taxonomically widespread group of wild birds. Stallknecht reports findings of influenza A in ninety species representing twelve orders of birds, primarily shorebirds and waterfowl (Stallknecht, 1997; Stallknecht and Shane, 1988). Despite the importance of avian influenza to human public health, the incidence and frequency of avian disease caused by the various subtypes have not been extensively documented. The best studied group, the ducks, show pronounced geographical and temporal variation in subtype frequency and overall carriage. These infections are typically asymptomatic, with higher incidence in juveniles (Webster et al., 1992).

Most knowledge of genetics and population biology of avian influenza comes from studies of domestic poultry. Transmission of influenza viruses to domestic poultry by infected migratory waterfowl as they encounter poultry farms is thought to introduce wild strains of avian influenza into domestic stock (Webster et al., 1997). Symptoms vary by viral subtype. Most avian viruses are carried asymptomatically and replicate in the cells lining the intestinal tract (Kawaoka et al., 1988), but a few subtypes (H5 and H7) can cause systemic disease ("fowl plague") with central nervous system involvement and rapid progression to death (Murphy and Webster, 1996). These highly pathogenic avian influenza strains can evolve as they replicate in domestic poultry from avirulent strains by means of a few point mutations in the HA gene (Horimoto and Kawaoka, 2001).

The evolutionary history of avian influenza can be inferred from phylogenetic reconstruction using sequence data. Phylogenetic analysis suggests that geography and host taxonomy play roles in generating the observed patterns of genetic variation in avian strains (Webster et al., 1992). Phylogenetic trees differ somewhat depending on the gene studied. For example, PB1 sequences appear to show that avian viruses isolated from North American waterfowl are most closely related to viruses from gulls. In contrast, trees constructed using the NP, M, and PA genes show a closer relationship between viruses from North American waterfowl and Old World waterfowl (Gorman et al., 1992). The differences between gene trees may be attributable to genetic reassortment, which occurs frequently between avian strains, or to the small number of sequences used in these analyses. One interesting result of these studies is that mutations are fixed into the influenza genome at a much slower rate in wild birds than in humans or swine. We return to the topic of evolutionary rates after summarizing influenza in swine, horses, and humans.

EVOLUTION IN SWINE AND HORSES

Influenza viruses cause acute respiratory disease in swine and are endemic in swine populations worldwide. The first documented influenza epidemic in swine occurred concurrently with the 1918 H1N1 pandemic in humans. The descendants of that outbreak in swine, referred to as "classical" swine H1N1 influenza viruses, are a major source of disease in swine in North America to this day. However, introductions of both avian and human strains into swine occur fairly frequently. In addition to the classical H1N1 viruses, currently circulating swine strains include more recently introduced avian H1N1 strains, human H1N1 and H3N2 strains, and reassortants among these strains (Murphy and Webster, 1996; Brown, 2000; Webby et al., 2000).

Historical records suggest that influenza has long infected horses, where it causes acute respiratory disease. In the past century the H3N8 and H7N7 subtypes have caused both limited and decades-long outbreaks. Outbreaks of equine influenza do not follow the seasonal patterns observed in humans; instead, they seem to be triggered by geographic movement of infected horses. For reviews, see Mumford and Chambers (1998) and Manuguerra et al. (2000).

HOST SPECIFICITY

Direct transmission of avian viruses to humans has been reported rarely and was believed to be highly restricted. This is because influenza viruses from humans and birds preferentially bind to different forms of the sialic acid receptor on host cells. Swine, which have both the cell receptors preferred by avian viruses and those preferred by human influenza viruses, were thus proposed as necessary intermediate hosts for the transmission of avian viruses to humans (Scholtissek, 1990). This hypothesis is consistent with the observation that many epidemics and pandemics appear to originate in Southeast Asia, where agricultural practices put ducks, swine, and humans in close contact (de Jong et al., 2000). However, no known human epidemics have been caused by swine-adapted influenza (Murphy and Webster, 1996). Even the so-called "swine flu" outbreak in 1976 among military personnel at Fort Dix, which prompted a mass vaccination program in the U.S., was ultimately a short-lived outbreak (Dowdle, 1997).

Several recent events suggest that receptor specificity may be only one factor in determining host specificity. In 1997, avian H5N1 viruses infected at least eighteen humans, six of whom died, in Hong Kong SAR (de Jong et al., 1997; Shortridge et al., 1998). The H5N1 avian influenza viruses responsible for this outbreak were reassortants between viruses from geese and either quail or teal which had been caged in close proximity in live poultry markets in Hong Kong (Guan et al., 1999). These were the first confirmed isolations of avian influenza viruses from humans with severe respiratory disease. These viruses apparently were not capable of transmission between humans (Katz et al., 1999; Bridges et al., 2002), and no further human H5N1 infections were seen after elimination of infected local poultry. These findings show that host receptor type is not a strict barrier to infection of humans by avian influenza viruses.

Evidence for direct infection of humans by avian viruses does not prove that swine are never involved in the transmission of avian influenza to

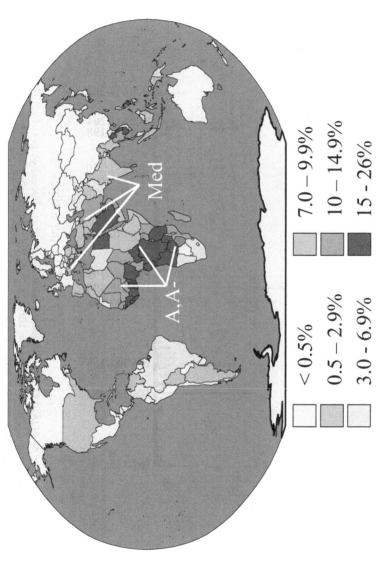

Figure 5.2. Global distribution and frequency of G6PD deficiency (adapted from Vulliamy et al., 1992) with the location of common G6PD alleles.

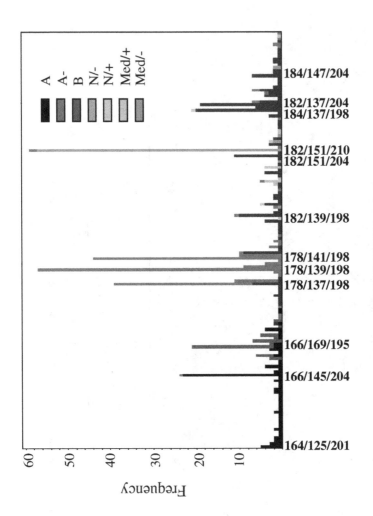

Figure 5.4. Relative frequencies of AC/AT/CTT microsatellite haplotypes on B chromosomes ($n = 183$), A chromosomes ($n = 90$), A- chromosomes ($n = 42$), Norm/*Bcl*II(−) chromosomes ($n = 188$), Norm/*Bcl*II(+) chromosomes ($n = 63$), Med/*Bcl*II(+) chromosomes ($n = 17$), Med/*Bcl*II(−) chromosomes ($n = 8$). Haplotypes are ordered by size of the AC, then the AT, then the CTT (data from Tishkoff et al., 2001).

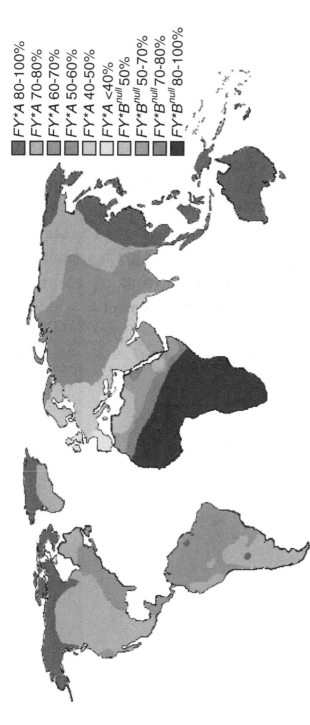

Figure 6.1. Distribution of *FY*A*, *FY*B*, and *FY*B^null^* alleles. In the regions of Africa where the *FY*B^null^* allele frequency drops below 80–100%, both *FY*A* and *FY*B* alleles are observed at low frequency. Outside of Africa, in regions where the *FY*A* allele frequency is below 80–100%, the *FY*B* allele is observed at frequencies of 1-*FY*A*. Adapted from Cavalli-Sforza *et al.* (59).

Legend:

- *FY*A* 80-100%
- *FY*A* 70-80%
- *FY*A* 60-70%
- *FY*A* 50-60%
- *FY*A* 40-50%
- *FY*A* <40%
- *FY*B^null^* 50%
- *FY*B^null^* 50-70%
- *FY*B^null^* 70-80%
- *FY*B^null^* 80-100%

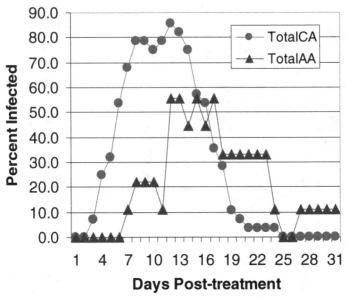

Figure 6.2. Summary of *P. knowlesi* infections in a cohort of malaria therapy patients (73). Caucasian (CA) and African American (AA) patients were compared based on the time to first blood-stage parasitemia, duration of blood-stage positivity, infectiousness to susceptible rhesus monkeys, and development of immunity to reinfection.

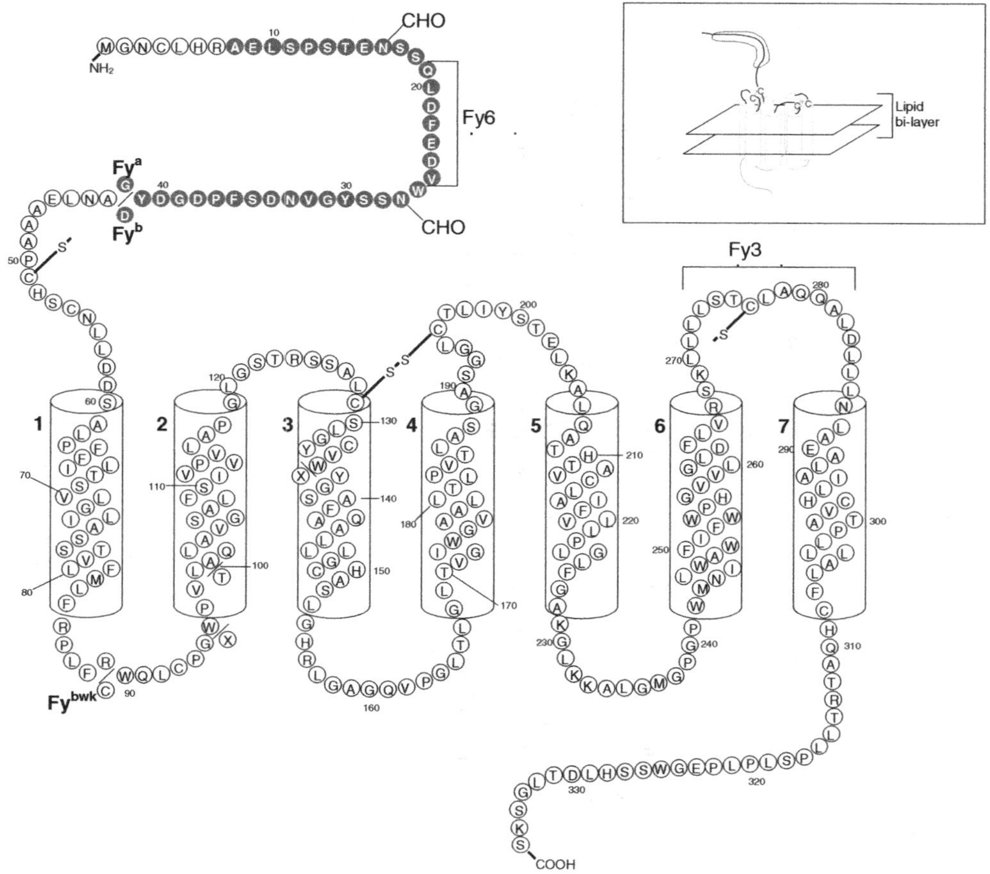

Figure 6.3.

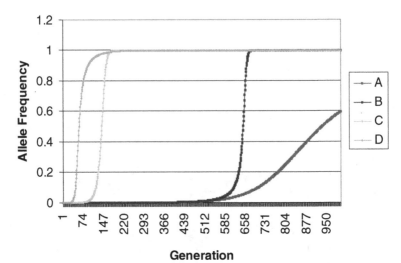

Figure 6.4. Models predicting the time frame over which the $FY*B^{null}$ may have evolved to fixation. For each simulation the initial $FY*B^{null}$ allele frequency was set to 0.00001. Assigned relative fitness scores for the $FY*B^{null}/B^{null}$, $FY*B/B^{null}$, and $FY*B/B$ genotypes were 0.376, 0.375, and 0.370 (model A, red line); 0.470, 0.375, and 0.370 (model B, blue line); 0.470, 0.400, and 0.370 (model C, green line): and 0.470, 0.460, and 0.370 (model D, gold line). The number of generations required to reach a frequency of 0.500 was 934 (model A), 650 (model B), 141 (model C), and 58 (model D). Models were based upon standard Hardy–Weinberg equilibrium and relative fitness equations (135).

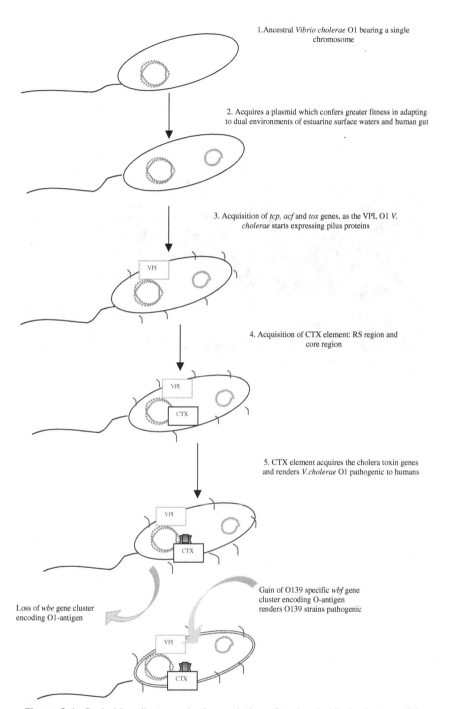

1. Ancestral *Vibrio cholerae* O1 bearing a single chromosome

2. Acquires a plasmid which confers greater fitness in adapting to dual environments of estuarine surface waters and human gut

3. Acquisition of *tcp, acf* and *tox* genes, as the VPI, O1 *V. cholerae* starts expressing pilus proteins

4. Acquisition of CTX element: RS region and core region

5. CTX element acquires the cholera toxin genes and renders *V.cholerae* O1 pathogenic to humans

Loss of *wbe* gene cluster encoding O1-antigen

Gain of O139 specific *wbf* gene cluster encoding O-antigen renders O139 strains pathogenic

Figure 8.1. Probable milestones in the evolution of toxigenic *Vibrio cholerae* O1 and O139.

Figure 8.2. Attachment of *V. cholerae* to copepods.

humans. It suggests, however, that a barrier to establishment in mammals may be the lack of efficient transmission between individuals. Birds generally harbor influenza viruses in their intestinal tract, not in their lungs. Thus avian viruses must adapt both to conditions in the mammalian respiratory tract and to airborne transmission. Dehydration during aerosol transmission among humans, for example, is a challenge not experienced during spread in feces and in the aquatic environments of waterfowl.

The live poultry markets in Southeast Asia constitute a natural setting for genetic reassortment between avian influenza viruses, and this setting also provides opportunities for these new strains to infect and adapt to humans. One fortunate outcome of the H5N1 outbreak has been greatly increased influenza surveillance in Hong Kong. This has allowed the detection of a number of additional reassortment events between avian influenza viruses of different subtypes along with studies of their pathogenicity in both poultry and mice, as reviewed by Hatta and Kawaoka (2002).

HUMAN INFLUENZA

Both influenza A and B cause acute, highly contagious respiratory illness in humans, with epidemics caused by influenza B viruses being less severe. Symptoms typically appear abruptly and can persist for one to two weeks. In adults, virus replication typically peaks at about 48 hours post-infection, declines slowly thereafter, with little viral shedding after six to eight days. Children may shed influenza virus for longer periods and at higher titers than adults. Influenza infections range in severity from asymptomatic to lethal. Death is most common in the elderly or those with compromised cardiovascular or immune systems and is typically associated with a secondary bacterial pneumonia or exacerbation of underlying health conditions. We return to influenza C, which causes neither epidemics nor pandemics, when we discuss relative rates of evolution among the different types of influenza.

Influenza epidemics are widespread outbreaks of disease that appear suddenly, persist for a few weeks, and then abruptly disappear. Epidemics are caused by influenza viruses that have evolved through antigenic drift from strains already circulating in humans. Such epidemics occur nearly annually and may be caused by one or more of the strains of influenza circulating in humans at a particular point in time. Attack rates during influenza epidemics are generally 10–20%, often being highest in young children (Cox and Subbarao, 2000). Influenza B, and influenza A H1N1 and H3N2, subtype viruses are currently circulating among humans.

Influenza pandemics, in contrast to epidemics, result from antigenic shift and are characterized by the rapid worldwide spread of a virus containing an HA and sometimes an NA to which humans have had no previous exposure. Pandemics result in high rates of morbidity and mortality, social disruption, and economic loss. After a few years, pandemic strains evolve to a lower level of virulence and subsequently appear as typical winter epidemic strains. However, the cumulative morbidity and mortality during inter-pandemic years due to epidemic influenza actually exceeds that of pandemics (Cox and Kawaoka, 1998).

We begin our discussion of human influenza by describing the introduction of influenza, through antigenic shift, into the human population in recent times. We then return to epidemics and trace the history of the most recent pandemic strains during their evolution within humans.

PANDEMIC INFLUENZA

Historical records describing influenza pandemics go back hundreds of years (Potter, 2001). There have been only two opportunities (1957 and 1968) to observe pandemics since a human influenza A virus was first isolated in 1933 (Smith et al., 1933). No pandemic has occurred since the advent of molecular biology, but sequencing of archived viruses has allowed us to partially reconstruct the genetic changes associated with the three most recent pandemics.

1918 H1N1 "Spanish Flu" Pandemic

The 1918 H1N1 pandemic occurred in waves of increasingly virulent disease starting in the spring of 1918 and continuing through the subsequent winter. This pandemic resulted in the deaths of between twenty and forty million people world-wide. Although called "Spanish flu," the geographic origin of this virus is controversial. The 1918 pandemic differed from previous pandemics, according to historical records, and from subsequent pandemics in two respects. First, it had a much higher per-case mortality rate, with deaths primarily occurring in young adults rather than among the elderly and infants. The second difference was the manner in which deaths occurred. Influenza-related mortality typically occurs a week or two post-infection and is associated with a secondary bacterial pneumonia or other complications. In 1918, many people died within just a few days from hemorrhagic pulmonary edema (Taubenberger et al., 2000).

The rapid and horrifying manner in which people died in 1918 has prompted much speculation about how this virus differed from other influenza strains. This has led Taubenberger, Reid, and colleagues to sequence H1N1 viruses preserved in the lung tissue of two army soldiers and in an Alaskan Inuit woman frozen in permafrost – all victims of the 1918 pandemic (Taubenberger et al., 1997; Reid et al., 1999). These ongoing studies have yet to reveal why this strain was so deadly. The HA and NA resemble the oldest available classical H1N1 swine influenza strains (from 1930) but share characteristics with modern avian H1N1 strains as well. Sequences from viruses isolated from waterfowl collected in 1917 and preserved in alcohol in the American Museum of Natural History have done little to resolve this mystery (Fanning et al., 2002).

The first known influenza outbreak in swine occurred in 1918, suggesting the hypothesis that swine served as intermediate hosts in the evolution of the 1918 human pandemic strain. However, influenza can spread from humans to swine as well (Brown, 2000), thus this swine outbreak could have originated in humans. The paucity of archived viral isolates from human, swine, and avian hosts may prevent us from ever determining the sequence of events resulting in the 1918 pandemic. Unfortunately, almost as little is known about the avian and swine viruses in circulation at the time of the subsequent 1957 and 1968 pandemics. This lack of knowledge speaks for the preservation of existing museum collections from which viral RNA can be extracted and for expanded surveillance of extant avian strains.

1957 H2N2 "Asian Flu" and 1968 H3N2 "Hong Kong Flu" Pandemics

The H2N2 pandemic began in February of 1957 in China and by summer had spread world-wide. A second wave of disease occurred in the winter of 1958. Total influenza-related excess mortality in the United States was estimated at about 70,000 (Noble, 1982), much less than the half-million Americans who died in 1918. The 1957 pandemic strain was a result of reassortment. An H1N1 strain circulating in humans – a descendant of the 1918 pandemic – had obtained avian H2, N2, and PB1 genes (Kawaoka et al., 1989). With the spread of the H2N2 strain, the parental H1N1 lineage disappeared from circulation in humans.

The H2N2 pandemic strain subsequently circulated as a typical winter epidemic strain until being displaced itself during the next pandemic. The 1968 H3N2 "Hong Kong flu" pandemic was caused by a human-adapted H2N2 virus that obtained avian H3 and PB1 genes (Kawaoka et al., 1989). This pandemic, which also emerged from Southeast Asia, resulted in only about half the mortality in the United States of the H2N2 pandemic (Noble, 1982), perhaps because the human population already had antibodies to the N2 protein.

With the emergence of the H2N2 and H3N2 pandemic strains (in 1957 and 1968, respectively), the previously circulating human influenza A strains soon became extinct. This serial replacement of one influenza A subtype by another suggested that only a single influenza A subtype could circulate in humans at one time. However, an event in 1997 proved this not to be so.

REAPPEARANCE OF H1N1

In 1977, twenty years after they had last circulated in humans, human H1N1 viruses reappeared in northern China (Raymond et al., 1986). These viruses were almost genetically and antigenically identical to H1N1 viruses isolated from humans in 1950 (Nakajima et al., 1978) and were most likely preserved during the intervening decades in a laboratory freezer or possibly in a frozen state in nature. The reemerged H1N1 strain spread rapidly but caused relatively mild disease (Cox and Regnery, 1996). Illness occurred almost exclusively among persons under twenty years of age because older individuals retained antibodies from previous exposure.

There is some question as to the extent to which pandemic viruses actually evolve to lower levels of virulence. A decreased severity of disease over time might also occur because prolonged circulation has resulted in a host population with antibodies that are broadly protective against the new subtype. However, evidence suggesting that pandemic strains can evolve towards milder forms comes from comparing the consequences of the infection of young adults by the 1918 H1N1 virus, which was often deadly, with the much milder disease caused by infection by descendants of that pandemic strain when it reemerged in 1977.

The H1N1 and H3N2 subtypes of influenza A have cocirculated in humans since 1977. Reassortment between the two subtypes has occurred repeatedly, but the resulting strains did not persist (Guo et al., 1992). In the 2001–2 influenza season, a new H1N2 reassortant spread across a wide geographic area (Gregory et al., 2002). Consistent with previous observations of reassortment between human-adapted strains (Cox and Bender, 1995), the resulting disease was not particularly severe. Whether this strain persists or replaces the currently circulating H3N2 and H1N1 strains remains to be seen.

To summarize, sequencing of archived viruses has shown that the three most recent influenza pandemics involved the cross-species transfer of either the entire avian influenza genome or the transfer of some avian influenza genes, including that for the HA, into viruses that infected humans. During this period (1918 to the present), influenza B has presumably circulated (it was first isolated in 1940) in humans along with either one or two subtypes of influenza A having HA alleles H1–H3. Surveys of antibody titers suggest that some of the HA alleles infecting humans in this century may have caused pandemics in the last century as well. These data were obtained by surveying antibodies present in different human age cohorts, a technique for reconstructing past disease outbreaks called seroarchaeology.

SEROARCHAEOLOGY

The nature and duration of the human immune response to influenza is only partially understood. A variety of immune mechanisms are likely to be involved. However, immunity appears to be subtype- and strain-specific and primarily mediated by antibodies against the surface glycoproteins. For unknown reasons, the first variant of a subtype encountered by an individual typically causes the strongest antibody response. Subsequent infections with related variants tend to reinforce the response to the first variant. Thus, the highest antibody titers in an age group appear to reflect the dominant antigens of the virus responsible for the childhood infections of the group. This phenomenon is called "original antigenic sin," (Francis, 1953).

In the mid-1900s, older people were screened for antibodies to a number of HA antigens. The results suggest that the unusually high mortality in young adults in the 1918 pandemic might be better described as an unusually low mortality rate among the middle-aged and elderly. Older people in 1918 appeared to be at least partially protected by antibodies presumably produced during exposure to an H1N1 virus in the late 1800s. Similarly, serum collected before the H2N2 or H3N2 subtypes entered the human population in 1957 and 1968, respectively, suggests that older human cohorts had been previously infected by these strains as well.

As summarized by Dowdle (1999), seroarchaeological studies suggest that the H1 and H3 hemagglutinin alleles (with varying N alleles) have each been introduced into humans at least twice in the recent past. The recent circulation of the H2 allele and the apparent recirculation of the H1 and H3 alleles do not guarantee, however, that viruses carrying other hemagglutinin subtypes cannot infect humans. Farmers in Southeast Asia have been reported to carry antibodies to a number of avian influenza subtypes not known to circulate in humans including the H5 allele recently involved in outbreaks of human illness in Hong Kong (Shortridge et al., 1998).

Pandemics occur as several waves of disease. The pandemic strains then evolve and reappear, primarily during winter epidemics. Gene sequencing did not become a standard part of the World Health Organization influenza surveillance program until the mid-1980s. Because there has not been a pandemic since 1968, there has been no opportunity to use molecular markers to study the genetic changes associated with successive waves of pandemic disease or to study the evolution of a novel strain during its initial adaptation to humans. However, over the past two decades, sequence data have become increasingly used to study antigenic drift, the ongoing evolution of strains currently causing epidemics in human populations.

EPIDEMIC INFLUENZA

Influenza is said to be epidemic when the incidence of influenza-related illness rises above a seasonal baseline. This concept can be illustrated by looking at the levels of influenza- and pneumonia-related (P & I) deaths in Figure 7.1. Winter epidemics occur almost yearly in both the north and south temperate zones, but influenza can be isolated from humans somewhere in the world year round. There is substantial temporal and geographic heterogeneity in influenza incidence. Thus Figure 7.1, which depicts P & I mortality data for the U.S., may not accurately portray influenza deaths worldwide.

It has long been known from immunological studies that more than one type or subtype of human influenza can circulate during an influenza season. Figure 7.2 shows the relative frequency of the three influenza strains circulating in humans (H1N1, H3N2, and B) in the U.S. since 1985. In many years one strain is predominant, but in the 2001–2 influenza season (October 1, 2001–September 30, 2002), for example, influenza A subtype H1N1 and influenza B viruses were reported at similar frequencies.

During short intervals of time there may appear to be a regular periodicity in the order in which the types or subtypes predominate, such as during the first 10 years shown in Figure 7.2. However, upon examination of longer periods of time, these patterns disappear (Figure 7.2). These data are again only for the U.S.; substantial geographic variation occurs. For example, during the 1985–6 season, influenza B was the predominant influenza virus reported in the U.S.; elsewhere in the world influenza A H3N2 viruses were most commonly reported.

Comparison of Figures 7.1 and 7.2 shows that although there is no exact correlation between disease severity (Figure 7.1) and the type or subtype of influenza in circulation (Figure 7.2), there is a correlation between the predominance of influenza A/H3N1 viruses and the severity of the season. For example, seasons with the highest peaks of excess P & I deaths (1993–4 and 1996–7) were those during which influenza A (H3N2)

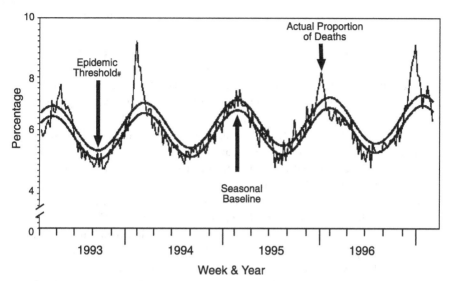

Figure 7.1. Weekly pneumonia and influenza (P&I) mortality as a percentage of all deaths in 122 cities in the United States from January 3, 1993 to April 5, 1977. The epidemic threshold is 1.645 standard deviations above the seasonal baseline. The expected seasonal baseline is projected using a robust regression procedure in which a periodic regression model is applied to observed percentages of death from P&I since 1983. From *Morbidity and Mortality Weekly Reports* 46(5):327.

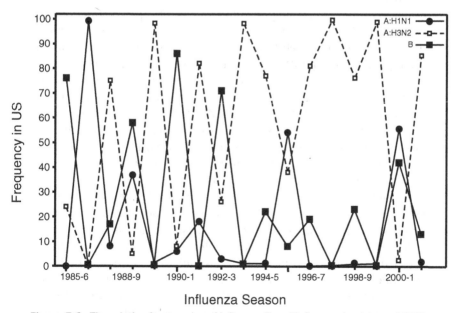

Figure 7.2. The relative frequencies of influenza B and influenza A subtypes H3N2 and H1N1 in the U.S.

viruses predominated. These figures also show, however, that the number of P & I deaths cannot be predicted solely from the incidence of a particular subtype. For instance, the incidence of influenza-related disease in 1994–5 barely exceeded the seasonal baseline but was at epidemic levels for over two months during 1996–7 (Figure 7.1). Yet the relative frequencies of the three strains circulating in the 1994–5 and 1996–7 influenza seasons were almost identical (Figure 7.2). To explain such variation both the accumulation of host immunity to the different circulating strains and the ongoing evolution of those strains must be considered. In this case, influenza A viruses antigenically similar to the H3N2 reference strain A/Johannesburg/33/94 circulated during both the 1994–5 and 1995–6 seasons. This presumably resulted in widespread immunity in the human population. The epidemic of 1996–7 was caused by an H3N2 virus that had evolved from an ancestor of the A/Johannesburg/33/94-like strains. These newly evolved viruses, related to the A/Wuhan/359/95 reference strain, were not effectively neutralized by existing antibodies against A/Johannesburg/33/94. This example illustrates the continual reinfection of humans by influenza viruses evolving through the process of antigenic drift.

ANTIGENIC DRIFT
Antigenic drift occurs rapidly in influenza. The RNA polymerase complex makes errors during replication of viral genes; one estimate is $10^{-5.5}$ mutations/site/replication (Webster and Laver, 1980). Because there is no proofreading mechanism for RNA viruses, mutations accumulate (Parvin et al., 1986). Over time, the accumulated replacements alter the shape and the electric charge of viral surface antigens, particularly on the distal surface and around the receptor binding site of the HA1 domain of hemagglutinin (Cox and Bender, 1995). These cumulative changes, many of which occur in known antibody binding sites (Wilson and Cox, 1990), eventually allow escape from antibodies raised by prior infection or vaccination (Wiley et al., 1981; Both et al., 1983; Nakajima et al., 1983).

Current methods of influenza vaccine production require that selection of the viral strains for inclusion in the vaccine be made about eight months prior to the beginning of the next influenza season. The goal of influenza surveillance is to track the evolution of existing lineages and to detect the emergence of novel antigenic variants that may cause future epidemics. In the past two decades, gene sequencing has become a routine part of influenza surveillance. Phylogenetic trees constructed using sequence data are a natural way to monitor the genetic changes associated with antigenic drift. Figure 7.3 shows phylogenies constructed using

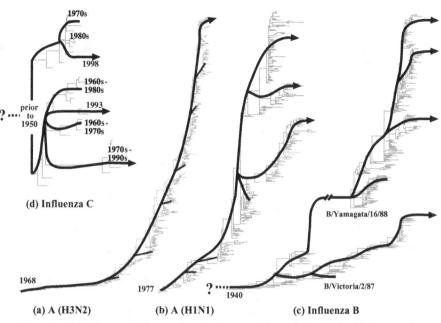

(d) Influenza C

(a) A (H3N2)   (b) A (H1N1)   (c) Influenza B

Figure 7.3. Phylogenetic trees showing the evolution of the HA1 of influenza A subtypes H3N2 (a) and H1N1 (b) and influenza B (c) and of the HE of influenza C (d). Dark lines are superimposed to indicate some of the major genetic lineages. As explained in the text, because of non-random sampling, the relative size of various branches is not an accurate reflection of the frequency of those genetic lineages in nature. Horizontal branch lengths indicate genetic distance; the length of vertical branches is simply a function of the number of sequences that need to be fit on the page. This explains some features of these trees, such as the sharp bend in the H3N2 tree around the late 1980s, the time at which sequencing became a standard tool in surveillance. Lines ending in arrows indicate extant lineages.

the HA1 domain of the hemagglutinin gene of human influenza A subtypes H3N2 (7.3a) and H1N1 (7.3b) and of influenza B (7.3c). Figure 7.3d was constructed using the HE gene of influenza C. Here we use a comparative approach to illustrate evolutionary features held in common by different strains of influenza and to highlight the ways in which they differ. These differences may provide insight into the forces shaping influenza evolution.

The main difference, highlighted by black lines superimposed over the main lineages in each tree, is variation among the trees in the degree of branching. With the exception of influenza C (7.3d), the trees are very similar in having relatively short side branches. These short side branches give these trees a very slender, linear look, as opposed to the bushy or

star-shaped trees seen for organisms having numerous long-lived cocirculating lineages, such as HIV.

The phylogenetic tree constructed using the gene sequence of the HA1
domain of H3N2 human influenza (Figure 7.3a) gives the most striking
example of linearity. H3N2 first infected humans in 1968. New lineages of
H3 can be seen to arise continually, and at any time, a number of closely
related lineages may cocirculate. However, all of these lineages except one
become extinct very quickly, on average within one to two years (Fitch
et al., 1997).

Figure 7.3b shows the evolution of the human H1N1 tree since its
reappearance in 1977. This tree is similar to the H3N2 tree (Figure 7.3a)
in having a linear shape with the exception of a split in the mid-1980s
and another in the mid-1990s (Cox et al., 1989; Nakajima et al., 2000).
Influenza B (Figure 7.3c) exhibits greater branching still, having diverged
into two lineages, each containing multiple cocirculating sublineages, in
the 1970s (Lindstrom et al., 1999). At the far end of the evolutionary
spectrum is influenza C (Figure 7.3d) which consists of a number of slowly
evolving, cocirculating lineages (Buonagurio et al., 1986; Matsuzaki et al.,
1994). Question marks are used to indicate the root of the influenza B and
C trees, as it is not known when they first infected humans.

It would be a mistake to equate the linear appearance of these trees
with the idea that each subsequent epidemic evolved from the previous
epidemic strain. In many cases epidemic strains evolve from small, linger-
ing populations of strains that are descendants of the ancestors of viruses
causing the most recent epidemic. One example is the A/Singapore/1/86-
like H1N1 viruses that caused the 1986–7 epidemic. These viruses did not
evolve from the previous epidemic strain, the A/Chile/1/83-like viruses,
but rather from an ancestor that had been detected earlier in China. The
emergence of new epidemics from small, isolated populations poses a challenge to the early detection necessary for vaccine strain selection, which
is based on antigenic, genetic, and epidemiological data.

Figure 7.3 illustrates the recent evolutionary patterns in influenza. The
processes that generate these patterns are not clear. Two factors that often
co-vary with the pattern of branching in phylogenetic trees of influenza
are

(1) the rate at which new mutations become fixed in populations and
(2) the length of time a strain has been evolving in humans.

However, it is not known whether this covariation indicates causal
relationships.

Published estimates of fixation rates vary depending on the time period involved, the method used, and the particular data examined. However, all suggest that H3 changes at about the same rate or possibly a little more rapidly than H1 ($4^{-5} \times 10^{-3}$ nucleotide substitutions/site/year). Influenza B HA evolves more slowly ($1^{-2} \times 10^{-3}$ nucleotide substitutions/site/year) than either H1 or H3 HA (Both et al., 1983; Raymond et al., 1986; Yamashita et al., 1988; Rota et al., 1990; Cox et al., 1993; Fitch et al., 1997). As can be seen in Figure 7.3d, influenza C HA has undergone comparatively little genetic change in recent years.

Several lines of evidence suggest that influenza viruses evolve more rapidly after introduction into a novel host. The rate in the natural host of the virus – waterfowl – is much slower than in humans (Webster et al., 1992). The fixation rate in swine, also a novel host for influenza A, is similar to the rates seen in humans (Scholtissek et al., 1993). The slower fixation rates reported for birds may be due to the shorter life span of birds and the relative lack of immune selection during influenza virus replication in birds. However, avian influenza viruses introduced into domestic poultry, which are not natural hosts of influenza, show fixation rates similar to those seen among human influenza viruses (Garcia et al., 1997).

Many other factors influence influenza evolution and thus may be responsible for the variation seen among the trees in Figure 7.3. For instance, there may be inherent differences in the underlying mutation rates of these strains, different levels of cross-immunity among strains, differences in rates of transmission, or differences in the functional constraints on the HA. Also, genes other than the HA may influence the patterns of HA evolution. For example, a virus with a particular HA genotype might outcompete a cocirculating virus because of differences in a gene other than the HA. Reassortment with other segments could also affect the evolution of the HA; this would not be evident from the HA trees. One way to learn more about the genetic changes resulting in the variation among the trees in Figure 7.3 would be to look more closely at the association between particular genetic changes and changes in antigenicity, thus identifying the targets of selection by the host immune system.

TARGETS OF SELECTION

Influenza surveillance based on HA1 nucleotide sequences provides more detail on evolutionary change in circulating strains than does traditional antigenic analysis. However, the usefulness of sequence data for epidemic prediction is limited because the relationship between genetic and antigenic variation is often unclear. For instance, in humans the evolution of a new "drift variant" appears to require at least four amino acid replacements occurring in at least two of the five antigenic sites in the HA1 of the H3N2 strain (Cox and Bender, 1995). However, there is a great deal of variation in this pattern from year to year. One way to more directly address the question of how genetic change relates to antigenic change is to use evolutionary hypotheses to try to anticipate the fitness of particular types of new mutant strains.

It has been shown that a small number of codons in the HA1 of influenza A subtype H3N2 have been under selection in the past to repeatedly change the amino acid they encode. These codons show an excess of non-synonymous (amino-acid changing) versus synonymous (silent) nucleotide substitutions, a pattern consistent with the assumption that host antibody pressure drives change in HA conformation. Analysis of phylogenetic trees representing the evolution of the HA1 of human H3N2 viruses through a short period of time (1987–97) showed that strains with more mutations in a set of eighteen codons with a history of having excess non-synonymous substitutions were the progenitors of successful new lineages in nine of eleven recent influenza seasons (Bush et al., 1999a, 1999b). A causal explanation for these results was suggested by the physical location of these codons in the HA1. Most occur in or near two important antibody binding sites, and some surround the receptor binding pocket as well.

If replacements at key amino acid positions truly explain differential fitness then additional replacements in these positions may also be advantageous in the future. Screening new strains for additional replacements at these particular sites may provide an early warning of potentially successful viral strains and thus prove a useful adjunct to current methods of influenza surveillance and vaccine strain selection.

BARRIERS TO STUDYING INFLUENZA EVOLUTION
The existing influenza sequence data are among the best available for studying the evolution of an infectious disease. However, there are problems with using these data to study influenza evolution and population biology. One problem is the presence of laboratory artifacts in the sequence data. Although cell culture is increasingly used, amplification of the virus by passage in embryonated hens' eggs was the standard laboratory practice for the culture of influenza viruses for many years. Egg-passage is still required of strains that will be used in the influenza vaccine in the U.S. Unfortunately, the HA1 of human influenza viruses evolves rapidly to adapt to replication in eggs (Robertson, 1993). The resulting sequences may thus contain replacements that either were not present or were at low frequency in the original viral sample. These laboratory artifacts often occur at sites involved in adaptation to humans as well as to eggs (Cox and Bender, 1995).

It is possible to estimate the proportion of amino acid replacements resulting from egg-passage by contrasting the number of replacements found in sequences in cell-passaged and egg-passaged isolates (Bush et al., 2000, 2001). In the HA1 domain of influenza A H3, egg-passage was associated with about 8% of amino acid replacements (Bush et al., 1999b). Unfortunately, in the absence of controls – viruses that have never been passaged – it is impossible to determine which replacements in a data set are artifacts.

Non-synonymous substitutions due to egg-passage should be eliminated from analyses seeking evidence of selection by the human immune system because these changes do not result from coadaptation with the human host. One way to minimize error due to these artifacts from selection studies is to discard changes assigned to the terminal branches of the trees when estimating substitution rates (Bush et al., 1999b, 2000, 2001). Studies that have failed to exclude egg-passaged replacements routinely find evidence for selection on codons for which there is no evidence of a selective advantage in humans (Huelsenbeck et al., 2001; Yang, 2000; Yang et al., 2000; Nielsen and Huelsenbeck, 2002).

Another barrier to studying selection in influenza is sampling bias. In the course of influenza surveillance, sequencing efforts are biased in favor of viruses that are antigenically distinguishable from known reference strains. Thus there is a strong bias towards sequencing viruses with non-synonymous substitutions in the HA, as these viruses appear antigenically novel in laboratory tests. This sampling bias causes error in evolutionary analyses that assume that an excess of non-synonymous substitutions reflects adaptive change accrued in response to host immune pressure. This bias is reduced if not entirely eliminated by the method described above for dealing with egg changes. This sampling bias also means that the sequences currently available in the public data bases (GenBank and the Los Alamos Influenza Database) are not representative of the frequencies of different types of influenza viruses in nature.

EFFECTS OF HUMAN ACTIVITY ON INFLUENZA EVOLUTION

The activities of humans have affected the evolution of influenza, but not to our advantage. Close confinement of various strains of fowl in live poultry markets provides conditions ripe for the formation of new reassortant viruses and their transmission to humans. In contrast, our intentional efforts to decrease disease burden have yet to affect its evolution; global vaccination rates are not high enough to exert selective pressure on currently circulating strains.

Antiviral drug-resistant strains of influenza have not circulated widely; however, viral strains resistant to adamantanes have been detected in individuals who had not been treated with these drugs (Ziegler et al., 1999). Furthermore, the limited susceptibility of some naturally occurring influenza viruses to the new class of neuraminidase-inhibiting drugs suggests that resistance to these compounds also might develop with widespread use; however, resistance to neuraminidase inhibitors appears to be less of a problem than resistance to adamantanes (Glezen and Couch, 1997).

UNANSWERED QUESTIONS
Influenza viruses are among the best studied of the causative agents of human diseases. Nonetheless, much remains to be learned. We do not understand either the factors limiting the introduction of avian viruses into humans or the initial stages in the adaptation of these avian viruses to their new host. Epidemics caused by drift variants account for substantial morbidity and mortality over time, yet very little is known about the types and amounts of genetic change needed to produce a new epidemic strain. Can the same antigenic variant arise simultaneously in more than one location because of similarities in immune pressure from the host?

Influenza epidemics occur during winters in the north and south temperate zones. It is not known whether influenza viruses migrate between the northern and southern hemisphere during respective winters or whether the viruses are seeded and perpetuated during the summer. One possible explanation for the seasonality of temperate zone outbreaks is the increased survival time of the virus in aerosol form with low humidity conditions that exist indoors in winter (Murphy and Webster, 1996; Glezen and Couch, 1997).

The ever-changing distribution of immunity in the human population probably contributes to the temporal order in which H1N1, H3N2, and influenza B have caused recent epidemics, but this remains an untested hypothesis. Current efforts toward understanding host, immunity in both humans and birds, along with expanded influenza surveillance in human and non-human hosts, will surely increase our understanding of antigenic shift and drift.

REFERENCES
Both, G. W., M. J. Sleigh, N. J. Cox, and A. P. Kendal. 1983. Antigenic drift in influenza virus H3 hemagglutinin from 1968 to 1980: multiple evolutionary pathways and sequential amino acid changes at key antigenic sites. *Journal of Virology* 48:52–60.
Bridges, C. B., W. Lim, J. Hu-Primmer, L. Sims, K. Fukuda, K. H. Mak, T. Rowe, W. W. Thompson, L. Conn, X. Lu, N. J. Cox, and J. M. Katz. 2002. Risk of influenza A (H5N1) infection among poultry workers, Hong Kong, 1997–1998. *Journal of Infectious Diseases* 185:1005–10.

Brown, I. H. 2000. The epidemiology and evolution of influenza viruses in pigs. *Veterinary Microbiology* 74:29–46.

Buonagurio, D. A., S. Nakada, W. M. Fitch, and P. Palese. 1986. Epidemiology of influenza C virus in man: multiple evolutionary lineages and low rate of change. *Virology* 153:12–21.

Bush, R. M., C. A. Bender, K. Subbarao, N. J. Cox, and W. M. Fitch. 1999a. Predicting the evolution of human influenza A. *Science* 286:1921–5.

Bush, R. M., W. M. Fitch, C. A. Bender, and N. J. Cox. 1999b. Positive selection on the H3 hemagglutinin gene of human influenza virus A. *Molecular Biology and Evolution* 16:1457–65.

Bush, R. M., C. B. Smith, N. J. Cox, and W. M. Fitch. 2000. Effects of passage history and sampling bias on phylogenetic reconstruction of human influenza A evolution. *Proceedings of the National Academy of Sciences USA* 97:6974–80.

Bush, R. M., W. M. Fitch, C. B. Smith, and N. J. Cox. 2001. Predicting influenza evolution: the impact of terminal and egg-adapted mutations. In *Options for the Control of Influenza IV* (Osterhaus, A. D. M. E., Ed.), pp. 147–53. Elsevier, Amsterdam.

Chen, W., P. A. Calvo, D. Malide, J. Gibbs, U. Schubert, I. Bacik, S. Basta, R. O'Neill, J. Schickli, P. Palese, P. Henklein, J. R. Bennink, and J. W. Yewdell. 2001. A novel influenza A virus mitochondrial protein that induces cell death. *Nature Medicine* 7:1306–12.

Cox, N., X. Xu, C. Bender, A. Kendal, H. Regnery, M. Hemphill, and P. Rota. 1993. Evolution of hemagglutinin in epidemic variants and selection of vaccine viruses. In *Options for the Control of Influenza II* (Hannoun, C., A. P. Kendal, H. D. Klenk, and F. L. Ruben, Eds.), pp. 223–30. Excerpta Medica, Amsterdam.

Cox, N., and H. Regnery. 1996. Global influenza surveillance: tracking a moving target in a rapidly changing world. In *Options for the Control of Influenza III* (Brown, L. E., Q. W. Hampson, and R. G. Webster, Eds.), pp. 591–8. Elsevier Science Publishers B. V. Amsterdam.

Cox, N. J., and C. A. Bender. 1995. The molecular epidemiology of influenza viruses. *Seminars in Virology* 6:359–70.

Cox, N. J., and Y. Kawaoka. 1998. Orthomyxoviruses: influenza. In *Topley and Wilson's Microbiology and Microbial Infections* (Mahy, B. W. J., and L. Collier, Eds.), pp. 385–433. Arnold, London.

Cox, N. J., and K. Subbarao. 2000. Global epidemiology of influenza: past and present. *Annual Review of Medicine* 51:407–21.

Cox, N. J., R. A. Black, and A. P. Kendal. 1989. Pathways of evolution of influenza A (H1N1) viruses from 1977 to 1986 as determined by oligonucleotide mapping and sequencing studies. *Journal of General Virology* 70:299–313.

de Jong, J. C., E. C. Claas, A. D. Osterhaus, R. G. Webster, and W. L. Lim. 1997. A pandemic warning? *Nature* 389:554.

de Jong, J. C., G. F. Rimmelzwaan, R. A. M. Fouchier, and A. Osterhaus. 2000. Influenza virus: a master of metamorphosis. *Journal of Infection* 40:218–28.

Dowdle, W. R. 1997. The 1976 experience. *Journal of Infectious Diseases* 176(Suppl. 1):S69–72.

Dowdle, W. R. 1999. Influenza A virus recycling revisited. *Bulletin of the World Health Organization* 77:820–8.

Fanning, T. G., R. D. Slemons, A. H. Reid, T. A. Janczewski, J. Dean, and J. K. Taubenberger. 2002. 1917 avian influenza virus sequences suggest that the 1918 pandemic virus did not acquire its hemagglutinin directly from birds. *Journal of Virology* 76:7860–2.

Fitch, W. M., R. M. Bush, C. A. Bender, and N. J. Cox. 1997. Long term trends in the evolution of H(3) HA1 human influenza type A. *Proceedings of the National Academy of Sciences USA* 94:7712–8.

Francis, T. 1953. Influenza, the new acquaintance. *Annals of Internal Medicine* 39:203–21.

Garcia, M., D. L. Suarez, J. M. Crawford, J. W. Latimer, R. D. Slemons, D. E. Swayne, and M. L. Perdue. 1997. Evolution of H5 subtype avian influenza A viruses in North America. *Virus Research* 51:115–24.

Glezen, W. P., and R. B. Couch. 1997. Influenza viruses. In *Viral Infections of Humans* (Evans, A. S., and R. A. Kaslow, Eds.), pp. 473–505. Plenum Medical Book Company, New York.

Gorman, O. T., W. J. Bean, and R. G. Webster. 1992. Evolutionary processes in influenza viruses: divergence, rapid evolution, and stasis. *Current Topics in Microbiology and Immunology* 176:75–97.

Gregory, V., M. Bennett, M. Orkhan, S. Al Hajjar, N. Varsano, E. Mendelson, M. Zambon, J. Ellis, A. Hay, and Y. Lin. 2002. Emergence of influenza A H1N2 reassortant viruses in the human population during 2001. *Virology* 300:1–7.

Guan, Y., K. F. Shortridge, S. Krauss, and R. G. Webster. 1999. Molecular characterization of H9N2 influenza viruses: were they the donors of the "internal" genes of H5N1 viruses in Hong Kong? *Proceedings of the National Academy of Sciences USA* 96:9363–7.

Guo, Y. J., F. G. Jin, P. Wang, M. Wang, and J. M. Zhu. 1983. Isolation of influenza C virus from pigs and experimental infection of pigs with influenza C virus. *Journal of General Virology* 64:177–82.

Guo, Y. J., X. Y. Xu, and N. J. Cox. 1992. Human influenza A (H1N2) viruses isolated from China. *Journal of General Virology* 73:383–7.

Hatta, M., and Y. Kawaoka. 2002. The continued pandemic threat posed by avian influenza viruses in Hong Kong. *Trends in Microbiology* 10:340–4.

Horimoto, T., and Y. Kawaoka. 2001. Pandemic threat posed by avian influenza A viruses. *Clinical Microbiology Reviews* 14:129–49.

Huelsenbeck, J. P., F. Ronquist, R. Nielsen, and J. P. Bollback. 2001. Bayesian inference of phylogeny and its impact on evolutionary biology. *Science* 294: 2310–4.

Katz, J. M., W. Lim, C. B. Bridges, T. Rowe, J. Hu-Primmer, X. Lu, R. A. Abernathy, M. Clarke, L. Conn, H. Kwong, M. Lee, G. Au, Y. Y. Ho, K. H. Mak, N. J. Cox, and K. Fukuda. 1999. Antibody response in individuals infected with avian influenza A (H5N1) viruses and detection of anti-H5 antibody among household and social contacts. *Journal of Infectious Diseases* 180:1763–70.

Kawaoka, Y., T. M. Chambers, W. L. Sladen, and R. G. Webster. 1988. Is the gene pool of influenza viruses in shorebirds and gulls different from that in wild ducks? *Virology* 163:247–50.

Kawaoka, Y., S. Krauss, and R. G. Webster. 1989. Avian-to-human transmission of the PB1 gene of influenza A viruses in the 1957 and 1968 pandemics. *Journal of Virology* 63:4603–8.

Lamb, R. A. 1989. Genes and proteins of the influenza viruses. In *The Influenza Viruses* (Krug, R. M., H. Fraenkel-Conrat, and R. R. Wagner, Eds.), pp. 1–88. Plenum Press, New York.

Lindstrom, S. E., Y. Hiromoto, H. Nishimura, T. Saito, R. Nerome, and K. Nerome. 1999. Comparative analysis of evolutionary mechanisms of the hemagglutinin and three internal protein genes of influenza B virus: multiple cocirculating lineages and frequent reassortment of the NP, M, and NS genes. *Journal of Virology* 73:4413–26.

Manuguerra, J. C., S. Zientara, C. Sailleau, C. Rousseaux, B. Gicquel, I. Rijks, and S. van der Werf. 2000. Evidence for evolutionary stasis and genetic drift by genetic analysis of two equine influenza H3 viruses isolated in France. *Veterinary Microbiology* 74:59–70.

Matsuzaki, Y., Y. Muraki, K. Sugawara, S. Hongo, H. Nishimura, F. Kitame, N. Katsushima, Y. Numazaki, and K. Nakamura. 1994. Cocirculation of two distinct groups of influenza C virus in Yamagata City, Japan. *Virology* 202:796–802.

Mounts, A. W., H. Kwong, H. S. Izurieta, Y. Ho, T. Au, M. Lee, C. Buxton Bridges, S. W. Williams, K. H. Mak, J. M. Katz, W. W. Thompson, N. J. Cox, and K. Fukuda. 1999. Case-control study of risk factors for avian influenza A (H5N1) disease, Hong Kong, 1997. *Journal of Infectious Diseases* 180:505–8.

Mumford, J. A., and T. M. Chambers. 1998. Equine influenza. In *Textbook of Influenza* (Nicholson, K. G., R. G. Webster, and A. J. Hay, Eds.), pp. 146–62. Blackwell Science, Oxford.

Murphy, B. R., and R. G. Webster. 1996. Orthomyxoviruses. In *Book Orthomyxoviruses* (Fields, B. N., D. M. Knipe, and P. M. Howley, Eds.), pp. 1397–445. Lippincott-Raven Publishers, Philadelphia.

Nakajima, K., U. Desselberger, and P. Palese. 1978. Recent human influenza A (H1N1) viruses are closely related genetically to strains isolated in 1950. *Nature* 274:334–9.

Nakajima, S., K. Nakajima, and A. P. Kendal. 1983. Identification of the binding sites to monoclonal antibodies on A/USSR/90/77 (H1N1) hemagglutinin and their involvement in antigenic drift in H1N1 influenza viruses. *Virology* 131:116–27.

Nakajima, S., F. Nishikawa, and K. Nakajima. 2000. Comparison of the evolution of recent and late phase of old influenza A (H1N1) viruses. *Microbiology and Immunology* 44:841–7.

Nielsen, R., and J. P. Huelsenbeck. 2002. Detecting positively selected amino acid sites using posterior predictive P-values. *Pacific Symposium on Biocomputing* pp. 576–88.

Noble, G. R. 1982. Epidemiological and clinical aspects of influenza. In *Basic and Applied Influenza Research* (Beare, A. S., Ed.), pp. 11–50. CDC Press, Boca Raton, FL.

Osterhaus, A., G. F. Rimmelzwaan, B. E. E. Martina, T. M. Bestebroer, and R. A. M. Fouchier. 2000. Influenza B virus in seals. *Science* 288:1051–3.

Parvin, J. D., A. Moscona, W. T. Pan, J. M. Leider, and P. Palese. 1986. Measurement of the mutation rates of animal viruses: influenza A virus and poliovirus type 1. *Journal of Virology* 59:377–83.

Peng, G., S. Hongo, Y. Muraki, K. Sugawara, H. Nishimura, F. Kitame, and K. Nakamura. 1994. Genetic reassortment of influenza C viruses in man. *Journal of General Virology* 75:3619–22.

Potter, C. W. 2001. A history of influenza. *Journal of Applied Microbiology* 91:572–9.

Raymond, F. L., A. J. Caton, N. J. Cox, A. P. Kendal, and G. G. Brownlee. 1986. The antigenicity and evolution of influenza H1 haemagglutinin, from 1950–1957 and 1977–1983: two pathways from one gene. *Virology* 148:275–87.

Reid, A. H., T. G. Fanning, J. V. Hultin, and J. K. Taubenberger. 1999. Origin and evolution of the 1918 "Spanish" influenza virus hemagglutinin gene. *Proceedings of the National Academy of Sciences USA* 96:1651–6.

Robertson, J. S. 1993. Clinical influenza virus and the embryonated hens egg. *Reviews in Medical Virology* 3:97–106.

Rota, P. A., T. R. Wallis, M. W. Harmon, J. S. Rota, A. P. Kendal, and K. Nerome. 1990. Cocirculation of two distinct evolutionary lineages of influenza type B virus since 1983. *Virology* 175:59–68.

Scholtissek, C. 1990. Pigs as 'mixing vessels' for the creation of new pandemic influenza A viruses. *Medical Principles and Practice* 2:65–71.

Scholtissek, C. 1998. Genetic reassortment of human influenza viruses in nature. *Textbook of Influenza* (Nicholson, K. G., R. G. Webster, and A. J. Hay, Eds.), pp. 120–5. Blackwell Science, Oxford.

Scholtissek, C., U. Schultz, S. Ludwig, and W. M. Fitch. 1993. The role of swine in the origin of pandemic influenza. *Options for the Control of Influenza II* (Hannoun, H. C., Ed.), pp. 193–201. Elsevier, Amsterdam.

Shortridge, K. F., N. N. Zhou, Y. Guan, P. Gao, T. Ito, Y. Kawaoka, S. Kodihalli, S. Krauss, D. Markwell, K. G., Murti, M. Norwood, D. Senne, L. Sims, A. Takada, and R. G. Webster. 1998. Characterization of avian H5N1 influenza viruses from poultry in Hong Kong. *Virology* 252:331–42.

Smith, W., D. H. Andrewes, and P. P. Laidlaw. 1933. A virus obtained from influenza patients. *Lancet* 2:66–8.

Snacken, R., A. P. Kendal, L. R. Haaheim, and J. M. Wood. 1999. The next influenza pandemic: lessons from Hong Kong, 1997. *Emerging Infectious Diseases* 5:195–203.

Stallknecht, D. E., 1997. Ecology and epidemiology of avian influenza viruses in wild bird populations: waterfowl, shorebirds, pelicans, cormorants, etc. *Fourth International Symposium on Avian Influenza* (Swayne, E. E., and R. D. Slemons, Eds.), pp. 61–7. American Association of Avian Pathologists, Athens, GA.

Stallknecht, D. E., and S. M. Shane. 1988. Host range of avian influenza virus in free-living birds. *Veterinary Research Communications* 12:125–41.

Suarez, D. L. 2000. Evolution of avian influenza viruses. *Veterinary Microbiology* 74:15–27.

Taubenberger, J. K., A. H. Reid, A. E. Krafft, K. E. Bijwaard, and T. G. Fanning. 1997. Initial genetic characterization of the 1918 "Spanish" influenza virus. *Science* 275:1793–6.

Taubenberger, J. K., A. H. Reid, and T. G. Fanning. 2000. The 1918 influenza virus: a killer comes into view. *Virology* 274:241–5.

Webby, R. J., S. L. Swenson, S. L. Krauss, P. J. Gerrish, S. M. Goyal, and R. G. Webster. 2000. Evolution of swine H3N2 influenza viruses in the United States. *Journal of Virology* 74:8243–51.

Webster, R. G., and W. G. Laver. 1980. Determination of the number of nonoverlapping antigenic areas on Hong Kong (H3N2) influenza virus hemagglutinin with monoclonal antibodies and the selection of variants with potential epidemiological significance. *Virology* 104:139–48.

Webster, R. G., W. J. Bean, O. T. Gorman, T. M. Chambers, and Y. Kawaoka. 1992. Evolution and ecology of influenza A viruses. *Microbiological Reviews* 56:152–79.

Webster, R. G., K. F. Shortridge, and Y. Kawaoka. 1997. Influenza: interspecies transmission and emergence of new pandemics. *Fems Immunology and Medical Microbiology* 18:275–9.

Wiley, D. C., I. A. Wilson, and J. J. Skehel. 1981. Structural identification of the antibody-binding sites of Hong Kong influenza haemagglutinin and their involvement in antigenic variation. *Nature* 289:373–8.

Wilson, I. A., and N. J. Cox. 1990. Structural basis of immune recognition of influenza virus hemagglutinin. *Annual Review of Immunology* 8:737–71.

Wood, J. M., D. Major, R. W. Newman, U. Dunleavy, C. Nicolson, J. S. Robertson, and G. C. Schild. 2002. Preparation of vaccines against H5N1 influenza. *Vaccine* 20(Suppl. 2):S84–7.

Xu, X., Y. Guo, P. Rota, M. Hemphill, A. P. Kendal, and N. Cox. 1993. Genetic reassortment of human influenza virus in nature. In *Options for the Control of Influenza II* (Hannoun, C., A. P. Kendal, H. D. Klenk, and F. L. Ruben, Eds.), pp. 203–7. Excerpta Medica, Amsterdam.

Yamashita, M., M. Krystal, W. M. Fitch, and P. Palese. 1988. Influenza B virus evolution: co-circulating lineages and comparison of evolutionary pattern with those of influenza A and C viruses. *Virology* 163:112–22.

Yang, Z. 2000. Maximum likelihood estimation on large phylogenies and analysis of adaptive evolution in human influenza virus A. *Journal of Molecular Evolution* 51:423–32.

Yang, Z., R. Nielsen, N. Goldman, and A. M. Pedersen. 2000. Codon-substitution models for heterogeneous selection pressure at amino acid sites. *Genetics* 155:431–49.

Ziegler, T., M. L. Hemphill, M. L. Ziegler, G. Perez-Oronoz, A. I. Klimov, A. W. Hampson, H. L. Regnery, and N. J. Cox. 1999. Low incidence of rimantadine resistance in field isolates of influenza A viruses. *Journal of Infectious Diseases* 180:935–9.

EIGHT

Free-Living to Freewheeling: The Evolution of *Vibrio cholerae* from Innocence to Infamy

Rita R. Colwell, Shah M. Faruque, and G. Balakrish Nair

INTRODUCTION

Toxigenic strains of the Gram-negative bacterium *Vibrio cholerae* cause the most severe form of dehydrating diarrhea known as cholera and have been responsible for at least seven distinct pandemics of cholera (Faruque et al., 1998*a*). Since the first recorded pandemic, which began in 1817, *V. cholerae* strains associated with different pandemics are assumed to have undergone phenotypic and genetic changes with time. For example, although the seventh pandemic was caused by the El Tor biotype of *V. cholerae*, the sixth and possibly earlier pandemics were caused by the classical biotype. The genetic changes in *V. cholerae* associated with epidemics and pandemics, however, were not fully appreciated until the development and application of molecular techniques to analyze strains.

Recent application of molecular approaches has enabled extraordinary progress in our understanding of the evolution of *V. cholerae*. Like other bacteria, *V. cholerae* can be assumed to have existed well before human evolution and therefore to have originated primarily as a free-living aquatic microorganism. The fact that over a period of a million years some clones (serogroups) of this organism have acquired the ability to colonize transiently the human intestine and become an efficient human pathogen reflects the progressive acquisition of the genetic capability to become pathogenic for humans. They could have acquired this capability by embracing another "life style," albeit accidentally, by carrying out a function in the human host that it performs either symbiotically or commensally for its nonhuman host but is, in effect, pathogenic to the human host, or alternatively, by finding a niche whereby the organism could amplify and perpetuate itself as effectively or more effectively, than it could in a free-living state. Virulence factors encoded on bacteriophages (and other mobile genetic elements) are known to allow the bacterium to enlarge its host range and increase its pathogenic potential by promoting evasion of human host immune defenses or providing mechanisms to breach human host structural barriers (Miao and Miller, 1999). What functions these genes might have in the natural environment with its nonhuman host are unknown. Nevertheless, lateral gene transfer has played an integral role in the evolution of bacterial genomes and in the diversification and specification of enteric and other bacteria (Ochman et al., 2000).

V. cholerae survives and persists under continuously changing environmental conditions in tidal estuaries and is therefore adapted to them. We know that *V. cholerae*, including strains of O1 (Colwell et al., 1977; Colwell, 1996) and O139 (Cholera Working Group, 1993; Nair et al., 1994*b*), exist as autochthonous natural inhabitants of riverine, coastal, and estuarine ecosystems, yet at the same time is pathogenic for humans. Here, we attempt to reconstruct salient events, particularly those that may have led to the evolution of *V. cholerae* from a free-living state to its freewheeling life style and to identify those molecular events that made this transition possible.

CLASSIFICATION

V. cholerae is a genetically diverse Gram-negative, polar monotrichous, oxidase positive, asporogenous species native to coastal ecosystems (Colwell et al., 1977). The organism is classified by biochemical tests and DNA homology studies and is further subdivided into serogroups based on the antigenicity of the somatic O-antigen. The O-antigen is heat-stable and shows enormous serological diversity. Of approximately 200 currently known O-antigen serogroups (Yamai et al., 1997), just two serogroups, O1 and O139, are associated with epidemic cholera. All strains that are identified as *V. cholerae* based on biochemical tests but do not agglutinate with O1 or O139 antisera are collectively referred to as the "non-O1 non-O139" (Nair et al., 1994*b*) or as the nonepidemic serogroups. Other formerly used but inaccurate names for this group were "noncholera vibrios" (NCVs) and nonagglutinable vibrios (NAGs). In-depth taxonomical studies, including biochemical and molecular tests and polyphasic numerical taxonomy, have proven that all strains of *V. cholerae*, independent of serogroup, are the same species (Citarella and Colwell, 1970; Colwell, 1973). In addition, 16S rRNA gene sequencing has shown no differences at the species level between *V. cholerae* O1 El Tor and classical strains or between serogroups of *V. cholerae* (Lipp et al., 2002). Although certain of the nonepidemic serogroups cause sporadic diarrheal disease (Aldova et al.,

1968; Dakin et al., 1974), most *V. cholerae* organisms probably do not infect humans. Some strains of *V. cholerae* have occasionally been isolated from extra-intestinal infections, including wounds, ear, sputum, urine, and cerebrospinal fluid (Morris and Black, 1985; Morris, 1990).

The O1 serogroup is divided into two biotypes, classical and El Tor, which can be differentiated in the laboratory by employing assays for hemolysis, hemagglutination, phage susceptibility, polymyxin B sensitivity, and the Voges Proskauer reaction. The more recent approach, however, is to use biotype-specific genes (such as *tcpA* and *rtxC*) to differentiate between the two biotypes (Nair et al., 2002). Each biotype can be further subdivided into two major serotypes, "Ogawa" and "Inaba." Ogawa strains produce the A and B antigens and a small amount of C antigen, whereas Inaba strains produce only the A and C antigens (Sakazaki and Donovan, 1984). A third serotype, Hikojima, contains all three antigens but is rare and unstable (Kelly, 1991).

The classical and El Tor biotypes of *V. cholerae* are closely related in their O-antigen biosynthetic genes (Manning et al., 1994; Yamasaki et al., 1999a), although these two biotypes differ in many other regions of their genome (Voss and Attridge, 1993; Beltran et al., 1999; Karaolis et al., 2001). Thus, O1 El Tor strains might have arisen following transfer of O1 antigen biosynthetic genes into a previously unknown environmental strain. Conversely, O139 and O1 El Tor strains are closely related in most parts of their genomes, but they carry different O-antigen genes, suggesting the transfer of O139-specific genes from an unknown donor into a recipient El Tor strain (Bik et al., 1995; Stroeher et al., 1995). Similar conclusions concerning transfer of genes have emerged from comparisons of serogroups and sequences of diagnostic housekeeping genes of nonepidemic isolates (Beltran et al., 1999). Recently, a new variety of *V. cholerae* O1, which appears to be a hybrid of the classical and El Tor biotypes, has been identified in hospitalized cases of acute diarrhea (Nair et al., 2002).

FREE-LIVING AQUATIC LIFE STYLE

The fact that *V. cholerae* is a naturally occurring member of the aquatic microbial community was not appreciated until Colwell et al. (1977) described the organism as an autochthonous inhabitant of riverine, estuarine, and coastal waters. Physicians and clinical workers who believed that the presence of *V. cholerae* in the aquatic environment was coincidental and dependent on contamination by infected individuals initially viewed this report with skepticism. At that time, it was believed by clinical workers that the endemicity of cholera was maintained by low-level continuous transmission through people with asymptomatic infection or mild disease and therefore transmission was essentially human-to-human. Now it is clear that *V. cholerae* has a life cycle consisting of two distinct phases, one of which includes a pathogenic life style and the other of which is outside of the human host in aquatic environs. Furthermore, *V. cholerae* can be found as free-swimming cells or attached to surfaces such as those provided by plants, filamentous green algae, copepods, crustaceans, and insects (Islam et al., 1994*a,b*; Colwell, 1996). Biofilm formation (Watnick and Kolter, 2000; Watnick et al., 2001) and entry into a viable but nonculturable state in response to nutrient deprivation (Xu et al., 1982; Colwell et al., 1985) are thought to be important in the life cycle of *V. cholerae*, facilitating persistence in its native aquatic habitat during interepidemic periods (Reidl and Klose, 2002). The genomic elements that the organism possesses to function in association with plankton and to occupy biofilm, on abiotic surfaces, have not yet been fully characterized.

Although *V. cholerae* is part of the normal estuarine flora, toxigenic strains are usually isolated from the environment in those areas where cholera patients abound. Environmental isolates from areas that are distant from regions of massive epidemics do not usually harbor the CT genes (Faruque et al., 1998*a*). Whether an epidemic or the acquisition of CT genes occurs first is also unknown. Thus, what is known is only that *V. cholerae* can be considered an autochthonous marine bacterium that colonizes and thrives in the human gut, causing infection, and spends the time between epidemics in its natural habitat, the estuary.

HUMAN PATHOGENIC LIFE STYLE
Development of the ability to infect and colonize the human gut apparently has provided enormous advantage to *V. cholerae*, in terms of rapid multiplication and dissemination, and thus can be considered a significant phase of its evolutionary history. In the host, the bacteria are virulent, whereby the organisms adhere to the intestinal epithelium, replicate, and cause disease. The acute watery diarrhea characteristic of cholera leads to rapid spread of the organisms, particularly during an epidemic. In order to cause disease, the ingested vibrios from contaminated water or food must pass the gastric acid barrier and colonize the small intestine. Once an infective dose of the bacteria is ingested, colonization of the intestine occurs with the help of the pilus colonization factor TCP that attaches to receptors on the gastric mucosa (Taylor et al., 1987). Colonization and multiplication lead to an increase in the concentration of vibrios on the mucosal surface as large as 10^7 to 10^8 cells per gram. With this large concentration

of vibrios closely attached to the mucosa, delivery of enterotoxin becomes highly efficient.

The pathogenesis of cholera is a complex process and involves a number of genes encoding virulence factors that aid the pathogen in its passage to the epithelium of the small intestine and then allow it to colonize the epithelium and produce the cholera toxin (CT), which disrupts ion transport by the intestinal epithelial cells. The major virulence genes include the CTX genetic element, which is the genome of a lysogenic bacteriophage, designated CTXΦ (Waldor and Mekalanos, 1996), that carries the genes encoding CT, and the TCP pathogenicity island, which carries genes for toxin-coregulated pilus (TCP). The structural features of the TCP pathogenicity island suggest that the TCP pathogenicity island may have originated from a bacteriophage which is now defective (Kovach et al., 1996; Karaolis et al., 1998). Like other phage-derived virulence genetic elements, the TCP pathogenicity island includes groups of virulence genes, a regulator of virulence genes, a transposase gene, specific (att-like) attachment sites flanking each end of the island, and an integrase with homology to a phage integrase gene. Thus, the major virulence gene clusters in *V. cholerae* appear to have the ability to propagate laterally and thereby disperse among different strains. The existence and role of other factors possibly responsible for colonization have also been investigated, including mannose–fucose–resistant cell-associated hemagglutinin (MFRHA), mannose-sensitive hemagglutinin (MSHA), and some outer membrane proteins (OMPs) of *V. cholerae* (Sengupta et al., 1992; Franzon et al., 1993; Jonson et al., 1994). Although some of these factors, including MFRHA, MSHA, and OMPs, are suspected to have a role in enhancing adhesion and colonization, based on tests in animal models, the exact role of these factors in the virulence of *V. cholerae* in humans is still uncertain. Again, these factors may have significance in the nonhuman host relationship and not primarily associated with human infection.

The colonized *V. cholerae* secretes CT into the host small intestine. CT is an oligomeric protein composed of one A subunit (28 kDa) and five B subunits (11.5 kDa each) (Finkelstein, 1983). Whereas the A subunit catalyzes NAD-dependent ADP-ribosylation, the B subunit binds to ganglioside GM1, the holotoxin receptor in the gut. The A subunit is transported into the cell where it activates adrenyl cyclase, which leads to a marked increase in cyclic AMP. This perpetuates an increase in chloride secretion in the crypt cells and inhibition of neutral sodium chloride absorption in the villus cells (Field et al., 1972). These alterations lead to a massive outpouring of fluid into the small intestine that exceeds the normal absorptive capacity of the bowel, resulting in watery diarrhea containing large amounts of sodium, chloride, bicarbonate, and potassium, but no protein or red or white blood cells. In the nonhuman host, this activity may well serve to assist in osmoregulation. In the human, however, the bacteria eventually detach and exit the host via profuse diarrhea. During the illness, there are high concentrations of *V. cholerae* in the stool (up to 10^8 CFU/s per ml of stool). It has recently been suggested that human colonization creates a hyperinfectious bacterial state that is maintained after dissemination and that may contribute to the epidemic spread of cholera (Merrell et al., 2002).

EVOLUTION OF PATHOGENIC *VIBRIO CHOLERAE*
Evolutionary processes that occur over longer periods of time are termed "macroevolution." The evolution of some clones of *V. cholerae* from the free-living state to a human pathogen might be viewed to be a process of macroevolution. That is, nontoxigenic strains of *V. cholerae* O1 and O139 may be the precursors of this evolutionary process. *V. cholerae* produces virulence factors that are directly involved in human infection and disease by acquiring transmissible elements involving phages and mobile elements that contribute to the process. In terms of evolution, determinants acquired by horizontal gene transfer from other organisms lead to evolution in significant leaps. Genetic recombination between divergent bacterial strains can be advantageous, especially in complex or variable environments, because it generates new genotypes more efficiently than does mutation alone (Mukhopadhyay et al., 2001).

There are two critical sequential steps in the evolution of pathogenic *V. cholerae*. First, strains have to acquire the genes encoding a pilus colonization factor, known as TCP. The TCP genes are part of a larger genetic element, the TCP pathogenicity island, alternatively referred to as the VPI. Secondly, the TCP positive strains are infected with and lysogenized by CTXφ (Waldor and Mekalanos, 1996; Mekalanos et al., 1997; Faruque et al., 1998*b,c*). Acquisition of TCP genes is a prerequisite for infection by CTX phage because TCP also acts as the receptor for CTX phage. In animal models, the intestinal milieu has been shown to be a site where strains can acquire these mobile elements efficiently (Waldor and Mekalanos, 1996; Lazar and Waldor, 1998). Where this would occur in the natural environment is not known. Acquisition of the two gene clusters appears to provide the marine bacterium the ability to colonize the human intestine and secrete a potent toxin resulting in severe diarrhea (Mekalanos, 1983; Taylor et al., 1987; Waldor and Mekalanos,

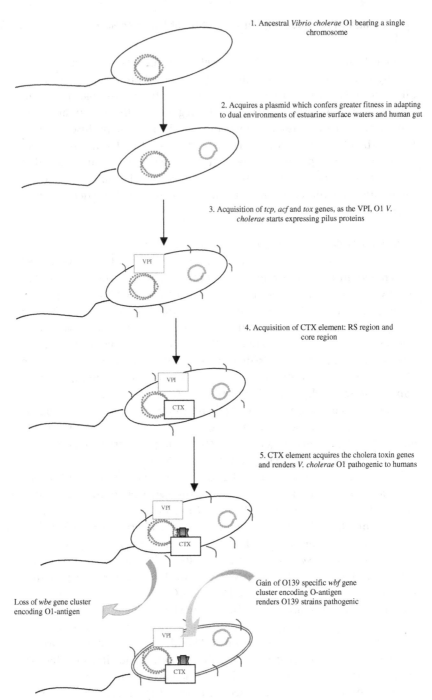

1. Ancestral *Vibrio cholerae* O1 bearing a single chromosome

2. Acquires a plasmid which confers greater fitness in adapting to dual environments of estuarine surface waters and human gut

3. Acquisition of *tcp, acf* and *tox* genes, as the VPI, O1 *V. cholerae* starts expressing pilus proteins

4. Acquisition of CTX element: RS region and core region

5. CTX element acquires the cholera toxin genes and renders *V. cholerae* O1 pathogenic to humans

Loss of *wbe* gene cluster encoding O1-antigen

Gain of O139 specific *wbf* gene cluster encoding O-antigen renders O139 strains pathogenic

Figure 8.1. Probable milestones in the evolution of toxigenic *Vibrio cholerae* O1 and O139. For a colour version of this figure, see www.cambridge.org/9780521126557.

1996; Karaolis et al., 1999). Such functions with respect to its nonhuman host would be interesting to know. These virulence gene clusters are absent from nonpathogenic strains. Interestingly, the specialized acquired genetic elements that confer the potential of pathogenicity to *V. cholerae* are themselves mosaics of genes from various sources. The potential for scrambling of functional modules within these elements and their transmission between strains may also speed bacterial adaptation to diverse or inconstant environments and contribute to the emergence of new, highly virulent strains of *V. cholerae* in at-risk human societies (Mukhopadhyay et al., 2001). Again, what advantage this gives the bacterium in nature needs to be determined. Therefore, *V. cholerae*, as well as its acquired genetic elements, is in a perpetual state of co-evolution. Probable events in the evolution of toxigenic *V. cholerae* are shown in Figure 8.1.

CTXΦ

A major transmissible element that *V. cholerae* has acquired is the filamentous temperate bacteriophage, designated CTXΦ. The genome of CTXΦ, among other genes, contains the *ctxAB* operon that encodes for cholera toxin, which essentially is responsible for many of the symptoms of cholera. CTXΦ was discovered by Waldor and Mekalanos (1996), who employed simple but elegant experiments to show that El Tor chromosomal CTX could be transferred to classical strains, where it was recovered as a plasmid (Waldor and Mekalanos, 1996). The CTXΦ genome can integrate into the *V. cholerae* genome to form a stable prophage, or it can replicate as a plasmid in isolates lacking an appropriate integration site. Lysogenic conversion of nontoxigenic strains of *V. cholerae* to toxigenicity by CTXΦ infection is perhaps the single most important event in the evolution of an innocuous strain into a fully pathogenic one (Heilpern and Waldor, 2000).

The CTXϕ genome is comprised of two functional regions: a 4.6-kb core region that includes *ctxAB* and a 2.4-kb region designated as RS2 (Waldor et al., 1997). Besides the genes encoding CT, the core region contains genes that are thought to encode the coat proteins (Psh, Cep, OrfU, and Ace) and a protein required for CTXϕ assembly (Zot) (Waldor and Mekalanos, 1996). The RS2 region encodes genes required for replication (*rstA*), integration (*rstB*), and regulation (*rstR*) of CTXϕ (Waldor et al., 1997). In El Tor *V. cholerae* isolates, the CTX prophage genome is often flanked by an additional 2.7-kb region, designated RS1 (Goldberg and Mekalanos, 1986), that is similar to RS2 but contains an additional gene, *rstC* (Waldor et al., 1997). CTXΦ DNA is generally found integrated at either one (El Tor) or two (classical) loci within the *V. cholerae* genome

(Mekalanos, 1983; Pearson et al., 1993; Waldor and Mekalanos, 1996). The organization of the core-encoded genes and the deduced amino acid sequences of their products (with the exception of *ctxAB*) resemble that of filamentous phages derived from a variety of bacterial species (Waldor and Mekalanos, 1996). However, unlike other filamentous phages and similar to integrating double-stranded DNA bacteriophages, CTXΦ encodes a repressor and forms lysogens (Kimsey and Waldor, 1998).

Recent studies on the various genes within the genome, of the CTXΦ have shown that the *ctxAB* operon does not constitute an integral part of the genome but has probably been acquired over the course of time. The percent GC content of the *ctxA* and the *ctxB* genes, 38% and 33%, respectively, is significantly different from that of the rest of the core region genes, suggesting that these genes evolved separately from the remainder of the phage genome and were acquired after the emergence of a precursor form of CTXΦ that lacked *ctxAB* (Boyd et al., 2000). Results of studies of the evolutionary history of CTXΦ are consistent with the hypothesis that a CTXΦ precursor that lacked *ctxAB* simultaneously acquired the toxin genes and their regulatory sequences (Boyd et al., 2000). Comparative nucleotide sequence analyses have revealed that CTXΦ derived from classical and El Tor *V. cholerae* isolates comprise two distinct lineages, within otherwise nearly identical chromosomal backgrounds, indicating that nontoxigenic precursors of the two *V. cholerae* O1 biotypes independently acquired distinct CTXΦ (Boyd et al., 2000).

Similarly, among the genes of the RS2 region of the CTXΦ, the *rstR* sequences in El Tor and classical CTXΦ are highly diverse, with less than 30% amino acid sequence similarity (Kimsey and Waldor, 1998; Kimsey et al., 1998; Davis et al., 1999). CTXΦ repressors of more than three different specificities are known, and strains carrying one prophage can often be infected or lysogenized by CTXΦ of other repressor specificities (Kimsey et al., 1998; Davis et al., 1999). Additional putative *rstR* prophage repressor genes have been found recently in non-O1 non-O139 strains isolated from the environment (Mukhopadhyay et al., 2001). The hypervariability of *rstR* and its distinct GC content suggest that CTXΦ variants obtained diverse *rstR* cassettes via horizontal gene transfer, followed by recombination, and reinforce the notion that *V. cholerae* is extremely diverse genetically.

TCP PATHOGENICITY ISLAND

Besides CTX phage, the other important genetic factor acquired by *V. cholerae* for conferring epidemic potential to the O1 and O139 serogroups is the TCP pathogenicity island, alternatively referred to as the *Vibrio* pathogenicity island (VPI). The TCP island of the El Tor biotype strain

N16961 is 41,272 bp and encodes 29 predicted proteins, whereas that of the classical biotype strain 395 is 41,290 bp (Karaolis et al., 2001). The structural features of the TCP island include the presence of groups of virulence genes, a regulator of virulence genes, a transposase gene, specific (*att*-like) attachment sites flanking each end of the island, and an integrase with homology to a phage integrase gene (Kovach et al., 1996; Karaolis et al., 1999). There are several gene clusters included within the TCP island. These include the *tcpA* gene encoding a type IV pilus known as TCP, an essential colonization factor (Taylor et al., 1987; Herrington et al., 1988); several genes that encode proteins involved in the regulation of virulence (DiRita et al., 1991; DiRita, 1992; Hase and Mekalanos, 1998); and numerous open reading frames with no known function (Karaolis et al., 2001). As remarkable examples of evolutionary co-adaptation, the CTXΦ virion uses TCP as a receptor during infection (Karaolis et al., 1999). Therefore, the TCP pathogenicity island is one of the initial genetic factors required for the emergence of epidemic strains of *V. cholerae*.

TCP mediates bacterial colonization of the intestine by facilitating microcolony formation via pilus-mediated bacterial interactions and perhaps direct attachment to the intestinal brush border (Kirn et al., 2000). The biogenesis and regulation of TCP production involve at least 15 genes encoded in the TCP gene cluster as well as several unlinked genes (Peek and Taylor, 1992; Kaufman et al., 1993; Manning, 1997; Taylor et al., 1988). Having established that the TCP island is an essential element for epidemic and pandemic strains, it has been postulated that the evolution of toxigenic from nontoxigenic strains must be a multistep process, the initial step of which would be the acquisition of the VPI. As with the sequence variability observed in type IV pili of several bacterial species (Blank et al., 2000), Boyd and Waldor (2002) have found extensive sequence variability in TcpA, the major subunit of TCP. Most likely this diversity is a reflection of diversifying selection in adaptation to the host immune response or to CTXφ susceptibility (Boyd and Waldor, 2002). According to Karaolis et al. (2001), the pool of variation in the central segment of the TCP island (containing the *tcp* gene cluster encoding proteins) has contributed to the emergence and evolution of the pathogenic forms by horizontal gene transfer and recombination exchange.

ENVIRONMENTAL STRAINS CARRYING VARIANTS OF CTXφ AND VPI

V. cholerae strains from the environment are more likely not to possess the CTXφ and the VPI gene clusters. However, recent studies have revealed that environmental strains of *V. cholerae* belonging to diverse serogroups sometimes carry virulence genes or their homologues (Chakraborty et al., 2000; Rivera et al., 2001; Mukhopadhyay et al., 2001). When environmental isolates of *V. cholerae* were examined for the presence of *ctxA, hlyA, ompU, stn/sto, tcpA, tcpI, toxR,* and *zot* genes, using multiplex PCR, the *toxR, hlyA,* and *ompU* genes were present in 100, 98.6, and 87.01% of the *V. cholerae* strains, respectively (Rivera et al., 2001). Three of four non-toxigenic *V. cholerae* O1 strains contained *tcpA* ET. Interestingly, among the isolates of *V. cholerae* non-Ol/non-O139, two had *tcpA* (classical), nine contained *tcpA* El Tor, three showed homology with both biotype genes, and four carried the *ctxA* gene. The *stn/sto* genes were present in 28.2% of the non-Ol/non-O139 strains, in 10.5% of the toxigenic *V. cholerae* O1, and in 14.3% of the O139 serogroups. The CTX genetic element and toxin-coregulated pilus El Tor (*tcpA* ET) gene were present in all toxigenic *V. cholerae* O1 and *V. cholerae* O139 strains examined in this study. Except for *stn/sto* genes, all of the genes studied occurred with high frequency in toxigenic *V. cholerae* O1 and O139 strains. This study indicated that surveillance of non-Ol/non-O139 *V. cholerae* in the aquatic environment, combined with genotype monitoring using *ctxA, stn/sto,* and *tcpA* genes, will be valuable in human health risk assessment.

Presence of virulence genes in environmental strains of *V. cholerae* has also been detected in several other studies (Chakraborty et al., 2000, 2001; Mukhopadhyay et al., 2001). The function of virulence gene homologues in the environment, however, is not sufficiently investigated. It has been suggested that in fresh water systems, the local ionic microenvironment can be controlled by *V. cholerae* O1 by use of toxin acting on other living cells. Some studies have illustrated the ability of *V. cholerae* to associate with a variety of aquatic organisms including crustaceans, zooplankton, phytoplankton, and algae (Islam et al., 1994a,b) (see Figure 8.2). The association prolongs survival and presumably the vibrios gain nutrients from the host. It appears that possession and conservation of virulence genes including colonization factors and cholera toxin may play crucial roles in the symbiotic association between *V. cholerae* and specific aquatic organisms. It is worth noting that under *in vitro* conditions, virulence genes are known to express more adequately at a lower temperature of 30°C, as opposed to 37°C, the latter being the physiological temperature of the mammalian host. Furthermore, the findings that virulence genes or their homologues are dispersed among environmental strains suggest the existence of a pool of virulence-associated genes or their homologues in the environment. This also suggests that virulence genes carried by clinical

Figure 8.2. Attachment of *V. cholerae* to copepods. For a colour version of this figure, see www.cambridge.org/9780521126557.

V. cholerae derive from environmental genes, and a crucial combination of different genes may be necessary for the origination of a potentially pathogenic strain. Selection mediated by aquatic flora or fauna cannot be ruled out.

COORDINATE REGULATION OF *V. CHOLERAE* GENES

Expression of several critical virulence genes in *V. cholerae* is coordinately regulated so that multiple genes respond in a similar fashion to environmental conditions (DiRita, 1992; Skorupski and Taylor, 1997). ToxR, a 32-kDa transmembrane protein is the master regulator, which is itself regulated by environmental signals, and initiates the activity of a cascading system of regulatory factors. The ToxR protein binds to a tandemly repeated 7-bp DNA sequence found upstream of the *ctxAB*, structural genes and increases transcription of *ctxAB*, resulting in higher levels of CT expression. Besides *ctxAB* genes, ToxR also regulates the expression of at least 17 distinct genes that constitute the ToxR regulon. These include the TCP colonization factor, the accessory colonization factor, outer membrane proteins OmpT and OmpU, and three other lipoproteins (Peterson and Mekalanos, 1988; Parsot et al., 1991; Hughes et al., 1994). Except for the *ctxAB* genes, genes in the ToxR, regulon are controlled through another regulatory factor, called ToxT, a 32-kDa protein, the expression of which is controlled by ToxR. ToxT encodes a member of the AraC family of bacterial transcription activators. The increased expression of the ToxT protein leads to activation of other genes in the ToxR regulon. Thus, ToxR, whose expression is controlled by environmental factors, remains at the top of the regulatory cascade that controls the expression of a number of other genes (Skorupski and Taylor, 1997).

The coordinated regulation of a number of genes through the *toxR* regulon demonstrates that the organism has developed a mechanism of sampling and responding to its environment. Hence, it seems obvious that *V. cholerae* is able to activate or inactivate a set of genes including those encoding colonization factors or toxins as an appropriate response to changing environmental conditions.

GENETICS OF THE CELL SURFACE LIPOPOLYSACCHARIDES
Different serogroups of *V. cholerae* carry distinct cell surface lipopolysaccharide antigens referred to as the O-antigens. The genes responsible for the synthesis of O-antigen are present in a cluster designated as the *wb** region. Whereas the serotype-specific genetic region of O1 strains is designated as *wbe*, that of the O139 strain is designated as *wbf*, and these regions comprise a large number of distinct genes or open reading frames. Genetic changes in the *wb* regions can lead to serotype conversion, and this may occur due to deletion or genetic recombination. For example, a large portion of DNA corresponding to the *wbe* region of O1 strains was found to be missing in O139 strains, and O139 strains were found to have acquired a unique DNA region (Bik et al., 1995, 1996; Sozhamannan et al., 1999). The emergence of the O139 epidemic strain of *V. cholerae* is assumed to have resulted from horizontal transfer of a fragment of DNA from an unknown donor into the region responsible for O-antigen biosynthesis of the seventh pandemic *V. cholerae* O1 El Tor strain (Stroeher et al., 1995; Bik et al., 1995; Comstock et al., 1996; Yamasaki et al., 1999*b*). It was later shown that the serogroup O139 resulted from a precise 22-kb deletion of the *wbe* (*rfb*) region of O1, with replacement by a 35-kb *wbf* region (*wbfA* through *wbfX*) encoding the O139 O-antigen (Comstock et al., 1996). Extensive analysis of O139 O-antigen biosynthesis genes showed that these genes have homology with that of several non-O1 serogroups with maximum homology to the O22 serogroup–specific genes (Dumontier and Berche, 1998; Yamasaki et al., 1999*b*). This indicated that the genes for O22-specific antigen could be the origin of the genes responsible for O139 biosynthesis genes. Molecular epidemiological studies support these findings and indicate that the O139 strains have genetic backbones very similar to those of the O1 El Tor Asian seventh pandemic strains (Berche et al., 1994; Johnson et al., 1994; Waldor and Mekalanos, 1994). However, unlike *V. cholerae* O1, serogroup O139 possesses a capsule distinct from the lipopolysaccharide antigens and has 3,6-dideoxyhexose (abequose or colitose), quinovosamine and glucosamine, and traces of tetradecanoic and hexadecanoic fatty acids (Johnson et al., 1994).

MICROEVOLUTION IN _VIBRIO CHOLERAE_

Epidemiological surveillance of cholera necessitated the development of new typing methods to differentiate among different clones of _V. cholerae_. Vibriophages have been used to type strains based on their susceptibility to different phages. _V. cholerae_ O1 strains of the same biotype and serotype can be differentiated into 146 phage types using 10 typing phages (Chattopadhyay et al., 1993), and a similar phage-typing scheme has been established for the O139 strains (Chakrabarti et al., 2000). However, until the development of new genetic typing methods, a standardized scheme for typing strains was limited in practice.

Multilocus enzyme electrophoresis (MEE) can distinguish between classical and El Tor strains (Wachsmuth et al., 1994; Freitas et al., 2002) and has grouped the toxigenic El Tor biotype strains of _V. cholerae_ O1 into four major clonal groups or electrophoretic types (ET) representing broad geographical areas which include the Australian clone (ET1), the Gulf Coast clone (ET2), the seventh pandemic clone (ET3), and the Latin American clone (ET4) (Chen et al., 1991; Salles and Momen, 1991; Wachsmuth et al., 1993).

More recently, several new genetic typing schemes, based on restriction fragment length polymorphisms of different genes, were developed. These included conserved rRNA genes as well as genes encoding virulence factors, e.g., the ctxAB operon. A standardized ribotyping scheme for _V. cholerae_ O1 (Popovic et al., 1993) and for O139 (Faruque et al., 2000) that can distinguish seven different ribotypes among classical strains, twenty ribotypes and subtypes among El Tor strains, and six distinct ribotypes among O139 strains has been effectively used to study the molecular epidemiology of cholera. Molecular analysis of epidemic isolates of _V. cholerae_ between 1961 and 1996 in Bangladesh revealed clonal diversity among strains isolated during different epidemics (Faruque et al., 1995, 1997_a_). These studies demonstrated the transient appearance and disappearance of more than six ribotypes among classical vibrios at least five ribotypes of El Tor vibrios, and three different ribotypes of _V. cholerae_ O139. Different ribotypes often showed different CTX genotypes resulting from differences in copy number of the CTX element and variations in the integration site of the CTX element in the chromosome (Faruque et al., 1997_a,b_).

Multilocus sequence typing (MLST), a method that is based on partial nucleotide sequences of multiple (usually around seven) housekeeping genes, has recently been shown to be a powerful technique for bacterial typing (Maiden et al., 1998). Housekeeping genes are preferred over virulence-associated genes, because an analysis of mutations (most of which are usually synonymous, given the strong selection against changes of the amino acid sequence in genes coding for proteins required for growth) in such genes is more likely to adequately reflect the phylogeny of strains. Results of genetic typing of *V. cholerae* strains indicate that there must have been continual emergence of new clones of toxigenic *V. cholerae* which replaced existing clones, possibly through natural selection involving unidentified environmental factors and immunity of the host population.

EVOLUTIONARY SIGNATURES IN THE WHOLE GENOME SEQUENCE

The complete genome of a representative *V. cholerae* O1 El Tor biotype strain N16961 was sequenced recently (Heidelberg et al., 2000). This sequence analysis is expected to represent an important step toward the molecular description of how this free-living environmental organism is both naturally occurring in the environment and a human pathogen by horizontal gene transfer. The impact of lateral gene transfer on bacterial evolution is underscored by the realization that foreign DNA can represent up to one-fifth of a given bacterial genome (Ochman et al., 2000). Even the most clonal bacteria, such as *E. coli*, are chimeras bearing chromosomal genes (DuBose et al., 1988) and portions of genes of different ancestries (Bowler et al., 1994). Much of the DNA of bacteria classified as *Escherichia coli* has been acquired relatively recently: more than 17% of the open reading frames of the *E. coli* K-12 genome was acquired in the last 100 or so million years from organisms with G + C ratios and codon usage patterns distinguishable from those of other strains of *E. coli* and closely related Enterobacteriaceae (Lawrence and Ochman, 1998). The genome of *V. cholerae* consists of two circular chromosomes (Trucksis et al., 1998), and the entire genome sequence of a biotype El Tor strain N16961 has recently been published (Heidelberg et al., 2000). The larger chromosome (chrI) was found to contain most of the genes involved in cell division, transcription, and translation. The major pathogenicity-associated elements, such as the VPI and the CTX prophage, were found to be located on this chromosome. The smaller chromosome has the majority of hypothetical genes, the functions of which are not yet well understood.

Recently, a distinct type of integron was discovered in the *V. cholerae* genome (Mazel et al., 1998). This element spans 126 kb, gathers at least 179 cassettes into a single structure termed a super-integron, and is located on the smaller chromosome (Rowe-Magnus and Mazel, 1999; Heidelberg et al., 2000). There are over a hundred reading frames that have duplicate copies in both the chromosomes.

Various features of the smaller chromosome led to the proposal that it could originally have been a megaplasmid captured by an ancestral *Vibrio* sp. Phylogenetic analysis shows that a gene near the putative replication origin of chr II, but not chr I, shows more affinity to plasmid cognates. Moreover, genes encoding ribosomal RNAs are found in chr I only. These differences, along with the fact that chr I contains most of the functional genes, are best explained by the "megaplasmid hypothesis." However, the GC content of the two chromosomes is almost the same, and this would mean that the two plasmids have been cohabitants for a significantly long evolutionary time. This finding is in conflict with the megaplasmid hypothesis, leading Waldor and RayChaudhuri (2000) to suggest that the small chromosome arose by excision from a single large ancestral chromosome. The presence of the two chromosomes could be evolutionarily advantageous for the organism given the fact that it has a complex life style involving the varied environments of the human gut and the estuarine habitat. Heidelberg et al. (2000) have also proposed that single chromosome cells may be generated in response to environmental signals, such that metabolic functions would continue but valuable nutrients would not be used up in replicating the cells. Such metabolically active, replication defective cells, called "drones," could aid the normal cells in bacterial biofilms or may be the viable but nonculturable cells of *Vibrio cholerae* described by Colwell and colleagues (Colwell and Grimes, 2000).

CONCLUDING REMARKS

There are still many intriguing questions concerning the evolution of vibrios. The number of *Vibrio* spp. recognized within the genus *Vibrio* has currently expanded to 55 (http://www.gbf.de/dsmz/bactnom/bactname. htm), of which at least 11 are known to be associated with human disease. From this plethora of vibrios, why was *V. cholerae* selected by nature to be both environmentally and clinically significant, becoming an efficient human pathogen? This relates to its genetic background, compared to other *Vibrio* spp., and would presumably include the amenability of *V. cholerae* to receive and thereby acquire foreign DNA. Only recently have a few serogroups of another *Vibrio*, namely *V. parahaemolyticus*, acquired pandemic potential, causing a pandemic that included eight countries (Okuda et al., 1997; Chowdhury et al., 2000; Matsumoto et al., 2000). As more whole genome sequences of *Vibrio* spp. are completed, answers to this and other questions will become available. At the time of writing this chapter, we are aware that the whole genome sequence of *V. parahaemolyticus* has just been completed. The other enigma is why, within the species *V. cholerae*, have only certain serogroups acquired the ability to cause cholera and to be associated with pandemics of cholera, despite the fact that, taxonomically, differences at the species level have not been observed among the serogroups that cause cholera and those that do not (Citarella and Colwell, 1970; Colwell, 1973). Answers to some of these perplexing questions will allow us to better understand the evolution of this enigmatic but fascinating organism.

ACKNOWLEDGMENTS

The ICDDR,B is supported by countries and agencies that share its concern for the health problems of developing countries. Current donors providing unrestricted support include the aid agencies of the governments of Australia, Bangladesh, Belgium, Canada, Japan, the Kingdom of Saudi Arabia, the Netherlands, Sweden, Sri Lanka, Switzerland, and the United States.

REFERENCES

Aldova, E., Laznickova, K., Stepankova, E., and Lietava, J. (1968). Isolation of nonag-glutinable vibrios from an enteritis outbreak in Czechoslovakia. *Journal of Infectious Diseases*, 118, 25–31.

Beltran, P., Delgado, G., Davarro, A., Trujillo, F., Selander, R. K., and Cravioto, A. (1999). Genetic diversity and population structure of *Vibrio cholerae*. *Journal of Clinical Microbiology*, 37, 581–90.

Berche, P., Poyart, C., Abachin, E., Lelievre, H., Vandepitte, J., Dodin, A., and Fournier, J. M. (1994). The novel epidemic strain O139 is closely related to the pandemic strain O1 of *Vibrio cholerae*. *Journal of Infectious Diseases*, 170, 701–4.

Bik, E. M., Bunschoten, A. E., Gouw, R. D., and Mooi, F. R. (1995). Genesis of the novel epidemic *Vibrio cholerae* O139 strain: evidence for horizontal transfer of genes involved in polysaccharide synthesis. *European Molecular Biology Organization Journal*, 14, 209–16.

Bik, E. M., Gouw, R. D., and Mooi, F. R. (1996). DNA fingerprinting of *Vibrio cholerae* strains with a novel insertion sequence element: a tool to identify epidemic strains. *Journal of Clinical Microbiology*, 34, 1453–61.

Blank, T. E., Zhong, H., Bell, A. L., Whittam, T. S., and Donnenberg, M. S. (2000). Molecular variation among type IV pilin (*bfpA*) genes from diverse enteropathogenic *Escherichia coli* strains. *Infection and Immunity*, 68, 7028–38.

Bowler, L. D., Zhang, Q. Y., Riou, J. Y., and Spratt, B. G. (1994). Interspecies recombination between the *penA* genes of *Neisseria meningitidis* and commensal *Neisseria* species during the emergence of penicillin resistance in *N. meningitidis*: natural events and laboratory simulation. *Journal of Bacteriology*, 176, 333–7.

Boyd, E. F., Moyer, K. E., Shi, L., and Waldor, M. K. (2000). Infectious CTXΦ and the *Vibrio* pathogenicity island prophage in *Vibrio mimicus*: evidence for recent horizontal transfer between *V. mimicus* and *V. cholerae*. *Infection and Immunity*, 68, 1507–13.

Boyd, E. F., and Waldor, M. K. (2002). Evolutionary and functional analyses of variants of the toxin-coregulated pilus protein TcpA from toxigenic *Vibrio cholerae* non-O1/non-O139 serogroup isolates. *Microbiology*, 148, 1655–66.

Chakrabarti, A. K., Ghosh, A. N., Nair, G. B., Niyogi, S. K., Bhattacharya, S. K., and Sarkar, B. L. (2000). Development and evaluation of a phage typing scheme for *Vibrio cholerae* O139. *Journal of Clinical Microbiology*, 38, 44–9.

Chakraborty, S., Mukhopadhyay, A. K., Bhadra, R. K., Ghosh, A. N., Mitra, R., Shimada, T., Yamasaki, S., Faruque, S. M., Takeda, Y., Colwell, R. R., and Nair, G. B. (2000). Virulence genes in environmental strains of *Vibrio cholerae*. *Applied and Environmental Microbiology*, 66, 4022–8.

Chakraborty, S., Garg, P., Ramamurthy, T., Thungapathra, M., Gautam, J. K., Kumar, C., Maiti, S., Yamasaki, S., Shimada, T., Takeda, Y., Ghosh, A., and Nair, G. B. (2001). Comparison of antibiogram, virulence genes, ribotypes and DNA fingerprints of *Vibrio cholerae* of matching serogroups isolated from hospitalised diarrhoea cases and from the environment during 1997–1998 in Calcutta. *India Journal of Medicine Microbiology*, 50, 879–88.

Chattopadhyay, D. J., Sarkar, B. L., Ansari, M. Q., Chakrabarti, B. K., Roy, M. K., Ghosh, A. N., and Pal, S. C. (1993). New phage typing scheme for *Vibrio cholerae* O1 biotype El Tor strains. *Journal of Clinical Microbiology*, 31, 1579–85.

Chen, F., Evins, G. M., Cook, W. L., Almeida, R., Bean, H. N., and Wachsmuth, I. K. (1991). Genetic diversity among toxigenic and non-toxigenic *Vibrio cholerae*-O1 isolated from the western hemisphere. *Epidemiology and Infection*, 107, 225–33.

Cholera Working Group. (1993). Large epidemic of cholera-like disease in Bangladesh caused by *Vibrio cholerae* O139 synonym Bengal. *Lancet*, 342, 387–90.

Chowdhury, N. R., Chakraborty, S., Ramamurthy, T., Nishibuchi, M., Yamasaki, S., Takeda, Y., and Nair, G. B. (2000). Molecular evidence of clonal *Vibrio parahaemolyticus* pandemic strains. *Emerging Infectious Diseases*, 6, 631–6.

Citarella, R. V., and Colwell, R. R. (1970). Polyphasic taxonomy of the genus *Vibrio*: polynucleotide sequence relationships among selected *Vibrio* species. *Journal of Bacteriology*, 104, 434–42.

Colwell, R. R. (1973). Genetic and phenetic classification of bacteria. *Advances in Applied Microbiology*, 16, 137–76.

Colwell, R. R. (1996). Global climate and infectious disease: the cholera paradigm. *Science*, 274, 2025–31.

Colwell, R. R., and Grimes, D. J. (2000). *Nonculturable Microorganisms in the Environment*. American Society of Microbiology Press, Washington, DC, 354 pp.

Colwell, R. R., Kaper, J., and Joseph, S. W. (1977). *Vibrio cholerae*, *Vibrio parahaemolyticus*, and other vibrios: occurrence and distribution in Chesapeake Bay. *Science*, 198, 394–6.

Colwell, R. R., Brayton, P. R., Grimes, D. J., Roszak, D. R., Huq, S. A., and Palmer, L. M. (1985). Viable but non-culturable *Vibrio cholerae* and related environmental pathogens in the environment: implication for release of genetically engineered microorganisms. *Bio/Technology*, 3, 817–20.

Comstock, L. E., Johnson, J. A., Michalski, J. M., Morris, J. G., Jr., and Kaper, J. B. (1996). Cloning and sequence of a region encoding a surface polysaccharide of

Vibrio cholerae O139 and characterization of the insertion site in the chromosome of *Vibrio cholerae* O1. *Molecular Microbiology*, 19, 815–26.

Dakin, W. P., Howell, D. J., Sutton, R. G., O'Keefe, M. F., and Thomas, P. (1974). Gastroenteritis due to non-agglutinable (non-cholera) vibrios. *Medical Journal of Australia*, 2, 487–90.

Davis, B. M., Kimsey, H. H., Chang, W., and Waldor, M. K. (1999). The *Vibrio cholerae* O139 Calcutta bacteriophage CTXΦ is infectious and encodes a novel repressor. *Journal of Bacteriology*, 181, 6779–87.

DiRita, V. J. (1992). Co-ordinate expression of virulence genes by ToxR in *Vibrio cholerae*. *Molecular Microbiology*, 6, 451–8.

DiRita, V. J., Parsot, C., Jander, G., and Mekalanos, J. J. (1991). Regulatory cascade controls virulence in *Vibrio cholerae*. *Proceedings of the National Academy of Sciences USA*, 88, 5403–7.

DuBose, R. F., Dykhuizen, D. E., and Hartl, D. L. (1988). Genetic exchange among natural isolates of bacteria: recombination within the *phoA* gene of *Escherichia coli*. *Proceedings of the National Academy of Sciences USA*, 85, 7036–40.

Dumontier, S., and Berche, P. (1998). *Vibrio cholerae* O22 might be a putative source of exogenous DNA resulting in the emergence of the new strain of *Vibrio cholerae* O139. *FEMS Microbiology Letters*, 164, 91–8.

Faruque, S. M., Roy, S. K., Alim, A. R. M. A., Siddique, A. K., and Albert, M. J. (1995). Molecular epidemiology of toxigenic *Vibrio cholerae* in Bangladesh studied by numerical analysis of rRNA gene restriction patterns. *Journal of Clinical Microbiology*, 33, 2833–8.

Faruque, S. M., Ahmed, K. M., Alim, A. R. M. A., Qadri, F., Siddique, A. K., and Albert, M. J. (1997*a*). Emergence of a new clone of toxigenic *Vibrio cholerae* biotype El Tor displacing *V. cholerae* O139 Bengal in Bangladesh. *Journal of Clinical Microbiology*, 35, 624–30.

Faruque, S. M., Ahmed, K. M., Siddique, A. K., Zaman, K., Alim, A. R. M. A., and Albert, M. J. (1997*b*). Molecular analysis of toxigenic *Vibrio cholerae* O139 Bengal isolated in Bangladesh between 1993 and 1996: evidence for the emergence of a new clone of the Bengal vibrios. *Journal of Clinical Microbiology*, 35, 2299–306.

Faruque, S. M., Albert, M. J., and Mekalanos, J. J. (1998*a*). Epidemiology, genetics, and ecology of toxigenic *Vibrio cholerae*. *Microbiology and Molecular Biology Reviews*, 62, 1301–14.

Faruque, S. M., Asadulghani, S. Saha, M. N., Alim, A. R. M. A., Albert, M. J., Islam, K. M. N., and Mekalanos, J. J. (1998*b*). Analysis of clinical and environmental strains of nontoxigenic *Vibrio cholerae* for susceptibility to CTXΦ: molecular basis for origination of new strains with epidemic potential. *Infection and Immunity*, 66, 5819–25.

Faruque, S. M., Asadulghani, S. Alim, A. R. M. A., Albert, M. J., Islam, K. M. N., and Mekalanos, J. J. (1998*c*). Induction of the lysogenic phage encoding cholera toxin in naturally occurring strains of toxigenic *Vibrio cholerae* O1 and O139. *Infection and Immununity*, 66, 3752–7.

Faruque, S. M., Saha, M. N., Asadulghani, Bag, P. K., Bhadra, R. K., Bhattacharya, S. K., Sack, R. B., Takeda, Y., and Nair, G. B. (2000). Genomic diversity among

Vibrio cholerae O139 strains isolated in Bangladesh and India between 1992 and 1998. *FEMS Microbiology Letters*, 84, 279–84.

Field, M., Fromm, D., Al-Awqati, Q., and Greenough, W. B., III. (1972). Effect of cholera enterotoxin on ion transport across isolated ileal mucosa. *Journal of Clinical Investigation*, 51, 796–804.

Finkelstein, R. A. (1983). Antigenic and structural variations in the cholera/coli family of enterotoxins. *Developments in Biological Standardization*, 53, 93–5.

Franzon, V. L., Barker, A., and Manning, P. A. (1993). Nucleotide sequence encoding the mannose-fucose-resistant hemagglutinin of *Vibrio cholerae* O1 and construction of a mutant. *Infection and Immununity*, 61, 3032–7.

Freitas, F. S., Momen, H., and Salles, C. A. (2002). The zymovars of *Vibrio cholerae*: multilocus enzyme electrophoresis of *Vibrio cholerae*. *Memórias do Instituto Oswaldo Cruz*, 97, 511–6.

Goldberg, I., and Mekalanos, J. J. (1986). Effect of a *recA* mutation on cholera toxin gene amplification and deletion events. *Journal of Bacteriology*, 165, 723–31.

Hase, C. C., and Mekalanos, J. J. (1998). TcpP protein is a positive regulator of virulence gene expression in *Vibrio cholerae*. *Proceedings of the National Academy of Sciences USA*, 95, 730–4.

Heidelberg, J. F., Eisen, J. A., Nelson, W. C., Clayton, R. A., Gwinn, M. L., Dodson, R. J., Haft, D. H., Hickey, E. K., Peterson, J. D., Umayam, L., Gill, S. R., Nelson, K. E., Read, T. D., Tettelin, H., Richardson, D., Ermolaeva, M. D., Vamathevan, J., Bass, S., Qin, H., Dragoi, I., Sellers, P., McDonald, L., Utterback, T., Fleishman, R. D., Nierman, W. C., White, O., Salzberg, S. L., Smith, H. O., Colwell, R. R., Mekalanos, J. J., Ventor, J. C., and Frasier, C. M. (2000). DNA sequence of both chromosomes of the cholera pathogen *Vibrio cholerae*. *Nature*, 406, 477–83.

Heilpern, A. J, and Waldor, M. K. (2000). CTXΦ infection of *Vibrio cholerae* requires the *tolQRA* gene products. *Journal of Bacteriology*, 182, 1739–47.

Herrington, D. A., Hall, R. H., Losonsky, G., Mekalanos, J. J., Taylor, R. K., and Levine, M. M. (1988). Toxin, toxin-coregulated pili and ToxR regulon are essential for *Vibrio cholerae* pathogenesis in humans. *Journal of Experimental Medicine*, 168, 1487–92.

Hughes, K. J., Everiss, K. D., Harkey, C. W., and Peterson, K. M. (1994). Identification of a *Vibrio cholerae* ToxR-activated gene (*tagD*) that is physically linked to the toxin coregulated pilus (*tcp*) gene cluster. *Gene*, 148, 97–100.

Islam, M. S., Drasar, B. S., and Sack, R. B. (1994*a*). The aquatic flora and fauna as reservoirs of *Vibrio cholerae*: a review. *Journal of Diarrhoeal Disease Research*, 12, 87–96.

Islam, M. S., Drasar, B. S., and Sack, R. B. (1994*b*). The aquatic environment as a reservoir of *Vibrio cholerae*: a review. *Journal of Diarrhoeal Disease Research*, 11, 197–206.

Johnson, J. A., Salles, C. A., Panigrahi, P., Albert, M. J., Wright, A. C., Johnson, R. J., and Morris, J. G., Jr. (1994). *Vibrio cholerae* O139 synonym Bengal is closely related to *Vibrio cholerae* El Tor but has important differences. *Infection and Immunity*, 62, 2108–10.

Jonson, G., Lebens, M., and Holmgren, J. (1994). Cloning and sequencing of *Vibrio cholerae* mannose-sensitive haemagglutinin pilin gene: localization of *mshA* within a cluster of type IV pilin genes. *Molecular Microbiology*, 13, 109–18.

Karaolis, D. K., Johnson, J. A., Bailey, C. C., Boedeker, E. C., Kaper, J. B., and Reeves, P. R. (1998). A *Vibrio cholerae* pathogenicity island associated with epidemic and pandemic strains. *Proceedings of the National Academy of Sciences USA*, 95, 3134–9.

Karaolis, D. K., Somara, S., Maneval, Jr., D. R., Johnson, J. A., and Kaper, J. B. (1999). A bacteriophage encoding a pathogenicity island, a type-IV pilus and a phage receptor in cholera bacteria. *Nature*, 399, 375–9.

Karaolis, D. K. R., Lan, R., Kaper, J. B., and Reeves, P. R. (2001). Comparison of *Vibrio cholerae* pathogenicity islands in sixth and seventh pandemic strains. *Infection and Immunity*, 69, 1947–52.

Kaufman, M. R., Shaw, C. E., Jones, I. D., and Taylor, R. K. (1993). Biogenesis and regulation of the *Vibrio cholerae* toxin-coregulated pilus: analogies to other virulence factor secretory systems. *Gene*, 126, 43–9.

Kelly, M. T. (1991). Pathogenic Vibrionaceae in patients and the environment. *Undersea Biomedical Research*, 18, 193–6.

Kimsey, H. H., and Waldor, M. K. (1998). CTXϕ immunity: application in the development of cholera vaccines. *Proceedings of the National Academy of Sciences USA*, 95, 7035–9.

Kimsey, H. H., Nair, G. B., Ghosh, A., and Waldor, M. K. (1998). Diverse CTXphis and evolution of new pathogenic *Vibrio cholerae*. *Lancet*, 352, 457–8.

Kirn, T. J., Lafferty, M. J., Sandoe, C. M., and Taylor, R. K. (2000). Delineation of pilin domains required for bacterial association into microcolonies and intestinal colonization by *Vibrio cholerae*. *Molecular Microbiology*, 35, 896–910.

Kovach, M. E., Shaffer, M. D., and Peterson, K. M. (1996). A putative integrase gene defines the distal end of a large cluster of *toxR*-regulated colonization genes in *Vibrio cholerae*. *Microbiology*, 142, 2165–74.

Lawrence, J. G., and Ochman, H. (1998). Molecular archaeology of the *Escherichia coli* genome. *Proceedings of the National Academy of Sciences USA*, 95, 9413–17.

Lazar, S., and Waldor, M. K. (1998). ToxR-independent expression of cholera toxin from the replicative form of CTXΦ. *Infection and Immunity*, 66, 394–7.

Lipp, E. K., Huq, A., and Colwell, R. R. (2002). Effects of global climate on infectious disease: the cholera model. *Clinical Microbiology Reviews*, 15, 757–70.

Maiden, M. C., Bygraves, J. A., Feil, E., Morelli, G., Russell, J. E., Urwin, R., Zhang, Q., Zhou, J., Zurth, K., Caugant, D. A., Feavers, I. M., Achtman, M., and Spratt, B. G. (1998). Multilocus sequence typing: a portable approach to the identification of clones within populations of pathogenic microorganisms. *Proceedings of the National Academy of Sciences USA*, 95, 3140–5.

Manning, P. A. (1997). The *tcp* gene cluster of *Vibrio cholerae*. *Gene*, 192, 63–70.

Manning, P. A., Stroeher, U. H., and Morona, R. (1994). Molecular basis for O-antigen biosynthesis in *Vibrio cholerae*: Ogawa-Inaba switching. In *Vibrio cholerae and Cholera: Molecular to Global Perspectives*, eds. I. K. Wachsmuth, P. A. Blake, and O. Olsvik, pp. 77–94. ASM Press, Washington, DC.

Matsumoto, C., Okuda, J., Ishibashi, M., Iwanaga, M., Garg, P., Ramamurthy, T., Wong, H. C., Depaola, A., Kim, Y. B., Albert, M. J., and Nishibuchi, M. (2000). Pandemic spread of an O3:K6 clone of *Vibrio parahaemolyticus* and emergence of related strains evidenced by arbitrarily primed PCR and *toxRS* sequence analyses. *Journal of Clinical Microbiology*, 38, 578–85.

Mazel, D., Dychinco, B., Webb, V. A., and Davies, J. (1998). A distinctive class of integron in the *Vibrio cholerae* genome. *Science*, 280, 605–8.

Mekalanos, J. J. (1983). Duplication and amplification of toxin genes in *Vibrio cholerae*. *Cell*, 35, 253–63.

Mekalanos, J. J., Rubin, E. J., and Waldor, M. K. (1997). Cholera: molecular basis for emergence and pathogenesis. *FEMS Immunology and Medical Microbiology*, 18, 241–8.

Merrell, D. S., Butler, S. M., Qadri, F., Dolganov, N. A., Alam, A., Cohen, M. B., Calderwood, S. B., Schoolnik, G. K., and Camilli, A. (2002). Host-induced epidemic spread of the cholera bacterium. *Nature*, 417, 642–5.

Miao, E. A., and Miller, S. I. (1999). Bacteriophages in the evolution of pathogen-host interactions. *Proceedings of the National Academy of Sciences USA*, 96, 9452–4.

Morris, J. G. (1990). Non-O group 1 *Vibrio cholerae*: a look at the epidemiology of an occasional pathogen. *Epidemiological Reviews* 12, 179–91.

Morris, J. G., Jr., and Black, R. E. (1985). Cholera and other vibrios in the United States. *New England Journal of Medicine*, 312, 343–50.

Mukhopadhyay, A. K., Chakraborty, S., Takeda, Y., Nair, G. B., and Berg, D. E. (2001). Characterization of VPI pathogenicity island and CTXΦ prophage in environmental strains of *Vibrio cholerae*. *Journal of Bacteriology*, 183, 4737–46.

Nair, G. B., Garg, S., Mukhopadhyay, A. K., Shimada, T., and Takeda, Y. (1994*a*). Laboratory diagnosis of *Vibrio cholerae* O139 Bengal, the new pandemic strain of cholera. *LabMedica International*, XI, 8–11.

Nair, G. B., Ramamurthy, T., Bhattacharya, S. K., Mukhopadhyay, A. K., Garg, S., Bhattacharya, M. K., Takeda, T., Shimada, T., Takeda, Y., and Deb, B. C. (1994*b*). Spread of *Vibrio cholerae* O139 Bengal in India. *Journal of Infectious Diseases*, 169, 1029–34.

Nair, G. B., Faruque, S. M., Bhuiyan, N. A., Kamruzzaman, M., Siddique, A. K., and Sack, D. A. (2002). New variants of *Vibrio cholerae* O1 biotype El Tor with attributes of the classical biotype from hospitalized patients with acute diarrhea in Bangladesh. *Journal of Clinical Microbiology*, 40, 3296–9.

Ochman, H., Lawrence, J. G., and Groisman, E. A. (2000). Lateral gene transfer and the nature of bacterial innovation. *Nature*, 405, 299–304.

Okuda, J., Ishibashi, M., Hayakawa, E., Nishino, T., Takeda, Y., Mukhopadhyay, A. K., Garg, S., Bhattacharya, S. K., Nair, G. B., and Nishibuchi, M. (1997). Emergence of a unique O3:K6 clone of *Vibrio parahaemolyticus* in Calcutta, India, and isolation of strains from the same clonal group from southeast Asian travellers arriving in Japan. *Journal of Clinical Microbiology*, 35, 3150–5.

Parsot, C., Taxman, E., and Mekalanos, J. J. (1991). ToxR regulates the production of lipoproteins and the expression of serum resistance in *Vibrio cholerae*. *Proceedings of the National Academy of Sciences USA*, 88, 1641–5.

Pearson, G. D. N., Woods, A., Chiang, S. L., and Mekalanos, J. J. (1993). CTX genetic element encodes a site-specific recombination system and an intestinal colonization factor. *Proceedings of the National Academy of Sciences USA*, 90, 3750–4.

Peek, J. A., and Taylor, R. K. (1992). Characterization of a periplasmic thiol: disulfide interchange protein required for the functional maturation of secreted virulence factors of *Vibrio cholerae*. *Proceedings of the National Academy of Sciences USA*, 89, 6210–4.

Peterson, K. M., and Mekalanos, J. J. (1988). Characterization of the *Vibrio cholerae* ToxR regulon: identification of novel genes involved in intestinal colonization. *Infection and Immunity*, 56, 2822–9.

Popovic, T., Bopp, C., Olsvik, O., and Wachsmuth, K. (1993). Epidemiologic application of a standardized ribotype scheme for *Vibrio cholerae* O1. *Journal of Clinical Microbiology*, 31, 2474–82.

Reidl, J., and Klose, K. E. (2002). *Vibrio cholerae* and cholera: out of the water and into the host. *FEMS Microbiology Reviews*, 26, 125–39.

Rivera, I. N. G., Chun, J., Huq, A., Sack, R. B., and Colwell, R. R. (2001). Genotypes associated with virulence in environmental isolates of *Vibrio cholerae*. *Applied and Environmental Microbiology*, 67, 2421–9.

Rowe-Magnus, D. A., and Mazel, D. (1999). Resistance gene capture. *Current Opinion in Microbiology*, 2, 483–8.

Sakazaki, R., and Donovan, T. J. (1984). Serology and epidemiology of *Vibrio cholerae* and *Vibrio mimicus*. In *Methods in Microbiology*, Vol. 16, ed. Y. Bergan, pp. 271–89. Academic Press, London.

Salles, C. A., and Momen, H. (1991). Identification of *Vibrio cholerae* by enzyme electrophoresis. *Transactions of the Royal Society of Tropical Medicine and Hygiene*, 85, 544–7.

Sengupta, D. K., Sengupta, T. K., and Ghose, A. C. (1992). Major outer membrane proteins of *Vibrio cholerae* and their role in induction of protective immunity through inhibition of intestinal colonization. *Infection and Immunity*, 60, 4848–55.

Skorupski, K., and Taylor, R. K. (1997). Control of the ToxR virulence regulon in *Vibrio cholerae* by environmental stimuli. *Molecular Microbiology*, 25, 1003–9.

Sozhamannan, S., Deng, Y. K., Li, M., Sulakvelidze, A., Kaper, J. B., Johnson, J. A., Nair, G. B., and Morris, J. G., Jr. (1999). Cloning and sequencing of the genes downstream of the *wbf* gene cluster of *Vibrio cholerae* serogroup O139 and analysis of the junction genes in other serogroups. *Infection and Immunity*, 67, 5033–40.

Stroeher, U. H., Jedani, K. E., Dredge, B. K., Morona, R., Brown, M. H., Karageorgos, L. E., Albert, J. M., and Manning, P. A. (1995). Genetic rearrangements in the *rfb* regions of *Vibrio cholerae* O1 and O139. *Proceedings of the National Academy of Sciences USA*, 92, 10,374–8.

Taylor, R. K., Miller, V. L., Furlong, D. B., and Mekalanos, J. J. (1987). Use of phoA gene fusions to identify a pilus colonization factor coordinately regulated with cholera toxin. *Proceedings of the National Academy of Sciences USA*, 84, 2833–7.

Taylor, R., Shaw, C., Peterson, K., Spears, P., and Mekalanos, J. (1988). Safe, live *Vibrio cholerae* vaccines? *Vaccine*, 6, 151–4.

Trucksis, M., Michalski, J., Deng, Y. K., and Kaper, J. B. (1998). The *Vibrio cholerae* genome contains two unique circular chromosomes. *Proceedings of the National Academy of Sciences USA*, 95, 14,464–9.

Voss, E., and Attridge, S. R. (1993). In vitro production of toxin-coregulated pili by *Vibrio cholerae* El Tor. *Microbial Pathogenesis*, 15, 255–68.

Wachsmuth, I. K., Evins, G. M., Fields, P. I., Olsvic, O., Popovic, T., Bopp, C. A., Wells, J. G., Cerrillo, C., and Blake, P. A. (1993). The molecular epidemiology of cholera in Latin America. *Journal of Infectious Diseases*, 167, 621–6.

Wachsmuth, I. K., Olsvik, O., Evins, G. M., and Popovic, T. (1994). Molecular epidemiology of cholera, In *Vibrio cholerae and Cholera: Molecular to Global Perspectives*, eds. K. Wachsmuth, P. A. Blake, and O. Olsvik, pp. 357–70. American Society of Microbiology, Washington, DC.

Waldor, M. K., and Mekalanos, J. J. (1994). *Vibrio cholerae* O139 specific gene sequences. *Lancet*, 343, 1366.

Waldor, M. K., and Mekalanos, J. J. (1996). Lysogenic conversion by a filamentous bacteriophage encoding cholera toxin. *Science*, 272, 1910–14.

Waldor, M. K., and RayChaudhuri, D. (2000). Bacterial genomics: Treasure trove for cholera research. *Nature*, 406(6795), 469–70.

Waldor, M. K., Rubin, E. J., Pearson, G. D. N., Kimsey, H., and Mekalanos, J. J. (1997). Regulation, replication, and integration functions of the *Vibrio cholerae* CTXΦ are encoded by regions RS2. *Molecular Microbiology*, 24, 917–26.

Watnick, P., and Kolter, R. (2000). Biofilm, city of microbes. *Journal of Bacteriology*, 182, 2675–9.

Watnick, P. I., Lauriano, C. M., Klose, K. E., Croal, L., and Kolter, R. (2001). The absence of a flagellum leads to altered colony morphology, biofilm development and virulence in *Vibrio cholerae* O139. *Molecular Microbiology*, 39, 223–35.

Xu, H. S., Roberts, N. C., Singleton, F. L., Attwell, R. W., Grimes, D. J., and Colwell, R. R. (1982). Survival and viability of nonculturable *Escherichia coli* and *Vibrio cholerae* in the estuarine and marine environment. *Microbial Ecology*, 8, 313–23.

Yamai, S., Okitsu, T., Shimada, T., and Katsube, Y. (1997). Distribution of serogroups of *Vibrio cholerae* non-O1 non-O139 with specific reference to their ability to produce cholera toxin, and addition of novel serogroups. *Kansenshogaku Zasshi*, 71, 1037–45. (In Japanese.)

Yamasaki, S., Garg, S., Nair, G. B., and Takeda, Y. (1999a). Distribution of *Vibrio cholerae* O1 antigen biosynthesis genes among O139 and other non-O1 serogroups of *Vibrio cholerae*. *FEMS Microbiology Letters*, 179, 115–21.

Yamasaki, S., Shimizu, T., Hoshino, K., Ho, S. T., Shimada, T., Nair, G. B., and Takeda, Y. (1999b). The genes responsible for O-antigen synthesis of *Vibrio cholerae* O139 are closely related to those of *Vibrio cholerae* O22. *Gene*, 237, 321–32.

NINE

Evolutionary Dynamics of *Daphnia* and Their Microparasites

Tom Little and Dieter Ebert

INTRODUCTION

Haldane (1949) was one of the first to speculate that the genetic outcome of parasitic interactions may have profound implications, in particular, that frequency-dependent selection on host and parasite polymorphisms can maintain genetic variation and promote sexual reproduction. Much theoretical work since then has supported his early insights: dynamic genetic polymorphisms or arms races are a common outcome of computer simulations of host–parasite coevolution. Recent theory now prompts us to ask precise questions about parasite-associated dynamics: Do host and parasite alleles cycle with regularity or as bursts followed by long periods of stasis? Do arms races result from directional selection on mutational input or does selection maintain alleles to antiquity (Hill et al., 1992; Hughes and Nei, 1992; Stahl et al., 1999)? The other major line of theory on host–parasite interactions, also foreshadowed by Haldane, concerns the evolution of virulence (the damage to the host caused by parasites), and this work has now developed explicit predictions concerning the conditions under which parasites ought to evolve towards doing greater or less harm to their hosts (Anderson and May, 1982; Ewald, 1983).

Probing these questions and predictions will shed light on the evolutionary significance of parasitism as envisioned by Haldane (1949) and others, because the exact tempo, mode, and outcome of selection matter if we are to properly evaluate the evolutionary significance of parasitism. However, empirical evidence lags behind theory, and it remains uncertain just how widespread and strong is genetic change associated with parasitic interactions in the wild. Gathering empirical evidence for these issues is no easy task. Ideally, the natural world would mirror that of theoretical studies, and hosts and parasites would possess just one or a few genes relevant for their engagement, and biologists would possess markers for these genes. Nature not been so permissive, but research using model systems have nevertheless made great progress (e.g., Burdon, Thrall, and Brown, 1999; Gemmill, Viney, and Read, 1997; Lively, Craddock, and Vrijenhoek, 1990; Schmid-Hempel, 1995).

As a starting point for research on coevolutionary issues, host–parasite coevolution is usefully defined as reciprocal natural selection on parasite infectivity and host resistance (Thompson, 1994). Thus, one approach to the study of coevolution is to look for evidence of natural selection on these traits, and to simplify the burden, evolutionary parasitologists have typically probed one-half of the coevolutionary coin at a time. With this approach, the study of coevolution differs little from the study of natural selection generally, in that one examines patterns of genetic variation in space and time to detect evidence of adaptive genetic change. We and our colleagues have used such a framework with a model system, the planktonic crustacean *Daphnia* (Cladocera: Crustacea) and its various bacterial, fungal, and microsporidial parasites. In this chapter we review what we have learned from empirical work on this system, which encompasses field studies, laboratory experiments, experimental evolution, and the use of a variety of genetic markers. We start with an outline of the biological features we exploit, and then we provide a summary of recent data, classified as to whether it illustrates genetic variation among populations, within populations, or through time.

THE *DAPHNIA*-MICROPARASITE SYSTEM

Daphnia are small (0.5–4.0 mm) ubiquitous fresh-water crustaceans that have a generation time of about one week so the evolution of populations (in the lab or the field) can be observed within practical time frames. The small size and ease of handling of *Daphnia* permits the collection of large samples from the wild and the maintenance of thousands of individuals in the lab. Thus, *Daphnia* have proven themselves model organisms in both the field and the lab and are practical for linking field to lab studies. Especially useful is the fact that *Daphnia* may reproduce either clonally or sexually and that this can be controlled by the experimenter. Sexual reproduction permits crossing experiments and studies of inheritance, and clonal reproduction allows us to replicate genotypes captured in the field, i.e., to "freeze" the population genetic structure from a given time point. This latter point is key, because it allows us to capture a living snapshot and use this in experiments which verify the cause(s) of field patterns.

Daphnia are attacked by a wide array of microparasites (sensu Anderson and May, 1981), with bacterial, fungal, and microsporidial parasites being particularly common (Green, 1974; Little and Ebert, 1999; Stirnadel and Ebert, 1997). Many of these parasites are known to affect the fitness of their victims substantially (Ebert, 1995; Green, 1974; Little and Ebert, 1999; Mangin, Lipsitch, and Ebert, 1995; Schwartz and Cameron, 1993; Stirnadel, 1994; Stirnadel and Ebert, 1997), with sites of infection including the gut, ovaries, fat cells, the muscles that control the swimming antennae, the epidermis, hemolymph, and the eggs. Parasite biomass within hosts is often considerable, and because *Daphnia* have a transparent carapace, it is possible to detect many infections without dissection (Ebert and Mangin, 1995; Ebert et al., 1996; Stirnadel and Ebert, 1997), although this is not possible with some of the parasites that infect the gut, and for viruses. Laboratory microcosm studies have shown that many of these parasites can strongly affect population sizes (Ebert, Lipsitch, and Mangin, 2000), as is generally accepted for most parasite–host interactions (Tompkins and Begon, 1999). In this chapter we focus on evidence that parasitic interactions affect population genetic structure.

HOST AND PARASITE GENETIC VARIATION AND RELATED PHENOMENA
As a starting point for detecting evolutionary dynamics in the *Daphnia*–microparasite system, numerous studies have tested for genetic variation that is related to host and pathogen traits. Genetic variation provides the fuel for evolutionary change, and all models of host–parasite coevolution assume its existence in one form or another. For the *Daphnia* system, tests for genetic variation have been carried out both among and within populations, and in some cases it has been possible to explicitly test if the variation is adaptive.

Among Population Variation
Analyses of variation among populations have revealed large differences in rates of parasitic attack (Little and Ebert, 1999; Stirnadel, 1994) and the presence of genetic variation in both hosts and parasites. For example, laboratory exposures of hosts and pathogens show that the spore dose required to establish infections can vary 1,000-fold depending on the particular host and parasite population studied (Little and Ebert, 2000a). Variation among populations in *Daphnia*–parasite interactions is, in some cases, a result of adaptation. In particular, studies of patterns among populations detected a local adaptation of parasites to their hosts using the host species *D. magna* and the horizontally transferred microsporidian

Glucoides intestinalis (formerly called *Pleistophora intestinalis*) (Ebert, 1994). Infectivity and virulence were higher in sympatric compared to allopatric host–parasite combinations, and both traits decreased with increasing geographic distance between the host and parasite populations, probably due to genetic isolation with distance (Ebert, 1994). Such local parasite adaptation indicates that parasites have evolved to maximize fitness on the hosts they commonly encounter and that they have a relative advantage in any arms race that might be occurring. Among-population studies using another parasite, the bacterium *Pasteuria ramosa*, of *D. magna*, did not show a clear pattern of local adaptation (Ebert, Zschokke-Rohringer, and Carius, 1998). However, in the *D. magna–P. ramosa* study, genetic variation for resistance among host genotypes within populations was very high and likely obscured any among-population pattern of adaptation, i.e., there was considerably greater specialization within than among populations.

Looking beyond *Daphnia*, reviews of other host–pathogen systems (Kaltz and Shykoff, 1998; Lajeunesse and Forbes, 2001) indicate that there are just as many studies that failed to show local adaptation as did show it. Undoubtedly, a great many factors can influence the occurrence of local adaptation, and the range of outcomes observed in studies of local adaptation indeed suggests that the biological factors particular to each system make generalizations about local adaptation difficult. For example, depending on the relative migration rates of host and parasite among populations (or possibly from seed banks (Little and Ebert, 2001)), one could observe either a pattern of local host adaptation or local parasite adaptation (Dybdahl and Lively, 1996; Ebert, 1994; Kaltz and Shykoff, 1998; Lively, 1989). Migration rates and gene flow are well studied, though inconclusive, in *Daphnia* (De Meester et al., 2002), whereas aspects of parasite movement remain unstudied, probably due to a lack of molecular markers. Alternatively, the occurrence of local adaptation may depend on the host range of each parasite (Lajeunesse and Forbes, 2001) with broader host range parasites being less likely than narrow host range parasites to show local adaptation to any particular host. Multiple hosts for *Daphnia* parasites are not common, and both *Daphnia* parasites that have been analyzed for local adaptation are directly horizontally transmitted to new hosts, although it is not absolutely certain if these interactions are as species-specific as they appear.

Within Population Variation

From the beginning of the molecular era, field studies of *Daphnia* populations have shown that clonal frequencies within populations may fluctuate wildly (Hebert, 1974a,b). The causes of these fluctuations were not

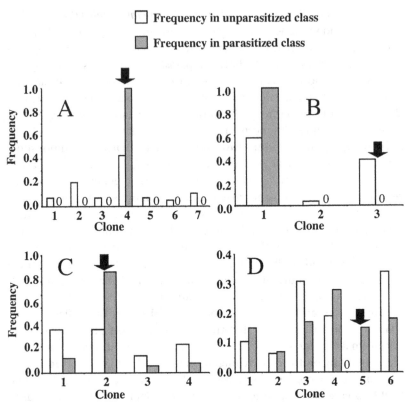

Figure 9.1. Examples of *Daphnia* populations for which the frequency of host genotypes (clones identified with allozymes) differed between the parasitized and unparasitized class of individuals. Arrows indicate particular host clones that were showing either very strong resistance or very high susceptibility in the field. (A) A population of *D. longispina* infected with *Thelohania sp.* (B) A population of *D. pulicaria* infected with *Thelohania sp.* (C) A population of *D. longispina* infected with microsporidian 1. (D) A population of *D. longispina* infected with microsporidian 4. Data are from Little and Ebert (1999).

established, but the dynamics were certainly consistent with parasite-driven genetic change. Little and Ebert (1999) showed the potential for these fluctuations to be caused by parasitic interactions by surveying allozyme variation in a range of *Daphnia* host and parasite species. In particular, they detected significant differences between the genotypic composition of parasitized and healthy *Daphnia* in 12 populations of 35 populations studied (Figure 9.1). This pattern indicates that the proportion of individuals infected commonly varies among host clones. Using allozymes to detect genetic variation for resistance within populations is possible because, in clonal organisms such as *Daphnia*, non-random associations between fitness-related loci and allozymes arise through the hitchhiking of allozymes with gene complexes that are under selection. A single bout of sexual reproduction will, however, considerably diminish these epistatic associations, and thus the utility of using allozymes to detect genetic variation due to parasitism (or any selective agent) depends critically on levels of sexual recruitment. In many *Daphnia* populations it appears that sexual recruitment is too frequent for allozyme studies such as that by Little and Ebert (1999) to be effective (Little and Ebert, 2000b). Thus, the overall impression from field studies of *Daphnia* is that genetic variation is widespread. *Daphnia* are not unique in this respect – virtually all studied host–pathogen associations appear to show considerable potential for parasite-mediated selection (reviewed in Little, 2002).

One aspect of within-population variation particularly emphasized by Haldane (1949) and later workers (Dybdahl and Lively, 1995) as a key mechanism driving frequency-dependent selection is parasite adaptation to common host clones. Briefly, parasite genotypes infective on the most common host genotype will be the most successful because they have the largest population of hosts available to them. Naively, one might predict that common host clones therefore tend to be over-parasitized. However, by virtue of being the target of parasite-mediated selection, common genotypes ought to decline in frequency. Particular host genotypes will thus oscillate between high and low frequency and also between being under-parasitized and over-parasitized. These oscillations make it difficult, or perhaps impossible, to determine if "over-infected" clones tend to be common ones (Dybdahl and Lively, 1995). Frank (1991) predicted that such temporal dynamics would generate a spatial pattern in which each population is in a different part of the cycle by chance alone. Thus, even in the presence of substantial clonal variation for infection, averaged over many populations, the clonal composition of the parasitized class should not differ from that of the random class. In agreement, Little and Ebert (1999) found significantly "over-parasitized" allozymic clones of *Daphnia* to span a full range of frequencies. Studies on other systems have concluded similarly (Dybdahl and Lively, 1995; Kaltz and Shykoff, 1998).

Laboratory studies have largely verified the evidence from molecular-based field work; under controlled environmental conditions, genetic variation for resistance within *Daphnia* populations is high (Ebert et al., 1998; Little and Ebert, 2000a, 2001). Moreover, studies have successfully linked laboratory results to field patterns. In particular, for many populations, individuals who were infected in field samples (but were subsequently cured with antibiotics and then raised in the laboratory) were more susceptible

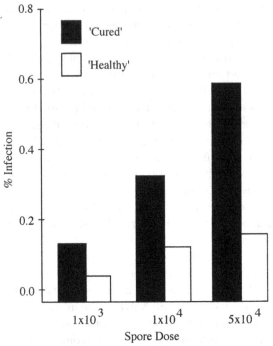

Figure 9.2. Experimental susceptibility of *D. magna* that were infected at the time of collection (but were "cured" with antibiotics prior to experimentation) was higher than that of hosts that were "healthy" at the time of collection. These experiments were conducted with up to three spore doses of *Pasteuria ramosa*. Data are from Little and Ebert (2000a).

than individuals who were healthy in field samples, under controlled conditions (Figure 9.2). Such a pattern unambiguously shows the effect of host genetic factors on the expression of disease in the wild. Not in all populations, however, was it clear that genetic factors influenced the expression of disease, and these deviations from the general pattern indicate the importance of environmental noise or genotype by environment interactions (Little and Ebert, 2000a).

Genetic variation on the parasite side has also been detected in abundance within populations of *D. magna* and the bacterium *P. ramosa* (Carius, Little, and Ebert, 2001). This parasite strain variation occurs simultaneously with host variation, thereby revealing significant host–genotype by parasite–genotype interactions, i.e., the susceptibility of a particular host genotype depends on which parasite genotype it encounters, and the infectivity of a particular parasite strain depends on which host genotype it encounters (see also Wedekind and Ruetschi, 2000). In addition, the range of specificities present within the populations studied by Carius et al.

(2001) was limited: there were no host strains resistant to all pathogen strains, or pathogen strains universally infective on all host strains. These two patterns – specificity and lack of universality – are both key assumptions in models for maintenance of genetic variability through frequency-dependent selection.

Moreover, these patterns shed light on the sort of genetic control of resistance present in this system. Much debate on the genetic control of interactions has centered on whether "matching allele" models or "gene-for-gene" systems are prevalent in nature. The patterns from the *D. magna/ P. ramosa* interaction are more in line with the matching allele model, especially given the lack of an overall superior genotype on either the host or parasite side. A matching allele system is more likely than a gene-for gene system to exhibit frequency-dependent dynamics and lead to the maintenance of sexual reproduction (Clay and Kover, 1996; Frank, 1996a,b; Parker, 1994, 1996). To what extent the results of Carius et al. (2001) can be generalized remains an important and pressing question, and studies on other systems are needed to verify whether the requirements for frequency-dependent dynamics are commonly met in the wild.

The genetic basis of genetic variation for infectivity or resistance has yet to be fully elucidated in most systems under study. It remains largely uncertain if resistance is conferred by single genes of large effect, a few major genes, or is polygenic (Sorci, Moller, and Boulinier, 1997). Are the genes involved strictly part of the innate immune system? Alternatively, are there major genes relevant to immunity because, for example, their constant presence in cell membranes prevents invasion in the first place? If the basis of resistance is polygenic, it could be due to a general efficiency at parasite clearance from the action of many enzyme pathways or to a complex behavioral trait. This latter possibility has been shown in *Daphnia*, providing the only clear understanding we presently have of the mechanisms behind resistance. *Daphnia magna* genotypes differ in their tendency to occupy the upper or lower portions of the water column. Those with a predilection for the deeper regions more often encounter parasite spores, which are abundant in pond sediments (Decaestecker, De Meester, and Ebert, 2002). Interestingly, however, parasite-mediated selection in favor of hosts that inhabit the upper regions of the water column is constrained because the risk of mortality due to visual hunting predators (mostly fish) is greater higher up. The study of Decaestecker et al. (2002) therefore provided a clear example of a trade-off encountered when organisms face multiple enemies.

Further efforts to identify the genetic basis of parasitic interactions in the *Daphnia* system are ongoing through the use of breeding experiments and molecular analyses. Gaps in the understanding of the genetic basis of resistance and infectivity (or adaptation generally) are widespread (Orr and Irving, 1997; Sorci et al., 1997). For example, considering all the arthropoda, it is essentially unknown which immune system genes might be the targets of diversifying pathogen-mediated selection. By analogy with plants and vertebrates, genetically diverse immune system genes would be those involved in the recognition of invading pathogens (Apanius et al., 1997; Stahl and Bishop, 2000), although in plants it appears that pathogen-attack molecules (specifially, chitinases) also show very high rates of diversity due to pathogen-mediated selection (Bishop, Dean, and Mitchell-Olds, 2000). Population genetic studies of the invertebrate immune system have so far not examined pathogen-recognition molecules. Genes that have been examined (all in *Drosophila*) include a number of cytosolic antimicrobial peptides (molecules in the final phase of the immune response that attack and destroy pathogens) (Clark and Wang, 1997; Lazzaro and Clark, 2001) and one of their transcription factors (Begun and Whitley, 2000). Results from these studies have been mixed – some genes show evidence of rapid evolution; others appear to be under purifying selection. No invertebrate immune-related gene has yet to reveal evidence of long-term persistence of alleles as has been observed for both MHC alleles in vertebrates and r-genes of plants. Clearly, much work needs to be done identifying genes involved in arthropod host–pathogen arms races, and we believe *Daphnia* should play a central role in such studies. The water flea may not be the host that immediately comes to mind for such studies; ideally, one would like a species for which extensive information on the genetic basis of host resistance is available. But obvious candidates like *Drosophila* and mosquitoes have a number of disadvantages. Most importantly, neither they nor any of their parasites can be cloned, and thus genotypes in the arms race cannot be extracted from the field and replicated for experimentation. In the case of *Drosophila*, there really are no established host–parasite systems, though parasitoids are well studied (e.g., Kraaijeveld and Godfray, 1999). For mosquitoes, we also lack much basic information about the ecology of their parasites in the field and, as is the case for malaria parasites, the parasites do not readily lend themselves to coevolutionary experiments or field manipulations. The major drawback of using *Daphnia* is of course the lack of knowledge at the molecular level, but efforts to accelerate *Daphnia* genomics are ongoing (Couzin, 2002; see also http://cgb.indiana.edu/genomics/projects/daphnia/).

EVOLUTIONARY DYNAMICS AND RELATED PHENOMENA
Showing genetic variation in host–parasite interactions, i.e., the potential for dynamics, has proven easier than actually showing a response to selection in almost all systems studied, including *Daphnia* (Little, 2002). This

is especially true for field studies, whereas the relatively simplified world of laboratory-based microcosms has shown more interpretable dynamic patterns. This is to be expected; the response to selection may be difficult to detect where genetic variance is small relative to the background of environmental variance, as is the case in noisy natural environments.

Host Dynamics

The first tests for genetic change over time due to parasites in the *Daphnia* system used molecular markers (allozymes) to test whether relatively over-parasitized host clones tended to decline in frequency whereas relatively uninfected hosts tended to rise in frequency (Little and Ebert, 1999). In some populations, gene frequency changes occurred in line with predictions based on genetic variation for resistance, but this was not so in other populations. However, those populations for which gene frequency changes did not appear to be caused by parasitism were those with low parasite prevalence and/or those containing a large number of rare clones such that statistical power to detect clonal fluctuations was weak. Thus, the overall impression from allozyme-based field studies was that parasites were indeed driving (sometimes extremely rapid) gene frequency changes and shifting populations towards genetic disequilibria (Figure 9.3).

However, testing of representative host population samples in the laboratory indicated that population mean resistance to the bacterium *P. ramosa* did not change markedly between (Little and Ebert, 2001) or within years (S. Mitchell, T. Little, and A. Read, personal communication, 2002), even though the populations studied possessed abundant genetic variation. For some populations and for some traits, small but significant increases in host resistance were detected between years, but the impression was hardly one of a rapid dynamic. Laboratory microcosm work (Capaul and Ebert, 2002), by contrast, showed strong parasite-mediated gene frequency changes, although this was for infection with two gut-infecting microsporidians (which reduce both mortality and fecundity by about 20%), and the (seemingly) more virulent *P. ramosa*, which usually renders hosts completely sterile, was not tested. The contrast between the microcosm results for the two microsporidians and the field-based work on *P. ramosa* is somewhat typical – the relatively simplified world of the microcosm yields results that seem meaningful, whereas studies that accommodate the full range of environmental uncertainties associated with the field yield results that would not have been expected based on naive interpretation of patterns of genetic variation for resistance.

Studies of a variety of trade-offs have shed light on responses to selection in the *Daphnia* system. First, it appears that resistance to one parasite species does not necessarily confer resistance to another (Capaul and

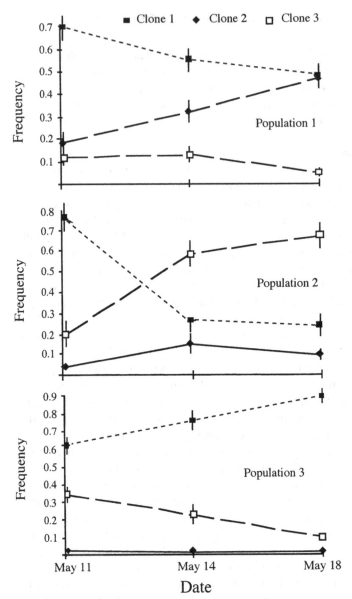

Figure 9.3. In many cases, *Daphnia* populations exhibit strong responses to parasite-mediated selection. Shown are data from three geographically close populations of *Daphnia pulex*. Each population was infected with a microsporidian and in each population "clone 1" (as identified by allozymes) was significantly overparasitized. In populations 1 and 2, clone 1 declined dramatically in frequency and the populations shifted towards genetic disequilibria. This was not so in population 3 where parasite prevalence was relatively low. Data are from Little and Ebert (1999).

Ebert, 2002; Decaestecker et al., 2003), although general resistance to a range of parasites does account for some hosts variation for resistance. The lack of strong genetic correlations in resistance indicates that selection by different parasites is independent, and this ought to provide considerable evolutionary flexibility for hosts faced with multiple enemies. It also appears that resistance to parasites is not traded-off against other fitness traits, for example, life-history traits or competitive ability (Capaul and Ebert, 2002; Little et al., 2001). Clonal organisms such as *Daphnia* provide a good opportunity to estimate fitness by measuring competitive ability. The preservation of genetic backgrounds via clonal reproduction permits the estimation of fitness in multi-generational competition assays which account for all relevant life-history variation so that even small fitness differences are amplified and detectable (Bell, 1997). Nevertheless, the isolates tested by Capaul and Ebert (2002) and by Little and Ebert (2001), which varied in resistance characteristics, did not differ in competitive ability in a way that was related to resistance. It may be that artificially selected (as opposed to the naturally occurring genotypes so far studied in *Daphnia*) lines offer the best opportunity to detect costs because their allocation to defense is sufficiently extreme that the traits against which resources have been traded-off can be detected.

Parasite Dynamics

One of the populations studied by Little and Ebert (2001) was tested for a change in the mean infectivity of the parasite (*Pasteuria ramosa*) population between years. As with the hosts in this population, these parasites had been the target of a study of strain variation for infectivity (Carius et al., 2001), and the potential for genetic change was also considerable. However, as with the hosts, there was little evidence of a strong dynamic. It is, again, possible that environmental noise and/or gene flow swamped adaptation and genetic change in the parasite population, although other explanations remain unexplored. For example, the parasite spore solutions that were used contained a mixture of parasite strains, so it is possible that interactions between parasite strains, as occurs in malarial infections (Taylor, Walliker, and Read, 1997), gave the spore solutions infective properties that did not reflect the natural situation. The incidence and dynamics of mixed infections in *Daphnia* are unstudied but potentially important issues. Future studies might investigate natural mixed infections by purifying strains obtained from single wild-caught individuals.

Experimental evolution in microcosms offers a method to test for parasite evolution over time in laboratory parasite populations without much of the noise of natural systems while retaining a reasonable measure of realism. Experimental evolution differs from selection experiments in at least two critical ways. Under experimental evolution, the experimenter does not select the traits of interest. Instead, the experiment imposes conditions upon populations of organisms and allows the organisms to find a solution by evolving towards a state that is better suited to the imposed conditions. By imposing particular conditions, one can make clear predictions about the course of evolution. The predictions used in studies of experimental evolution of *Daphnia* parasites are based on conventional theory regarding the evolution of virulence (Anderson and May, 1982; Ewald, 1983). Briefly, the theory assumes that a correlation between transmission and host survival (and therefore parasite survival) is based on an evolutionary trade-off for the parasite. If parasites have a low level of reproduction, this will have little impact on host longevity, but this results in a low level of transmission and fitness for the parasite. At the other extreme, a high level of parasite reproduction results in high transmission, but the period of transmission may be brief due to the reduced life span of the heavily diseased host. The evolutionary optimum for the parasite is to balance virulence and reproduction such that its fitness (i.e., transmission rate) is maximized over the course of infection. This optimum is influenced by, among other factors, expected life span of the host, within-host competition between parasite genotypes, host density, and the effectiveness of the host immune response (which is possibly related to host genotype).

Ebert and Mangin (1997) used experimental evolution to study the effect of changes in host demography on the virulence of the microsporidian *G. intestinalis*. In particular they predicted that parasite evolution due to increased host mortality should select for higher virulence and this ought to be evident as a faster parasite growth rate. Ebert and Mangin (1997) imposed various levels of extrinsic host mortality on infected host populations but culling. They observed, contrary to initial predictions, an increase in virulence in parasites from host populations with the longest life span but not in those infecting short-lived hosts. However, Ebert and Mangin (1997) provided evidence that within-host competition among parasite genotypes was greater in long-lived hosts. Within-host competition is thought to drive the evolution of higher virulence (Nowak and May, 1994), and thus this study of experimental evolution indicates the strength of parasite competition relative to the effect of host life span. In addition, this study confirmed that higher levels of parasite reproduction underpinned greater virulence (for discussion, see Ebert, 1998; Hochberg, 1998).

CONCLUSIONS AND FUTURE DIRECTIONS

Two major lines of theoretical endeavor into the genetics of host–parasite coevolution are

(a) modeling of frequency-dependent dynamics, which largely extends Haldane's (1949) verbal arguments about the evolutionary aspects of antagonist interactions, and

(b) modeling of the evolution of virulence, which principally concerns the trade-off for the parasite between transmission and optimal virulence, and the conditions under which virulence optima vary.

Empirical works on the *Daphnia*–microparasite system has principally tested predictions stemming from these two lines of inquiry.

Revealed is the widespread occurrence of genetic variation for host resistance and parasite infectivity in natural populations. Field studies have shown that certain genotypes are associated with disease, and laboratory work has verified that genetic variation is indeed an important determinant of infection. Evolutionary dynamics or a response to selection, be it in parasites or hosts, has been evident in some studies and under some conditions but not in others. Why this is so is not yet certain, but elucidating the tempo and mode of parasite-associated genetic change should be a priority. For some theories, it can be argued that if dynamics are sufficiently common and powerful enough to generate the evolutionary phenomena attributed to them, then rapid fluctuations in gene frequencies should be observable, both in the field and in real time. In particular, theoretical results that support the notion that coevolutionary interactions can select for sexual reproduction (the Red Queen hypothesis) usually require rapid allele frequency fluctuations (Howard and Lively, 1994; May and Anderson 1983; Peters and Lively, 1999).

We see the need for the following studies on factors which influence evolutionary dynamics:

(1) Studies of constraints on selection due to environmental factors. How large is the environmental variance around infection patterns in the field? How important are genotype (G) by environment (E) interactions? The most obvious environmental factor that could be used to explore G × E interactions is temperature. *Daphnia* certainly experience large temperature fluctuations on a daily and monthly basis, and the importance of interactions between temperature and infection in other arthropods has been shown (e.g., Blanford and Thomas, 2001). "Migration" from host or parasite "seed" banks (*Daphnia* produce

resting stages called epphipia, and many of their parasites also have resting stages) may also have a critical influence on evolutionary dynamics.

(2) Elucidating the genetic basis of resistance and infectivity should be a priority. Whether these traits are controlled by one, a few, or many genes has crucial bearing on the evolutionary significance of parasitism. In addition to knowing the number of genes involved, identifying those genes will be important. It is helpful that genome projects and functional analyses of immune-system genes are elucidating the genetic basis of arthropod host–pathogen interactions at an astonishing rate. However, it will also be critical to perform population genetic analyses (e.g., analyses of rates of synonymous to nonsynonymous nucleotide substitution) to establish which genes have been the target of diversifying pathogen-mediated selection. Variation due to arms races and other diversifying modes of pathogen-mediated selection underpins theory on the evolution of virulence. In particular, variation in infection rates and disease symptoms is attributable, in part, to genetic variation arising due to pathogen-mediated diversification. Therefore, to identify host genetic polymorphism involved in arms races in natural populations is to identify genes that determine virulence in the wild.

(3) Most of our ideas about *Daphnia* epidemiology stem from extrapolations from a particular set of models (Anderson and May, 1982; May and Anderson, 1983, 1990). We have little data on the factors that influence the spread of parasites in natural populations. Why do parasites disappear from populations despite the continual presence of susceptible hosts? Does mortality limit the transmission of parasites? How does competition among hosts and among parasites influence epidemics? Generally, what governs transmission dynamics? We require data bearing on these epidemiological questions to fully understand evolutionary dynamics.

In conclusion, although the evolutionary significance of parasitism remains uncertain, model systems such as the *Daphnia*–microparasite interaction have supplied a large measure of support for the notion that parasites have a strong influence on the genetic structure of host populations and that hosts in turn rapidly influence parasite adaptation. Much work remains to be done, but more than 50 years after Haldane first outlined his theory of disease and evolution, interaction with biological enemies remains an appealing explanation for the paradox of sexual reproduction and the maintenance of genetic variation in the wild.

REFERENCES

Anderson, R. M., and May, R. M. (1981). The population dynamics of microparasites and their invertebrate hosts. *Philosophical Transactions of the Royal Society, London, Series B* **291**, 451–524.

Anderson, R. M., and May, R. M. (1982). Coevolution of hosts and parasites. *Parasitology* **85**, 411–26.

Apanius, V., Penn, D., Slev, P. R., Ruff, L. R., and Potts, W. K. (1997). The nature of selection on the major histocompatibility complex. *Critical Reviews in Immunology* **17**, 179–224.

Begun, D. J., and Whitley, P. (2000). Adaptive evolution of relish, a Drosophila NF-kappa B/I kappa B protein. *Genetics* **154**, 1231–8.

Bell, G. (1997). *Selection: The Mechanism of Evolution*. Chapman & Hall, Florence, KY.

Bishop, J. G., Dean, A. M., and Mitchell-Olds, T. (2000). Rapid evolution in plant chitinases: Molecular targets of selection in plant-pathogen coevolution. *Proceedings of the National Academy of Sciences USA* **97**.

Blanford, S., and Thomas, M. B. (2001). Adult survival, maturation, and reproduction of the desert locust *Schistocerca gregaria* infected with the fungus Metarhizium anisopliae var acridum. *Journal of Invertebrate Pathology* **78**, 1–8.

Burdon, J. J., Thrall, P. H., and Brown, A. H. D. (1999). Resistance and virulence structure in two *Linum marginale-Melampsora* host-pathogen metapopulations with different mating systems. *Evolution*, 704–16.

Capaul, M., and Ebert, D., in press. Parasite mediated experimental evolution in *Daphnia magna* populations. *Evolution*.

Carius, H.-J., Little, T. J., and Ebert, D. (2001). Genetic variation in a host–parasite association: Potential for coevolution and frequency dependent selection. *Evolution* **55**, 1136–45.

Clark, A. G., and Wang, L. (1997). Molecular population genetics of Drosophila immune system genes. *Genetics* **147**, 713–24.

Clay, K., and Kover, P. (1996). The Red Queen hypothesis and plant/pathogen interactions. *Annual Review of Phytopathology* **34**, 29–50.

Couzin, J. (2002). NSF's ark draws alligators, algae, wasps. *Science* **297**, 1638–9.

De Meester, L., Gomez, A., Okamura, B., and Schwenk, K. (2002). The Monopolization Hypothesis and the dispersal-gene flow paradox in aquatic organisms. *Acta Oecologica-International Journal of Ecology* **23**, 121–35.

Decaestecker, E., De Meester, L., and Ebert, D. (2002). In deep trouble: Habitat selection constrained by multiple enemies in zooplankton. *Proceedings of the National Academy of Sciences USA* **99**, 5481–5.

Decaestecker, E., Vergote, A., Ebert, D., and De Meester, L., (in press). Evolutionary flexibility in a one host–multiple parasite system. *Evolution*.

Dybdahl, M. F., and Lively, C. M. (1995). Host-parasite interactions: Infection of common clones in natural populations of a freshwater snail (*Potamopyrgus antipodarum*). *Proceedings of the Royal Society of London, Series B* **260**, 99–103.

Dybdahl, M. F., and Lively, C. M. (1996). The geography of coevolution: Comparative population structures for a snail and its trematode parasite. *Evolution* **50**, 2264–75.

Ebert, D. (1994). Virulence and local adaptation of a horizontally transmitted parasite. *Science* **265**, 1084–6.

Ebert, D. (1995). The ecological interactions between a microsporidian parasite and its host *Daphnia magna. Journal of Animal Ecology* **64**, 361–9.

Ebert, D. (1998). Infectivity, dose effects, and the genetic correlation between within-host growth and parasite virulence. *Evolution* **52**, 1869–71.

Ebert, D., and Mangin, K. (1995). The evolution of virulence: When familiarity breeds death. *Biologist* **42**, 154–6.

Ebert, D., and Mangin, K. L. (1997). The influence of host demography on the evolution of virulence of a microsporidian gut parasite. *Evolution* **51**, 1828–37.

Ebert, D., Rainey, P., Embley, T. M., and Scholz, D. (1996). Development, life cycle, ultrastructure and phylogenetic position of *Pasteuria ramosa* Metchnikoff 1888: Rediscovery of an obligate endoparasite of *Daphnia magna* Straus. *Philosophical Transactions of the Royal Society, London, Series B* **351**, 1689–1701.

Ebert, D., Zschokke-Rohringer, C. D., and Carius, H. J. (1998). Within- and between-population variation for resistance of *Daphnia magna* to the bacterial endoparasite *Pasteuria ramosa. Proceedings of the Royal Society of London, Series B Science* **265**, 2127–34.

Ebert, D., Lipsitch, M., and Mangin, K. L. (2000). The effect of parasites on host population density and extinction: Experimental epidemiology with Daphnia and six microparasites. *American Naturalist* **156**, 459–77.

Ewald, P. W. (1983). Host-parasite relations, vectors, and the evolution of disease severity. *Annual Review of Ecology and Systematics* **14**, 465–85.

Frank, S. A. (1991). Spatial variation in coevolutionary dynamics. *Evolutionary Ecology* **5**, 193–214.

Frank, S. A. (1996a). Problems inferring the specificity of plant-pathogen genetics. *Evolutionary Ecology* **10**, 323–5.

Frank, S. A. (1996b). Statistical properties of polymorphism in host-parasite genetics. *Evolutionary Ecology* **307**–17.

Gemmill, A. W., Viney, M. E., and Read, A. F. (1997). Host immune status determines sexuality in a parasitic nematode. *Evolution* **51**, 393–401.

Green, J. (1974). Parasites and epibionts of Cladocera. *Transactions of the Zoological Society of London* **32**, 417–515.

Haldane J. B. S. (1949). Disease and evolution. *La Ricerca Scientifica* **19**, 2–11.

Hebert, P. D. N. (1974a). Enzyme variability in natural populations of *Daphnia magna*. II. Genotypic frequencies in permanent populations. *Genetics* **77**, 323–34.

Hebert, P. D. N. (1974b). Enzyme variability in natural populations of *Daphnia magna*. III. Genotypic frequencies in intermittent populations. *Genetics* **77**, 335–41.

Hill, A. V. S., Kwiatkowski, D., McMIchael, A. J., Greenwood, B. M., and Bennett, S. (1992). Maintenance of Mhc polymorphism – Reply. *Nature* **355**, 403–403.

Hochberg, M. E. (1998). Establishing genetic correlations involving parasite virulence. *Evolution* **52**, 1865–8.

Howard, R. S., and Lively, C. M. (1994). Parasitism, mutation accumulation and the maintenance of sex. *Nature* **367**, 554–7.

Hughes, A. L., and Nei, M. (1992). Maintenance of Mhc polymorphism. *Nature* **355**, 402–3.

Kaltz, O., and Shykoff, J. A. (1998). Local adaptation in host-parasite systems. *Heredity* **81**, 361–70.

Kraaijeveld, A. R., and Godfray, H. C. J. (1999). Geographic patterns in the evolution of resistance and virulence in Drosophila and its parasitoids. *American Naturalist* **153**, S61–S74.

Lajeunesse, M. J., and Forbes, M. R., in press. Host range and local parasite adaptation. *Proceedings of the Royal Society of London, Series B*.

Lazzaro, B. P., and Clark, A. G. (2001). Evidence for recurrent paralogous gene conversion and exceptional allelic divergence in the Attacin genes of *Drosophila melanogaster. Genetics* **159**, 659–71.

Little, T. J. (2002). The evolutionary significance of parasitism: Do parasite-driven genetic dynamics occur *ex silico? Journal of Evolutionary Biology* **15**, 1–9.

Little, T. J., Carius, H.-J., Sakwinska, O., and Ebert, D., in press. Competitiveness and life-history characteristics of *Daphnia* with respect to susceptibility to a parasite. *Journal of Evolutionary Biology*.

Little, T. J., and Ebert, D. (1999). Associations between parasitism and host genotype in natural populations of *Daphnia* (Crustacea: Cladocera). *Journal of Animal Ecology* **68**, 134–49.

Little, T. J., and Ebert, D. (2000a). The cause of parasitic infection in natural populations of *Daphnia*: The role of host genetics. *Proceedings of the Royal Society of London, Series B* **267**, 2037–42.

Little, T. J., and Ebert, D. (2000b). Sex, linkage disequilibrium and patterns of parasitism in three species of cyclically parthenogenetic *Daphnia* (Cladocera: Crustacea). *Heredity* **85**, 257–65.

Little, T. J., and Ebert, D. (2001). Temporal patterns of genetic variation for resistance and infectivity in a *Daphnia*-microparasite system. *Evolution* **55**, 1146–52.

Lively, C. M. (1989). Adaptation by a parasitic trematode to local populations of its snail host. *Evolution* **43**, 1663–71.

Lively, C. M., Craddock, C., and Vrijenhoek, R. C. (1990). Red Queen hypothesis supported by parasitism in sexual and clonal fish. *Nature* **344**, 864–6.

Mangin, K. L., Lipsitch, M., and Ebert, D. (1995). Virulence and transmission mode of two microsporidia in *Daphnia magna. Parasitology* **111**, 133–42.

May, R. M., and Anderson, R. M. (1983). Epidemiology and genetics in the coevolution of parasites and hosts. *Proceedings of the Royal Society, London, Series B* **219**, 281–313.

May, R. M., and Anderson, R. M., (1990). Parasite-host coevolution. *Parasitology* **100**, S89–S101.

Nowak, M. A., and May, R. M. (1994). Superinfection and the evolution of parasite virulence. *Proceedings of the Royal Society, London, Series B* **255**, 81–9.

Orr, H. A., and Irving, S. (1997). The genetics of adaptation: The genetic basis of resistance to wasp parasitism in *Drosophila melanogaster. Evolution* **51**, 1877–85.

Parker, M. A. (1994). Pathogens and sex in plants. *Evolutionary Ecology* **8**, 560–84.

Parker, M. A. (1996). The nature of plant-parasite specificity. *Evolutionary Ecology* **10**, 319–22.

Peters, A. D., and Lively, C. M. (1999). The red queen and fluctuating epistasis: A population genetic analysis of antagonistic coevolution. *American Naturalist* **154**, 393–405.

Schmid-Hempel, P. (1995). Parasites and social insects. *Apidologie* 26, 255–71.

Schwartz, S. S., and Cameron, G. N. (1993). How do parasites cost their hosts? Preliminary answers from trematodes and *Daphnia obtusa*. *Limnology and Oceanography* 38, 602–12.

Sorci, G., Moller, A. P., and Boulinier, T. (1997). Genetics of host-parasite interactions. *Trends in Ecology & Evolution* 12, 196–200.

Stahl, E. A., and Bishop, J. G. (2000). Plant-pathogen arms races at the molecular level. *Current Opinion in Plant Biology* 3, 299–304.

Stahl, E. A., Dwyer, G., Mauricio, R., Kreitman, M., and Bergelson, J. (1999). Dynamics of disease resistance polymorphism at the Rpm1 locus of Arabidopsis. *Nature* 400, 667–71.

Stirnadel, H. A. (1994). *The Ecology of Three Daphnia Species – Their Microparasites and Epibionts*. Master's thesis, University of Basel, Basel, Switzerland.

Stirnadel, H. A., and Ebert, D. (1997). The ecology of three *Daphnia* species – Their microparasites and epibionts. *Journal of Animal Ecology* 66, 212–22.

Taylor, L. H., Walliker, D., and Read, A. F. (1997). Mixed-genotype infections of the rodent malaria *Plasmodium chabaudi* are more infectious to mosquitoes than single-genotype infections. *Parasitology* 115, 121–32.

Thompson, J. N. (1994). *The Coevolutionary Process*. Chicago University Press, Chicago.

Tompkins, D. M., and Begon, M. (1999). Parasites can regulate wildlife populations. *Parasitology Today* 15, 311–13.

Wedekind, C., and Ruetschi, A. (2000). Parasite heterogeneity affects infection success and the occurrence of within-host competition: An experimental study with a cestode. *Evolutionary Ecology Research* 2, 1031–43.

TEN

Human Susceptibility to Visceral Leishmaniasis (*Leishmania donovani*) and to Schistosomiasis (*Schistosoma mansoni*) Is Controlled by Major Genetic Loci

A. Dessein, B. Bucheton, L. Argiro, N. M. A. Elwali, V. Rodrigues, C. Chevillard, S. Marquet, H. Dessein, S. H. El-Safi, and L. Abel

SUSCEPTIBILITY TO SEVERE DISEASE DURING AN OUTBREAK OF VISCERAL LEISHMANIASIS IN A SUDANESE VILLAGE

Visceral leishmaniasis (VL) is caused by protozoan parasites of the *Leishmania* genus that are transmitted to humans by infected sand-flies (Figure 10.1a). Parasites rapidly invade host phagocytes and multiply inside phagolysosomes. Clinical disease is primarily due to the uncontrolled multiplication of the parasite in many organs including the spleen and the liver; clinical symptoms include recurrent fever, considerable splenomegaly, hepatomegaly, and adenopathy (1). Death is certain if the patient is left untreated. Violent outbreaks of VL have occurred in regions of eastern Africa (2) (Kenya, Sudan) and in India (Bihar). VL is endemic in South America and in the Mediterranean basin. VL is caused by three *Leishmania* species: *Leishmania donovani*, *L. chagasi/ infantum*, and *L. archibaldi* (*L. donovani* being the most pathogenic) (3). To identify the principal risk factors in VL, we carried out a five-year longitudinal study on 1,600 subjects from a village located on the Sudanese–Ethiopian border. The study was initiated in 1995 when the number of VL cases had just begun to rise in the village (4). Within five years, 28% of the population had been affected by VL. Most of these VL patients were treated and cured. Unfortunately, a small percentage either failed to respond to treatment or were not treated because of the difficulties involved in reaching them during the rainy season.

The immunological evaluation of one-third of the village population showed that, although they had not developed VL, healthy subjects had been exposed to infection, as 80% of them exhibited *Leishmania*-specific antibodies (Ab) (5). Individuals of less than 30 years, who were born in the

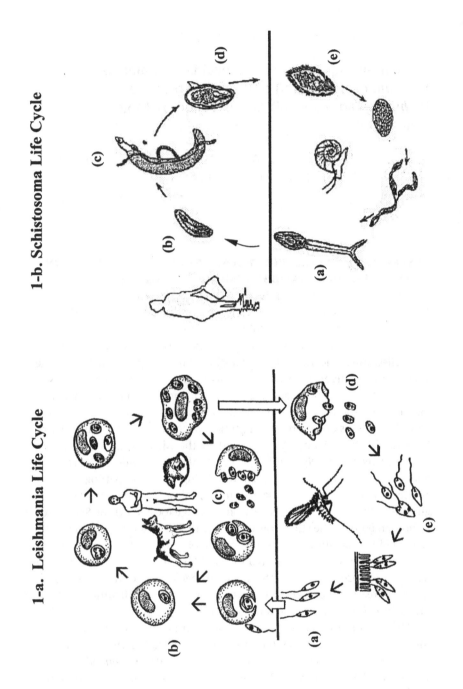

1-a. Leishmania Life Cycle

1-b. Schistosoma Life Cycle

242

village, had to have been exposed to the infection during this epidemic. The older subjects could have been exposed to VL during a previous outbreak that occurred in the area sóme 40 years ago or before settling in the village. As exposure did not differ much according to age or gender, it is likely that transmission occurred in the village itself and not in the forests near the village. Consistent with this hypothesis, large numbers of *Phlebotomus orientalis*, the vector of *Leishmania donovani* in Sudan, were found inside huts in the village (6). We looked for eco-epidemiological factors that could account for phenotypic heterogeneity (VL, asymptomatic) in this population: the presence of certain animals near households and certain types of vegetation were found to increase the risk of VL. The most important risk factors were the ethnic and family origins (7), which suggests that hereditary factors are important determinants in the development of disease. The study population comprised three ethnic groups: the Aringas, the Haoussas, and the Fellatas. The incidence of VL was higher in the Aringas than in the two other ethnic groups at all times during the outbreak (Figure 10.2). Interestingly, the incidence of VL was strongly age-dependent in the Aringas but not in the other two groups. This suggests that these ethnic groups harbor specific susceptibility factors as these tribes live close to each other in the village and probably experienced similar conditions of exposure to infection (7). Furthermore, young Aringas were especially susceptible, suggesting that children harbor some specific susceptibility factors. This view is also supported by the observation that adults developed VL late in the outbreak, whereas almost 20% of

---

Figure 10.1. (a) Leishmania Life Cycle. When the sandfly feeds, metacyclic promastigotes (**a**) are delivered into the vertebrate host where they are phagocytosed by macrophages. The promastigotes are then converted into amastigotes (**b**) within the phagolysosome and undergo multiple rounds of binary fission. The infected macrophages lyse (**c**) and release the amastigotes, which are taken up by other macrophages, thus reinitiating the replication cycle. Infected macrophages are also ingested by the sandfly (**d**). Amastigotes are released and convert to promastigotes in the gut of the sandfly (**e**) where they replicate by binary fission and migrate to the foregut. Biochemical changes in the promastigote surface and slight morphological changes occur as the parasite stops replicating and develops into infective metacyclic promastigotes. (b) Schistosoma Life Cycle. Infection occurs when cercariae (**a**) penetrate the skin while the human, the definitive host, is in the water. The cercaria loses its tail and becomes a schistosomule (**b**), which moves through various tissues to the mesenteric veins where it matures into an adult worm (**c**). Females release eggs (**d**), which are eliminated in the feces. When it comes into contact with water, the egg hatches and liberates a miracidium (**e**), which infects the mollusc vector (*Biophalaria glabrata*).

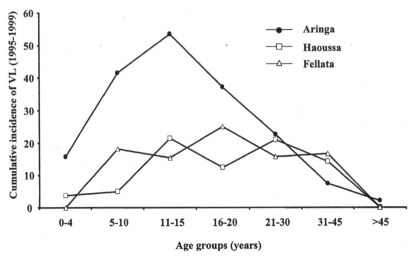

Figure 10.2. The incidence of VL during an outbreak in Barbar El Fugara varied according to ethnic group. Cumulative incidence of VL per age group during the outbreak period among the population living in the central district of the village: 426, 133, and 101 subjects belonged to the Aringa, Haoussa, and Fellata ethnic groups, respectively. The cumulative incidence of VL was higher in the Aringas than among the Haoussas and Fellatas. Among the Aringas, children and teenagers were the most affected, whereas among the Haoussas and Fellatas, all age groups had a similar risk of developing the disease.

children and teenagers were infected just two to three years after the start of the outbreak (Figure 10.3). We shall see that the results of the genetic analysis also support the view that early and late VL (during the outbreak) might be due to different susceptibility factors. In addition, some subjects developed VL one or two years after infection had been detected by serological tests (5).

SUSCEPTIBILITY TO VISCERAL LEISHMANIASIS IS DETERMINED BY A MAJOR SUSCEPTIBILITY LOCUS

The analysis of the distribution of VL cases in large pedigrees failed to reveal a Mendelian pattern of segregation, and a segregation analysis performed on these families did not find a simple Mendelian model fitting the distribution of VL cases (Bucheton *et al.*, unpublished data).

Nevertheless, studies in mice convincingly demonstrated that the susceptibility of inbred strains of mice to infection by *L. donovani* is controlled by the *Ish* locus (8, 9), which contains *Nramp1* (10), in the early phase of the infection, and by the *HLA* locus (11) in the later phase. The *Nramp1*

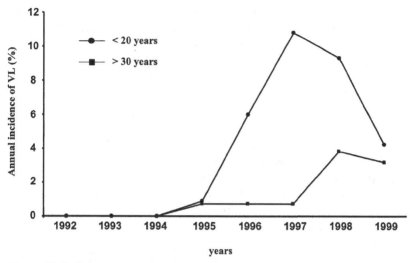

Figure 10.3. Younger subjects developed VL early in the outbreak, whereas older adults developed the disease toward the end of the outbreak. Annual incidence of VL during the outbreak for the population of the central district of Barbar El Fugara for subjects aged less than 20 years (396 subjects) and for subjects aged more than 30 years (165 subjects).

product is a divalent cation transporter found in the membranes of intracellular vesicular compartments (12, 13). It is thought to regulate microbicidal enzyme activities in phagolysosomes where mycobacteria and *Leishmania* multiply (14). Allelic forms of the human orthologue or *NRAMP1* (15) have been shown to be associated with increased susceptibility to human tuberculosis (16) and leprosy (17, 18), and a *NRAMP1* promoter polymorphism has been shown to be associated with different levels of mRNA *in vitro* (19). This is reviewed in another chapter of this book.

For these reasons, we tested whether the *NRAMP1* gene region was linked to VL in our multicase families. The analysis was carried out for *NRAMP1* together with four other regions containing candidate genes (20). The results show evidence for linkage ($p = 0.01$) between VL and the *NRAMP1* region. These data, however, require confirmation in a larger population sample to increase the statistical significance.

To conduct a more general search for a susceptibility locus, we performed a linkage analysis on affected siblings and screened the whole patient genome using microsatellite markers providing a 10–20 cM linkage map. Linkage was tested by the maximum-likelihood binomial method that allows the inclusion of more than two affected siblings from the same family and does not need to decompose sib-ships with more than two affected into constitutive sib-pairs (21). The study was performed in two steps: first, the genomes of 106 affected siblings belonging to 38 families were scanned. Three chromosomal regions, including the *NRAMP1* gene region, yielded suggestive lod-scores (lod-score > 0.83, p < 0.025) in this first scan and were further analyzed in 63 additional siblings (25 families) using 13 additional markers. The lod-scores of two of these three regions did not increase after the second analysis, whereas the lod-score of the third region reached statistical significance (lod-score = 3.5, $p = 3 \times 10^{-5}$). Our epidemiological observations suggested that the susceptibility factors that determine VL early in the outbreak differ from those that determine late VL cases, thus we repeated the analysis on early VL cases only. This restricted the study to 52 informative families and increased the lod-score to 3.9 ($p = 10^{-5}$). Finally, we tested gene–gene interactions between this susceptibility locus and the two other chromosomal regions with suggestive lod-scores. A conditional analysis (22), assuming negative gene–gene interactions (i.e., performed on the subgroup of 30 families whose lod-scores were negative at the susceptibility locus), suggested the existence of a second susceptibility locus. These data must be confirmed, however, on a larger patient sample. Thus, the linkage analysis provided strong evidence for a susceptibility locus and suggested the existence of a second locus on a distinct chromosome. If we can confirm that genetic heterogeneity plays a role in the control of VL in this population, this result might explain why we failed to define a simple Mendelian model by segregation analysis. Furthermore, as our study was restricted to the most susceptible ethnic group (the Aringa), it would be interesting to determine whether either of these two susceptibility loci play a role in susceptibility to VL in the less susceptible tribes (the Fellata and Haoussa). This will require a larger sample of subjects belonging to both tribes.

The observation that a strong genetic effect accounts for a large number of VL cases in Barbar's outbreak suggests that host genetics play an important role in outbreaks, determining to a large extent which subjects are at risk of severe disease. In the absence of treatment, most VL cases die. Because these are generally children or young teenagers, VL likely exerts a strong negative selective pressure on the susceptibility alleles in exposed populations. In this regard, the Haoussa and Fellata tribes had experienced a VL outbreak before this one and were less affected by the present outbreak than the Aringa, who immigrated to the village more recently. This observation suggests that previous exposure might have selected resistant genotypes in Fellata and Haoussa tribes. Comparing susceptibility allele frequencies in these populations and in populations from which these tribes originated could test this hypothesis.

GENETIC CONTROL OF *SCHISTOSOMA MANSONI* INFECTION
LEVELS BY GENES ON CHROMOSOME 5q31-q33

Schistosomiasis is caused by various species of schistosomes that have different geographical distributions, vectors, and tissue tropisms and cause different diseases. *Schistosoma mekongi, S. japonicum,* and *S. mansoni* have a tropism for the mesenteric veins and the portal veins and cause intestinal and hepatosplenic diseases; *S. hematobium* lives in the vessels of the urinary system and causes diseases of the urinary tract that culminate in hydronephrosis. All schistosomes can be lethal. The mortality rate associated with *S. hematobium* is, however, underestimated, as it often occurs after urinary infections in subjects with advanced hydronephrosis in both kidneys. Patients with hepatic schistosomiasis often die following esophageal bleeding and abdominal ascites. Schistosomes are 1–2 cm flat worms that live in the vascular system of their vertebrate hosts, where the female worms lay large numbers of eggs. The eggs must find their way out through the intestinal or bladder walls to reach fresh water, where they hatch, liberating a swimming larva that infests freshwater snails. The parasite multiplies asexually in the snail, producing thousands of cercarial larvae. Cercariae are emitted into the water, where they swim in search of the skin of their vertebrate host. They penetrate skin in a few minutes and then migrate to the vascular system where they develop into adult schistosomes in 5 to 6 weeks (the *Schistosoma* life cycle is presented in Figure 10.1b). Schistosome eggs that remain in the host tissues (intestine, bladder, ureter, liver) cause inflammation that may lead to severe clinical disease such as hydronephrosis, splenomegaly, portal hypertension, varices, and ascites in certain subjects (23, 24).

It was long thought that severe infections were mostly due to epidemiological factors, in particular the exposure of the patient to infected waters. We and others have shown that this is not the case in endemic areas and that severe infections are in fact due to the individual's capacity to resist infection rather than the level of exposure to infective cercariae (25–28). We and Hagan also demonstrated that resistance to infection is probably mediated by a Th2-T cell response involving IgE antibodies (29–32). It has also been demonstrated that protective immunity develops more slowly with age and that IgG4 antibodies counterbalance the action of IgE, probably by competing for the same antigenic determinants on the parasite (29, 33, 34). Most interestingly, subjects who are highly susceptible to infection were found to be clustered in certain families, whereas resistant individuals were concentrated in others (28). This suggested that a hereditary factor, possibly a genetic one, affects significantly the degree of immune resistance to schistosome infection. Thus, we tested the hypothesis that a single susceptibility locus controls infection levels. This analysis was performed taking into account the effects of other covariables that had previously been found to affect infection (28). We tested whether a genetic model with a major gene was better than a sporadic model that included only environmental factors age, and gender as explanatory variables to account for the phenotype (infection levels) distribution in families (35). The whole study required only 400 subjects. The results showed that the observed data were best explained by the effects of a major codominant gene (SM1), and the frequency of the susceptibility allele was calculated to be 0.22. According to this model, 6–8% of study subjects were homozygous susceptible, 32% were heterozygous, and 60–62% were homozygous resistant (35). The effects of this locus accounted for more than half of the variation in infection levels, which is considerable. Thus, infection levels are probably under strong genetic control, and this control is simple because all the heritability components of infection levels could be accounted for by a single major locus. To confirm the existence of this gene, we mapped it by genotyping 256 genetic markers, evenly distributed on the human chromosomes, in 153 subjects belonging to 11 families. This rather small sample of informative families was sufficient to detect a strong linkage between the major gene and markers of the 5q31q33 region (36). Thus, for the first time for an infectious disease, a wide genome scan was used to map a susceptibility gene. This result was confirmed by Muller-Myhsok et al. (37) who showed that high rates of infection by S. mansoni in a Senegalese population was also determined by gene(s) in the 5q31q33 chromosomal region.

The 5q31q33 region contains a cluster of cytokines that play a central role in the immune response including the Th2/Th1 differentiation. These cytokines include the granulocyte–macrophage colony stimulating factor (CSF2), several interleukins (IL-13, IL-3, IL-4, IL-5, IL-9, IL-12), the interferon regulatory factor 1 (IRF1), and the colony stimulating factor-1 receptor (CSF-1R). This cluster also contains gene(s) controlling total serum IgE levels and familial hypereosinophilia (38–40). Thus, based on the results of immunological studies on these subjects, we felt confident that the susceptibility genes were one of those obvious candidates. Nevertheless, all our attempts to demonstrate this point have failed. This is probably because the genetic control does not involve only one gene and because the control at this locus is more complex than expected; our small sample size is insufficient to resolve this complexity. It was not possible to increase the size of the sample because the implementation of mass control programs made it difficult to define the infection phenotypes of a large sample of additional individuals. We are now studying individuals from other countries where such studies are feasible, which should make it possible to re-evaluate this question.

PHENOTYPIC HETEROGENEITY IN PERIPORTAL FIBROSIS IN SUBJECTS INFECTED BY *S. MANSONI*

As stated above, the eggs cause the disease associated with schistosome infections. Eggs trapped in the host tissue secrete a number of substances, such as proteolytic enzymes that are toxic for the surrounding cells. Damaged endothelial cells release inflammatory substances, as do activated platelets. These mediators initiate an inflammatory reaction that, in experimental models, is massively infiltrated by eosinophils (41). Macrophages, T lymphocytes, and B lymphocytes are also present in the inflammation. This periovular reaction, which persists for several weeks due to the resistance of the egg shell to immune attack, is referred to as a granuloma (23, 24). The early stage of the granuloma is inflammatory-necrotic. It then becomes a fibrotic granuloma. These changes are regulated by a variety of cytokines and lipid-derived mediators (42, 43) produced by local tissue cells, such as hepatocytes, endothelial cells, and Kupfer cells, and by inflammatory cells, such as eosinophils, monocytes/macrophages, T-lymphocytes, and tissue mast cells (44–46).

In mice and humans with acute infections, the granuloma is down-regulated four to six weeks after the beginning of oviposition. This down-regulation probably involves the anti-inflammatory cytokine IL-10 (43, 47). The cytokine profile of the periovular cellular reaction is Th1-like in the early stages and Th2-like in the later stage. At least two important regulatory events affect the granuloma; one regulates the progression from the inflammatory stage to the fibrotic stage, and the other occurs when the initially acute infection becomes chronic. It is thought that either one or both of these regulatory mechanisms causes abnormal fibrosis. Fibrosis is the abnormal accumulation of extracellular matrix proteins (ECMP) into a dense and cross-linked network in a tissue. These ECMP (e.g., laminin, collagen, connectin) are produced by stellate cell (Itocell-derived) myofibroblasts. Stellate cells are activated by substances such as PGDF, which are released by damaged hepatocytes, endothelial cells, and activated platelets. Myofibroblasts are regulated by a variety of cytokines such as IFN-γ, which inhibits their multiplication and the production of ECMP (48–53). Conversely, IL-13, TGF-β, and IL-4 stimulate fibroblast division and ECMP production (54, 55). Thus, IFN-γ is anti-fibrogenic, whereas TGF-β and IL-13 are fibrogenic. Tissue fibrosis results from the excessive production of ECMP and/or the insufficient turnover of the fibrotic tissue,

Figure 10.4. Hepatic vascular trees of a liver from a healthy subject who died in a car accident (left) and of a liver from a subject who died from severe PPF due to *Schistosoma mansoni* infection (right). The hepatic vascular tree was injected with acrylic blue liquid resin. After polymerization, tissues were digested. Courtesy of Dr. E. Chapadeiro, Faculty of Medicine Uberaba, Brazil.

which depends on the action of metalloprotease (MP)/MP inhibitors (MPI), the synthesis and activities of which are also regulated by the above-mentioned cytokines, i.e., IFN-γ stimulates the production of MP and inhibits the synthesis of MPI (56).

The clinical presentation of schistosome infections in a population living in an endemic area is heterogeneous; most subjects do not exhibit marked clinical symptoms, whereas 5–20% of the population infected by *S. mansoni* may be affected by a marked hepatosplenomegaly, abdominal ascites, and bleeding from esophageal varices. These severe clinical symptoms usually result in death in countries where patients cannot be treated properly at the hospital; death is the consequence of an elevated portal blood hypertension caused by a massive periportal fibrosis (PPF) (Symmers fibrosis) (57). In these subjects, the fibrosis invades the space of Disse and the periportal space, occludes many branches of the portal vein, and eventually twists major branches of this vein, dramatically reducing the flow of blood through the portal system (Figure 10.4) (44).

To investigate risk factors in Symmers fibrosis, we conducted a study on the population of a Sudanese village in the Gezira region. This area is extensively irrigated and the village is surrounded by a dense network of water canals that are populated by *Biophalaria glabrata* snails, a vector of *S. mansoni*. This parasite is endemic in the whole province of Gezira. No mass treatments have been carried out for 30 to 40 years and the prevalence of infection is greater than 70% in the study village (57). This high prevalence can be explained by the fact that the domestic water supply is not correctly treated and by the fact that almost all of the inhabitants of the village are farmers who work in the irrigated fields where they are frequently in contact with infected water.

To evaluate PPF, abdominal ultrasonography was performed on the whole population (750 subjects). PPF was observed in a large number of subjects and evidence of severe disease (PPF + portal hypertension) (grade III fibrosis) was recorded in 2.5% of the study population. Twenty-eight percent exhibited no signs of PPF and 59% had either mild fibrosis or focal periportal inflammation (grade I disease). Interestingly, advanced PPF was only observed in adults, whereas grade I disease was observed in subjects of all ages and in a large number of children (57). This indicates that disease is not the result of a linear cumulative process and that some critical factors are probably responsible for the progression of disease to the severe clinical stages. As experimental models of schistosomiasis and fibrosis caused by other etiological agents showed that certain cytokines regulate the production and the degradation of ECMP, we evaluated cytokine production by the blood mononuclear cells in our study subjects. We studied all of the cytokines involved in ECMP regulation, except TGF-β. Only IFN-γ and IL-1β were found to be associated with severe fibrosis (PPF with portal hypertension). A multivariate analysis of the cytokine data, including other epidemiological covariates such as age, gender, and infection level, indicated that IFN-γ is the only cytokine strongly associated with disease (58). The risk of severe PPF was about ten-fold higher in subjects with the lowest IFN-γ levels than in high IFN-γ producers (median IFN-γ levels were used as a cutoff value to define low and high producers). These data also suggested that high TNF-α producers are at higher risk of disease than low TNF-α producers (58).

Analysis of the ultrasound data indicated that severe PPF was more prevalent in certain families. As an example, 30% of the individuals of a large pedigree presented advanced or severe fibrosis, whereas the prevalence of the disease in the whole population was 8%. Furthermore, the risk of PPF was increased in children born from parents with the disease, whereas the clinical phenotypes of the parents varied independently of each other. These observations suggested the existence of a genetic component in the control of PPF (57). This hypothesis was tested by segregation analysis.

SEVERE HEPATIC FIBROSIS IN PATIENTS WITH *S. MANSONI* INFECTION IS CONTROLLED BY A MAJOR LOCUS (SM2) THAT IS CLOSELY LINKED TO THE INTERFERON-γ RECEPTOR (*IFNGR1*) GENE

Segregation analysis was performed using regressive logistic models (59). These models specify a regression relationship between the probability of a person being affected (i.e., having a severe fibrosis) and a set of explanatory variables including major genotype, phenotypes of parents, and other co-variates.

The hypothesis of no familial dependence was rejected. A child with an affected parent was four-fold more likely to be affected than a child with an unaffected or an unknown parent. In the presence of parent–offspring dependence, there was evidence for a codominant major gene (SM2). Both the recessive and the dominant hypotheses for the major gene effect were rejected. Parent–offspring dependence residual from the codominant major gene was not significant. The frequency of the deleterious allele A was estimated at 0.162; consequently, the respective proportions of AA, Aa, and aa subjects were 0.03, 0.27, and 0.70.

To localize SM2, all informative families with multiple cases of severe fibrosis (8 families including 112 individuals) were genotyped, and linkage analysis was conducted in four candidate regions:

(1) the 5q31q33 region where SM1 and several candidate genes are located (see above);
(2) the HLA–TNF region (6p21) containing the HLA locus and the TNF-α and TNF-β genes;
(3) the 12q15 region including the IFN-γ gene and a gene controlling total serum IgE levels; and
(4) the 6q22q23 region containing the IFNGR1 gene. No maximum lod-score (Zmax) values exceeded +1 with any markers of regions 5q31q33 and 6p21 and 12q15. In contrast, significant Zmax values were observed in region 6q22q23 with D6S310 (Zmax = +2.81 at θ = .0) and the FA1 intragenic marker (Zmax = +1.80 at θ = .0) (59).

A combined segregation–linkage analysis was performed with the D6S310 marker to investigate further our linkage results with the 6q22q23 region. The results of this analysis showed that the hypothesis of no linkage with D6S310 could be rejected. According to the model accounting for linkage with D6S310, the maximum-likelihood estimate of θ was 0.0 and the lod-score was +3.11. The parameters of SM2 (penetrance and allelic frequency) estimated in this combined segregation–linkage analysis were very close to those estimated in our previous segregation analysis (described above). Figure 10.5 shows the penetrance with duration of exposure for AA subjects and Aa males as predicted by the model accounting for linkage. For AA males the penetrance is almost complete after 12 years, and for AA females the penetrance is almost complete after 17 years, whereas for Aa males the penetrance is 0.73 after 20 years of exposure. For

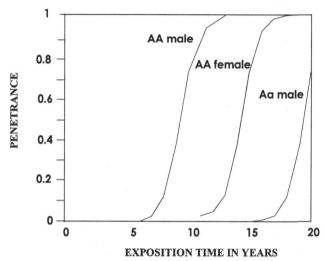

Figure 10.5. Penetrance of genotypes at the SM2 locus in the Al Taweel population (A is the high morbidity allele). The genetic model, defined by segregation analysis, indicates that 50% of the penetrance of the gene is reached after 9.14 years of residence in the area for AA males and after 14 years for AA females. The penetrance remains low after 20 years of exposure for other subjects (see, for example, Aa males). Nevertheless, with a sufficiently long period of residence, all heterozygous males are likely to develop the disease. Consequently, in this population, 30% (3% of homozygous and 27% of heterozygous) of males could potentially develop severe schistosomiasis if left untreated. The estimated penetrance of SM2 strongly depends on gender; the penetrance of fibrosis was lower in females than in males.

aa males the penetrance reaches 0.02 after 20 years of exposure and is lower than 0.01 before; for Aa and aa females the penetrance remains lower than 0.001 after 20 years of exposure.

This result shows that the major locus controlling fibrosis is not linked to chromosome 5q31q33 and maps to another region. It also demonstrates that anti-disease immunity and anti-infection immunity are controlled by distinct major genes. Obviously our results do not rule out an interaction between *SM1* and *SM2*. It is reasonable to postulate that disease development is accelerated in *SM2* subjects predisposed to high infections. However, the low and rather uniform levels of infection in our study population did not allow us to search for *SM1* and to evaluate possible interactions between *SM1* and *SM2*. Our linkage results obtained with two microsatellite markers of the *HLA–TNF* region also indicate that the major locus *SM2* is very unlikely to be located within this region, whereas associations have been reported between

(1) some HLA class I alleles (A1 and B5) and hepatosplenomegaly in Egypt (60–62) and
(2) an HLA class II allele (DQB1*0201) and biopsy-confirmed hepatic schistosomiasis in Brazil (63). However, it should be stressed that these results do not exclude the possibility that additional polymorphisms such as specific HLA antigens play a role. Such polymorphisms could be detected by carrying out an appropriate association study in this population.

It is too early to speculate on the mechanism of action of *SM2*. However, given the results of multipoint linkage analysis which mapped the susceptibility locus close to the *IFNGR1* gene, polymorphism(s) in the IFN-γ receptor gene may account for increased susceptibility to severe fibrosis. This hypothesis is consistent with various reports showing the strong anti-fibrogenic activity of IFN-γ as discussed above (50, 51).

CONCLUSION

One interesting finding from our studies is that susceptibility to parasites in humans living in endemic areas or exposed to an epidemic is largely determined by the genetic make-up of the host, despite the well known plasticity of parasite genomes that allow genetically different pathogens to merge and to escape host defenses raised against antigenically different individuals. Clearly, schistosomes and leishmania do not show the same genetic variability as trypanosomes and plasmodia, and it remains to be determined whether this conclusion also applies to these rapidly evolving pathogens.

Our studies also indicate that infections and diseases caused by parasites are controlled by major genetic effects that involve a small number of genetic loci. The demonstration that severe infection and advanced hepatic diseases in schistosome infections segregate in families as a Mendelian trait is a striking illustration of this fact. This is good news, as it was feared that genetic control would involve a large number of genes with small effects that would be difficult to analyze. This conclusion may not apply to all infected populations, and it is obvious that our choice to reduce genetic heterogeneity among our study populations as much as possible has simplified the genetic analysis.

The strong genetic control of schistosome infection was quite unexpected. We undertook our study to investigate the factors that accounted for the large variance in infection levels, with the idea that most of this variance was produced by heterogeneous living conditions including exposure to infected water. This was the explanation given by most epidemiologists when we started our study. The surprise came when we observed that above a certain level of exposure, which was reached by most subjects living in the study area, infections were strongly affected by human immune susceptibility to this parasite. The same observation was made by Butterworth *et al.* (25) and by Hagan *et al.* (29) working in Africa. This was a turning point. When we observed clusters of severe infections in certain families, it was clear that host genetics were probably playing an important role in the control of schistosome infection.

It would be a mistake to translate control by a major genetic locus into control by a single gene. Our data do not rule out the existence of more than one susceptibility gene at each major locus (genetic heterogeneity). Furthermore, it is even more likely that allelic heterogeneity will be the rule even if control is exerted by a single gene at the major locus. This complexity probably explains why we have not yet been able to identify the susceptibility gene(s) at the *SM1* and *SM2* loci (Figure 10.6). Indeed, our study populations have been much too small to make it possible to analyze these loci by SNP mapping. It has not been possible to increase the size of our study population in Brazil, as discussed above. Finally, these findings raise the important issue of whether the susceptibility gene(s) are the same in different populations. On the one hand, one may speculate that selective pressure will reduce the number of possible "genetic solutions." For example, the number of mutations altering the expression level of the genes encoding Th1/Th2 cytokines is probably limited by the fact that these cytokines are critical in various infections that often affect the same individuals: Th2 cytokines are vital for limiting host invasion by schistosomes, whereas they facilitate infections by *Leishmania*. Thus, in populations such as the Sudanese populations in which we conducted our studies, not all alterations of *IL-4*, *IL-13*, or *IFN-γ* genes are possible. On the other hand, it is clear that mutations in many different genes will have the same "global" effect. For example, mutations in *IFN-γ* and its receptor could have the same consequences on fibrosis. This may also (in part) apply to Stat4, which is the transcription factor that regulates IFN-γ regulated genes. This has been well illustrated in studies on mycobacterial infections (64). Thus, heterogeneity is certainly the rule in genetic control of susceptibility to infections, but the presence of many pathogens in the same populations has probably exerted major constraints on the selection of resistant/susceptibility alleles.

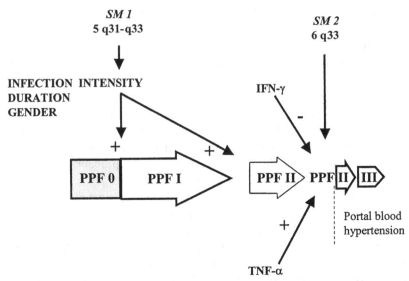

Figure 10.6. Summary of the genetic control of susceptibility to human schistosomiasis. The disease is initiated by eggs trapped in liver sinusoids and amplified by antigens released by adult worms. The first stage, perivascular inflammation, is associated with the deposition of small amounts of ECMP (grade I (FI) according to the WHO scale). In some subjects, ECMP accumulates around the secondary branches of the portal vein as long stretches of fibrosis (stage II); in some subjects, ECMP accumulates further and the fibrosis extends to the periphery of the organ, occludes certain secondary branches, and twists major branches; the gallbladder is also affected (stage III). Portal hypertension (PH), as detected by abnormal portal and splenic vein diameters, is observed in some FII subjects and in all FIII subjects. *SM2* accelerates the progression from FII to FIII + PH or to FII + PH. The protective role of IFN-γ is indicated by the association between FIII and low IFN-γ production and by various studies showing that IFN-γ decreases ECMP production and increases ECMP degradation. Other factors important in fibrosis are the duration of infection, the intensity of infection, and gender (males are more susceptible than females). The intensity of infection is probably most important during the early stage of the disease (FI) and is controlled by *SM1*, which probably acts on Thelper-cell differentiation.

Finally, it is quite stimulating that the available genetic tools have made it possible to analyze the genetic control of complex phenotypes in multifactorial parasitic diseases. It is particularly striking that the first successful genome scan to be performed in an infectious disease involved a parasitic infection (schistosomiasis). Finally, the demonstration of the strong effect of a major locus in a visceral leishmaniasis outbreak is extremely encouraging for further studies on the mechanisms of pathogenesis and vaccination in this lethal disease. It may also change our way of thinking about the key factors involved in the initiation and termination of outbreaks in epidemics.

ACKNOWLEDGMENTS

Financial assistance for this work was provided by the Institut National de la Santé et de la Recherche Médicale, the World Health Organization (ID096546), the European Economic Community (TS3CT940296, IC18CT970212), Scientific and Technical Cooperation with Developing Countries (IC18CT980373), the French Ministère de la Recherche et des Techniques (PRFMMIP), the Conseil Général Provence Alpes Côte d'Azur and the Conseil Régional Provence Alpes Côte d'Azur. BB and CC are supported by fellowships from the French Ministère de la Recherche et des Techniques, from the Conseil Général PACA, and from the Fondation pour la Recherche Médicale.

REFERENCES

1. Dessein AJ, Chevillard C, Marquet S, Henri S, Hillaire D, Dessein H. 2001. Genetics of parasitic infections. *Drug Metab. Dispos.* 29(4 Pt 2):484–8.
2. Seaman J, Mercer AJ, Sondorp E. 1996. The epidemic of visceral leishmaniasis in western Upper Nile, southern Sudan: Course and impact from 1984 to 1994. *Int. J. Epidemiol.* 25(4):862–71.
3. Desjeux P. The increase in risk factors for leishmaniasis worldwide. 2001. *Trans. R. Soc. Trop. Med. Hyg.* 95:239–43.
4. El-Safi SH, Bucheton B, Kheir MM, Musa HA, El-Obeid M, Hammad A, and Dessein A. 2002. Epidemiology of visceral leishmaniasis in Atbara River area, Eastern Sudan: The outbreak of Barbar El Fugara village (1996–1997). *Microb. Infect.* 14:1439–47.
5. Bucheton B, El-Safi SH, Hammad A, Kheir MM, Eudes N, Mergani A, and Dessein A. 2002. Anti-*Leishmania* antibodies in an outbreak of visceral leishmaniasis in Eastern Sudan: High antibody responses occur in resistant subjects and are not predictive of disease. *Trans. R. Trop. Med. Hyg.* (in press).
6. Lambert J, and 7 colleagues. 2002. The sand fly fauna in the visceral leishmaniasis focus of Gedaref (Atbara River area, Eastern Sudan). *Ann. Trop. Med. Parasitol.* 96(6):631–6.
7. Bucheton B, Kheir MM, El-Safi SH, Hammad A, Mergani A, Mary C, Abel L, and Dessein A. 2002. The interplay between environmental and host factors during an outbreak of visceral leishmaniasis in Eastern Sudan. *Microb. Infect.* 14:1449–57.
8. Bradley DJ, Taylor BA, Blackwell J, Evans EP, Freeman J. 1979. Regulation of *Leishmania* populations within the host. III. Mapping of the locus controlling susceptibility to visceral leishmaniasis in the mouse. *Clin. Exp. Immunol.* 37(1):7–14.
9. Blackwell JM. 1996. Genetic susceptibility to leishmanial infections: Studies in mice and man. *Parasitology* 112(Suppl):S67–S74.

10. Vidal S and 9 colleagues. 1995. The Ity/Lsh/Bcg locus: Natural resistance to infection with intracellular parasites is abrogated by disruption of the Nramp1 gene. *J. Exp. Med.* 182(3):655–66.

11. Blackwell J, Freeman J, Bradley D. 1980. Influence of H-2 complex on acquired resistance to *Leishmania donovani* infection in mice. *Nature* 283(5742): 72–4.

12. Searle S, Bright NA, Roach TI, Atkinson PG, Barton CH, Meloen RH, and colleagues. 1998. Localisation of Nramp1 in macrophages: Modulation with activation and infection. *J. Cell. Sci.* 111(Pt 19):2855–66.

13. Biggs TE, Baker ST, Botham MS, Dhital A, Barton CH, Perry VH. 2001. Nramp1 modulates iron homoeostasis in vivo and in vitro: Evidence for a role in cellular iron release involving de-acidification of intracellular vesicles. *Eur. J. Immunol.* 31(7):2060–70.

14. Blackwell JM, Goswami T, Evans CA, Sibthorpe D, Papo N, White JK, and 5 colleagues. 2001. SLC11A1 (formerly NRAMP1) and disease resistance. *Cell Microbiol.* 3(12):773–84.

15. Liu J, Fujiwara TM, Buu NT, Sanchez FO, Cellier M, Paradis AJ, and colleagues. 1995. Identification of polymorphisms and sequence variants in the human homologue of the mouse natural resistance-associated macrophage protein gene. *Am. J. Hum. Genet.* 56(4):845–53.

16. Bellamy R, Ruwende C, Corrah T, McAdam KP, Whittle HC, Hill AV. 1998. Variations in the NRAMP1 gene and susceptibility to tuberculosis in West Africans. *N. Engl. J. Med.* 338(10):640–4.

17. Abel L, Sanchez FO, Oberti J, Thuc NV, Hoa LV, Lap VD, and colleagues. 1998. Susceptibility to leprosy is linked to the human NRAMP1 gene. *J. Infect. Dis.* 177(1):133–45.

18. Greenwood CM, Fujiwara TM, Boothroyd LJ, Miller MA, Frappier D, Fanning EA, and colleagues. 2000. Linkage of tuberculosis to chromosome 2q35 loci, including NRAMP1, in a large aboriginal Canadian family. *Am. J. Hum. Genet.* 67(2):405–16.

19. Searle S, Blackwell JM. 1999. Evidence for a functional repeat polymorphism in the promoter of the human NRAMP1 gene that correlates with autoimmune versus infectious disease susceptibility. *J. Med. Genet.* 36(4):295–9.

20. Bucheton B, Abel L, Kheir MM, Mirghani A, El-Safi S, Chevillard C, Dessein A. 2003. Genetic control of visceral leishmaniasis in a Sudanese population: Candidate gene testing indicates a linkage to the *NRAMP1* gene. *Genes. Immun.* 4:104–109.

21. Abel L, Muller-Myhsok B. 1998. Robustness and power of the maximum-likelihood-binomial and maximum-likelihood-score methods, in multipoint linkage analysis of affected-sibship data. *Am. J. Hum. Genet.* 63(2):638–47.

22. Xu J, Meyers DA, Ober C, Blumenthal MN, Mellen B, Barnes KC, and 9 colleagues. 2001. Genomewide screen and identification of gene-gene interactions for asthma-susceptibility loci in three U.S. populations: Collaborative study on the genetics of asthma. *Am. J. Hum. Genet.* 68(6):1437–46.

23. Von Lichtenberg F. 1962. Host response to eggs of *S. mansoni*. I. Granuloma formation in the unsensitized laboratory mouse. *Am. J. Pathol.* 41:711–31.

24. Warren KS, Domingo EO, Cowan RB. 1967. Granuloma formation around schistosome eggs as a manifestation of delayed hypersensitivity. *Am. J. Pathol.* 51(5):735–56.

25. Butterworth AE, Capron M, Cordingley JS, Dalton PR, Dunne DW, Kariuki HC, and 9 colleagues. 1985. Immunity after treatment of human schistosomiasis mansoni. II. Identification of resistant individuals, and analysis of their immune responses. *Trans. R. Soc. Trop. Med. Hyg.* 79(3):393–408.

26. Dessein AJ, Begley M, Demeure C, Caillol D, Fueri J, dos Reis MG, and 3 colleagues. 1988. Human resistance to *Schistosoma mansoni* is associated with IgG reactivity to a 37-kDa larval surface antigen. *J. Immunol.* 140(8):2727–36.

27. Abel L, Dessein A. 1991. Genetic predisposition to high infections in an endemic area of *Schistosoma mansoni* [editorial]. *Rev. Soc. Bras. Med. Trop.* 24(1):1–3.

28. Dessein AJ, Couissinier P, Demeure C, Rihet P, Kohlstaedt S, Carneiro-Carvalho D, and 4 colleagues. 1992. Environmental, genetic and immunological factors in human resistance to *Schistosoma mansoni*. *Immunol. Invest.* 21(5):423–53.

29. Hagan P, Blumenthal UJ, Dunn D, Simpson AJ, Wilkins HA. 1991. Human IgE, IgG4 and resistance to reinfection with *Schistosoma haematobium*. *Nature* 349(6306):243–5.

30. Rihet P, Demeure CE, Bourgois A, Prata A, Dessein AJ. 1991. Evidence for an association between human resistance to *Schistosoma mansoni* and high anti-larval IgE levels. *Eur. J. Immunol.* 21(11):2679–86.

31. Couissinier P, Dessein AJ. 1995. *Schistosoma*-specific helper T cell from subjects resistant to infection by *Schistosoma mansoni* are ThO/2. *Euro. J. Immunol.* 25:2295–302.

32. Rodriguez V, Piper K, Couissinier-Paris P, Bacelar O, Dessein H, Dessein AJ. 1999. Genetic control of Schistosome infections by SM1 locus of the 5q31-q33 region is linked to differentition of type 2 helper T lymphocytes. *Infect. Immun.* 67, 4689–92.

33. Rihet P, Demeure CE, Dessein AJ, Bourgois A. 1992. Strong serum inhibition of specific IgE correlated to competing IgG4, revealed by a new methodology in subjects from a *S. mansoni* endemic area. *Eur. J. Immunol.* 22(8):2063–70.

34. Demeure CE, Rihet P, Abel L, Ouattara M, Bourgois A, Dessein AJ. 1993. Resistance to *Schistosoma mansoni* in humans: Influence of the IgE/IgG4 balance and IgG2 in immunity to reinfection after chemotherapy. *J. Infect. Dis.* 168(4):1000–8.

35. Abel L, Demenais F, Prata A, Souza AE, Dessein A. 1991. Evidence for the segregation of a major gene in human susceptibility/resistance to infection by *Schistosoma mansoni* [see comments]. *Am. J. Hum. Genet.* 48(5):959–70.

36. Marquet S, Abel L, Hillaire D, Dessein H, Kalil J, Feingold J, and 2 colleagues. 1996. Genetic localization of a locus controlling the intensity of infection by *Schistosoma mansoni* on chromosome 5q31-q33. *Nat. Genet.* 14(2):181–4.

37. Muller-Myhsok B, Stelma FF, Guisse-Sow F, Muntau B, Thye T, Burchard GD, and 2 colleagues. 1997. Further evidence suggesting the presence of a locus, on human chromosome 5q31-q33, influencing the intensity of infection with *Schistosoma mansoni* [letter; comment]. *Am. J. Hum. Genet.* 61(2):452–4.

38. Marsh DG, Neely JD, Breazeale DR, Ghosh B, Freidhoff LR, Ehrlich-Kautzky E, and colleagues. 1994. Linkage analysis of IL4 and other chromosome 5q31.1 markers and total serum immunoglobulin E concentrations. *Science* 264(5162):1152–6.

39. Martinez FD, Solomon S, Holberg CJ, Graves PE, Baldini M, Erickson RP. 1998. Linkage of circulating eosinophils markers on chromosome 5q. *Am. J. Respir. Crit. Care Med.* 158:1739–44.

40. Rioux JD, Stone VA, Daly MJ, Cargill M, Green T, Nguyen H, and colleagues. 1998. Familial eosinophiliz maps to the cytokine gene cluster on human chromosomal region 5q31-q33. *Am. J. Hum. Genet.* 63:1086–94.

41. Warren KS. 1977. Modulation of immunopathology and disease in schistosomiasis. *Am. J. Trop. Med. Hyg.* 26(6 Pt 2):113–9.

42. Chensue SW, Warmington KS, Ruth J, Lincoln PM, Kunkel SL. 1994. Cross-regulatory role of interferon-gamma (IFN-gamma), IL-4 and IL-10 in schistosome egg granuloma formation: The in vivo regulation of activity and inflammation. *Clin. Exp. Immunol.* 98(3):395–400.

43. Wynn TA, Morawetz R, Scharton-Kersten T, Hieny S, Morse HC, 3rd, Kuhn R, and colleagues. 1997. Analysis of granuloma formation in double cytokine-deficient mice reveals a central role for IL-10 in polarizing both T helper cell 1- and T helper cell 2-type cytokine responses in vivo. *J. Immunol.* 159(10):5014–23.

44. Grimaud JA, Borojevic R. 1977. Chronic human schistosomiasis mansoni. Pathology of the Disse's space. *Lab. Invest.* 36(3):268–73.

45. Gressner AM, Bachem MG. 1995. Molecular mechanisms of liver fibrogenesis – A homage to the role of activated fat-storing cells. *Digestion* 56(5):335–46.

46. Poli G. 2000. Pathogenesis of liver fibrosis: Role of oxidative stress. *Mol. Aspects Med.* 21(3):49–98.

47. Hoffmann KF, Cheever AW, Wynn TA. 2000. IL-10 and the dangers of immune polarization: Excessive type 1 and type 2 cytokine responses induce distinct forms of lethal immunopathology in murine schistosomiasis. *J. Immunol.* 164(12):6406–16.

48. Jimenez SA, Freundlich B, Rosenbloom J. 1984. Selective inhibition of human diploid fibroblast collagen synthesis by interferons. *J. Clin. Invest.* 74(3):1112–6.

49. Duncan MR, Berman B. 1985. Gamma interferon is the lymphokine and beta interferon the monokine responsible for inhibition of fibroblast collagen production and late but not early fibroblast proliferation. *J. Exp. Med.* 162(2):516–27.

50. Czaja MJ, Weiner FR, Eghbali M, Giambrone MA, Eghbali M, Zern MA. 1987. Differential effects of gamma-interferon on collagen and fibronectin gene expression. *J. Biol. Chem.* 262(27):13,348–51.

51. Czaja MJ, Weiner FR, Takahashi S, Giambrone MA, van der Meide PH, Schellekens H, and 2 colleagues. 1989. Gamma-interferon treatment inhibits collagen deposition in murine schistosomiasis. *Hepatology* 10(5):795–800.

52. Rockey DC, Chung JJ. 1994. Interferon gamma inhibits lipocyte activation and extracellular matrix mRNA expression during experimental liver injury: Implications for treatment of hepatic fibrosis. *J. Invest. Med.* 42(4):660–70.

53. Mallat A, Preaux AM, Blazejewski S, Rosenbaum J, Dhumeaux D, Mavier P. 1995. Interferon alfa and gamma inhibit proliferation and collagen synthesis of human Ito cells in culture. *Hepatology* 21(4):1003–10.

54. Roberts AB, Sporn MB, Assoian RK, Smith JM, Roche NS, Wakefield LM, and 2 colleagues. 1986. Transforming growth factor type beta: Rapid induction of fibrosis and angiogenesis in vivo and stimulation of collagen formation in vitro. *Proc. Natl. Acad. Sci. USA* 83(12):4167–71.

55. Tiggelman AM, Boers W, Linthorst C, Sala M, Chamuleau RA. 1995. Collagen synthesis by human liver (myo)fibroblasts in culture: Evidence for a regulatory role of IL-1 beta, IL-4, TGF beta and IFN gamma. *J. Hepatol.* 23(3):307–17.

56. Tamai K, Ishikawa H, Mauviel A, Uitto J. 1995. Interferon-gamma coordinately upregulates matrix metalloprotease (MMP)-1 and MMP-3, but not tissue inhibitor of metalloproteases (TIMP), expression in cultured keratinocytes. *J. Invest. Dermatol.* 104(3):384–90.

57. Mohamed-Ali Q, Elwali NE, Abdelhameed AA, Mergani A, Rahoud S, Elagib KE, and 3 colleagues. 1999. Susceptibility to periportal (Symmers) fibrosis in human *Schistosoma mansoni* infections: Evidence that intensity and duration of infection, gender, and inherited factors are critical in disease progression. *J. Infect. Dis.* 180(4):1298–306.

58. Henri S, Chevillard C, Mergani A, Paris P, Gaudart J, Camilla C, and colleagues. 2002. Cytokine regulation of periportal fibrosis in humans infected with *Schistosoma mansoni*: IFN-gamma is associated with protection against fibrosis and TNF-alpha with aggravation of disease. *J. Immunol.* 169(2):929–36.

59. Dessein AJ, Hillaire D, Elwali ENMA, Marquet S, Mohamed-Ali Q, Mirghani A, and colleagues. 1999. Severe hepatic fibrosis in *Schistosoma mansoni* infection is controlled by a major locus that is closely linked to the interferon-g receptor gene. *Am. J. Hum. Genet.* 68:709–21.

60. Salam EA, Ishaac S, Mahmoud AA. 1979. Histocompatibilty-linked susceptibility for hepatosplenomegaly in human schistosomiasis mansoni. *J. Immunol.* 123(4):1829–31.

61. Abaza H, Asser L, el Sawy M, Wasfy S, Montaser L, Hagras M, and 3 colleagues. 1985. HLA antigens in schistosomal hepatic fibrosis patients with haematemesis. *Tissue Antigens* 26(5):307–9.

62. Abdel-Wahab MF, Esmat G, Narooz SI, Yosery A, Struewing JP, Strickland GT. 1990. Sonographic studies of school children in a village endemic for *Schistosoma mansoni*. *Trans. R. Soc. Trop. Med. Hyg.* 84(1):69–73.

63. Secor WE, del Corral H, dos Reis MG, Ramos EA, Zimon AE, Matos EP, and colleagues. 1996. Association of hepatosplenic schistosomiasis with HLA-DQB1*0201. *J. Infect. Dis.* 174(5):1131–5.

64. Casanova JL, Abel L. 2002. Genetic dissection of immunity to mycobacteria: The human model. *Annu. Rev. Immunol.* 20:581–620.

PART FOUR

GENETIC AND EVOLUTIONARY CONSIDERATIONS

The Evolution of Pathogen Virulence in Response to Animal and Public Health Interventions

Andrew F. Read, Sylvain Gandon, Sean Nee, and
Margaret J. Mackinnon

INTRODUCTION

Pathogen evolution poses the critical challenge for infectious disease management in the twenty-first century. As is already painfully obvious in many parts of the world, the spread of drug-resistant and vaccine-escape (epitope) mutants can impair and even debilitate public and animal health programs. But there may also be another way in which pathogen evolution can erode the effectiveness of medical and veterinary interventions. Virulence- and transmission-related traits are intimately linked to pathogen fitness and are almost always genetically variable in pathogen populations. They can therefore evolve. Moreover, virulence and infectiousness are the target of medical and veterinary interventions. Here, we focus on vaccination and ask whether large-scale immunization programs might impose selection that results in the evolution of more-virulent pathogens.

The word virulence is used in a variety of ways in different disciplines. We take a parasite-centric view as follows. We use "disease severity" (morbidity and/or mortality) to mean the harm to the host following infection. Disease severity is thus a phenotype measured at the whole-organism (host) level that is determined by host genes, parasite genes, environmental effects, and the interaction between those factors. One component of this is virulence, a phenotypic trait of the pathogen whose expression depends on the host. Thus, virulence is the component of disease severity that is due to pathogen genes, and it can be measured only on a given host. We assume no specificity in the interaction between host and pathogen (more-virulent strains are always more virulent, whatever host they infect).

In "Malaria Virulence," we return to that assumption, and the evidence that there are more-or less-virulent pathogens within species.

We begin by asking how natural selection might shape virulence, and in light of that, we ask how widespread immunization might alter selection on virulence. We are deeply skeptical that there will be any simple generalizations about virulence evolution that will apply across all infectious diseases, and so we ground our argument in the biology of malaria, with which we are most familiar. Nonetheless, widespread vaccination has been used for a range of human and animal diseases for much of the twentieth century and our arguments may apply to some of these. In the second half of this chapter, we therefore consider infectious diseases other than malaria.

NATURAL SELECTION ON VIRULENCE

A large number of both host and pathogen genes are responsible for disease severity, and in almost all cases these genes will be interacting with each other and a very large number of environmental determinants such as host age, condition, and previous exposure. Understanding the nature of selection on genetic determinants of disease severity is clearly a mammoth task. But it is certainly possible to understand the evolution of traits under environmental and polygenic control (for example, body size, life span, and resistance), so we see no reason in principle why it should not be possible to make progress analyzing components of disease severity, such as virulence. In the future, it should become possible to analyze the natural selection on particular phenotypes contributing to overall virulence, such as pathogen replication rates and immune evasion. The extent to which analysis of individual components will be useful will depend on their contribution to variation in virulence and in turn to disease severity.

There is now a large body of theoretical work examining the evolution of virulence. The classical framework we use is the best developed theoretically, but, most importantly, it has the great merit that its assumptions and predictions are empirically testable. Nevertheless, we recognize that a challenge for the future is to incorporate such complexities as host genetic diversity and coevolution. The idea is that excessively virulent pathogen mutants are eliminated by natural selection because they kill their hosts and therefore themselves. Excessively avirulent mutants also have low fitness because they are more rapidly cleared from their hosts or they fail to maximize their output of transmission propagules. Natural selection should therefore optimize the balance between the costs of virulence (host death and/or morbidity) and the benefits (immune evasion and host resource extraction) (May and Anderson 1983; Bull 1994; Ewald

1994; Frank 1996; Ebert 1999; Read et al. 1999; Day 2001; Dieckmann et al. 2002). Empirical support for this trade-off framework comes from a number of animal and plant models (Bull et al. 1991; Bull and Molineux 1992; Herre 1993; Ebert and Mangin 1997; Fenner and Fantini 1999; Messenger et al. 1999). Its usefulness in a medical context has recently been questioned (Ebert and Bull 2003), but it seems to us that its utility in any disease context will depend on the extent to which the disease in question fulfills the assumptions involved in the framework. In particular, are there virulent and avirulent parasites in a population? Second, are the fitness costs and benefits of virulence as assumed by the trade-off model? These are disease-specific issues.

MALARIA VIRULENCE

Parasite Genetics?

Malaria disease severity is determined by a complex interaction of host and parasite genetics as well as factors such as previous exposure and socioeconomics (Mbogo et al. 1999; Mackinnon et al. 2000; Phillips 2001; Greenwood and Mutabingwa 2002; Miller et al. 2002). Malaria has of course provided some of the best examples of host genes involved in disease severity, and there is currently great enthusiasm about exploiting information from the human genome project to identify further such genes. However, it is very striking that despite the vast sums of money involved, there has been little attempt to use the tools of quantitative genetics to determine what proportion of the variation in disease outcome is actually due to host factors. In the only such study we know of, a pedigree analysis of malarial disease severity in human populations in Sri Lanka, as little as 10% of the variance was due to host additive genetic factors, and perhaps as much as another 10% was due to nonadditive genetic factors (Mackinnon et al. 2000). Environmental factors such as previous disease episodes, distance from water bodies, and housing type accounted for up to another 20%. The remaining unexplained variance could be due to unmeasured environmental effects or noise. It could also be due to parasite genetic variation. It is true that polymorphisms in many of the human genes most associated with malaria disease are rare or do not exist in Sri Lanka, so an analysis elsewhere may demonstrate a greater role for host genetics. Nonetheless, this early study does make clear that the evolutionary processes determining malarial disease severity need not involve only the human genome.

   Plasmodium would be an extraordinarily unusual pathogen if it was not variably virulent. So far as we are aware, wherever virulence variation has been looked for in other pathogen species, it has been found (e.g., Sibley and Boothroyd 1992; Lipsitch and Moxon 1997; Ebert 1998; Fenner and Fantini 1999; Pandey and Igarashi 2000; Ochman and Moran 2001; Read and Taylor 2001). Certainly there are substantial differences in virulence between *Plasmodium* species. However, direct evidence of virulence variation within species is actually rather limited. This is probably because the obligate sexuality of *Plasmodium* generates high rates of recombination in all but very low transmission situations, so that virulent and avirulent strains are impossible to recognize in the field. Nonetheless, some candidate virulence determinants in human *Plasmodium* species have been proposed (Marsh and Snow 1997; Hayward et al. 1999; Preiser et al. 1999; Miller et al. 2002), and overrepresentation of some parasite alleles among severe malaria cases has been reported (Engelbrecht et al. 1995; Robert et al. 1996; Kun et al. 1998; Ariey et al. 2001). Experimentally demonstrating the involvement of particular genes has proved difficult, not least because virulence cannot be assayed *in vitro*.

Nonetheless, a variety of indirect data point to genetic variation in *Plasmodium* virulence. First, deliberate infections of people demonstrated strain differences in virulence (James et al. 1936; Covell and Nicol 1951), and there were strain differences both within and between geographic regions in the speed with which infections became life-threatening (Figure 11.1) and in putative virulence determinants such as growth rates (Gravenor et al. 1995). Second, various phenotypes believed to be encoded by parasite genes are correlated with disease severity (Marsh 1992; Miller et al. 2002), such as *in vitro* proliferation rates (Figure 11.2, Chotivanich et al. 2000), rosetting (Carlson et al. 1990; Rowe et al. 1997), and cell selectivity (Simpson et al. 1999). Crucially, genetic variation for virulence has been readily uncovered in animal models of malaria (Yoeli et al. 1975; Cox 1988; Mackinnon and Read 1999a). For instance, in our laboratory, genetically distinct clones of *Plasmodium chabaudi* differ in their virulence in mice, measured as anemia, weight loss, and mortality (Mackinnon and Read 1999a; Ferguson et al., in press). These clone differences are consistent across a range of conditions such as dose (Timms et al. 2001), host sex (Mackinnon, personal communication, 2002), host immune status (Buckling and Read 2001; Mackinnon and Read 2003), drug treatment (Buckling et al. 1997), mosquito passage (Mackinnon, personal communication, 2002), the presence of competing clones (Taylor et al. 1998; Timms 2001), and host genotype (Mackinnon et al. 2002).

Taken together, then, it seems extremely likely that there is genetic variation for virulence in human *Plasmodium* populations. With the near

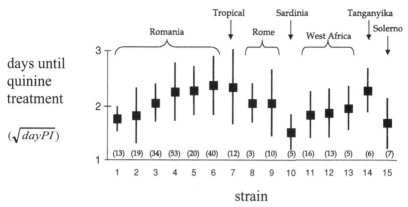

Figure 11.1. Days until quinine therapy became necessary for malaria-naive patients deliberately infected with different strains of *P. falciparum* as therapy for neurosyphilis in the Horton Hospital, UK, during the 1930s and 1940s. Quinine was administered when attending physicians considered the patient's life to be at risk. Data are available for a total of 250 patients infected with one of 19 strains; plotted points are for the 15 strains for which three or more patients were infected. The strains came from seven geographic regions, as labeled above. Sample sizes are given in brackets; plotted points are means ±1 s.e.m. The vertical axis is the square root of days post-infection (PÍ). Strains from various geographic regions differed significantly in how fast they induced life-threatening disease ($F_{6, 241} = 2.31$, p < 0.035), and there were also significant differences between strains within geographic regions ($F_{12, 241} = 3.32$, p < 0.0001). Data sources are described by Gravenor et al. (1995); data are from M. Gravenor (personal communication, 2002).

completion of the malaria genome project, and advances in bioinformatics, it is now possible to look for genetic differences in malaria parasites from patients with severe or mild disease, and there is every reason to think that particular parasite genes will be identified in the near future. But, from our perspective, a key issue is the fitness consequences of virulence for malaria parasites.

Selection for Virulence?

It is going to be extraordinarily difficult to obtain (ethically) data on the fitness costs and benefits of virulence for a human disease. However, it has proved relatively easy to identify fitness benefits of virulence in our rodent model for malaria, *P. chabaudi*, which shares many of the life-history characteristics of the most virulent of the human species, *P. falciparum*. In the absence of host death, more-virulent clones are cleared less rapidly,

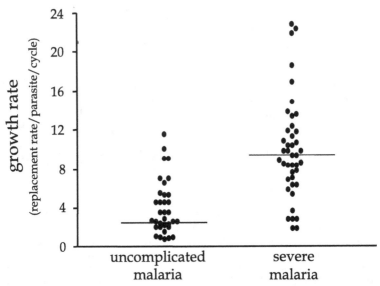

Figure 11.2. *In vitro* replication rates of *P. falciparum* parasites isolated from patients in hospitals in Thailand with life-threatening or mild malaria. Each dot represents parasites from a single patient. Redrawn from Chotivanich et al. (2000).

produce more gametocytes (transmission stages), and transmit better to mosquitoes (Mackinnon and Read 1999a, 2003; Ferguson et al., in press). Importantly for our argument below, the same advantages to virulence accrue in semi-immune hosts (Figure 11.3). Moreover, as others have found in malaria and many other systems (Ebert 1998), serial passage by syringe leads to increasing virulence (Mackinnon and Read 1999b). This implies a within-host fitness advantage: presumably, virulent variants arise that have an advantage in the race for the syringe. Thus, we have evidence of within- and between-host (transmission) advantages to virulence.

Selection against Virulence?

If virulence only enhanced fitness, ever more-virulent malaria parasites would be favored by natural selection. There are a number of possible fitness costs to virulence. Conventional wisdom is that excessively virulent pathogens kill their hosts and therefore themselves. However, it is remarkably difficult to get quantitative data on the fitness costs of pathogen-induced host death in any system. Clearly, if death is extremely rapid, and occurs before transmission propagules are produced, then pathogen fitness will be zero. But diseases are rarely that lethal. In our mouse malaria

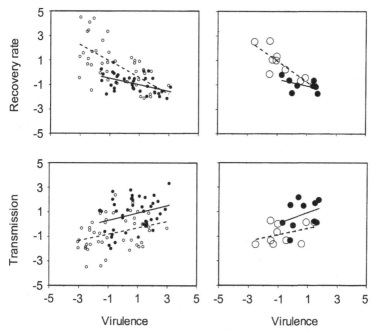

Figure 11.3. In the absence of host death, *P. chabaudi* virulence is associated with less-rapid clearance and enhanced transmission. Graphs show the phenotypic (cross-mouse, left) and genetic (cross-clone, right) relationships across immunologically naive (solid symbols and lines) and semi-immunized (open symbols, broken lines) mice. Numerical values are first principal components of traits associated with virulence (anemia and weight loss), transmission (gametocytemia), and recovery (duration of infection, rate of clearance of parasites). Mice were immunized with live parasites and cleared by chemotherapy four days later. Reproduced with permission from Mackinnon and Read (2003).

model, risk of death is at its greatest if the mice fail to control the initial proliferation of asexual parasites. Most transmission stages appear after this, so that in one of our experiments, host death resulted in a 75% reduction in transmission potential (Mackinnon et al. 2002). In humans, recent analysis of malaria therapy data suggests that failure to control initial proliferation is also a major determinant of risk of severe disease and death for *P. falciparum* (Molineaux et al. 2001), and here too, the bulk of gametocyte production occurs after the first wave of parasitemia (James et al. 1932, 1936). Thus, it seems likely that host death will impact pathogen fitness in the field. Whether this is sufficient to offset the fitness gains of virulence is very unclear. In a selection experiment with *P. chabaudi*, in which we mimicked a mortality cost of virulence by preventing parasites from 50–75% of the most virulent infections from proceeding to the next host, virulence increased nonetheless (Mackinnon and Read 1999b). Thus, this intensity of host-level selection was insufficient to prevent the evolution of more-virulent parasites. Case fatality rates in the section of human populations responsible for the bulk of malaria transmission are hard to estimate but they may be somewhere between 10% and 1% or even less (Snow et al. 1999; Trape et al. 2002). However, the key issue is not mortality risk *per se* but rather how an increased risk of host death is traded off against the increased transmission benefits of virulence in the absence of host death. To determine fitness functions for a human pathogen is a formidable challenge.

There are other possible costs of virulence (Ebert 1998; Day 2001). For instance, enhanced transmission to mosquitoes could be associated with reduced transmission to the next vertebrate host if strains more infectious to mosquitoes are also more lethal to them because of their greater parasite burdens. Malaria parasites can kill mosquitoes (Ferguson and Read 2002a,b), but we found no evidence that more-virulent infections in mice were more lethal to the mosquitoes they infected (Ferguson et al., in press). Another cost of virulence may be an inability to successfully infect a range of host genotypes. Pathogens that achieve high virulence on one host genotype might, for instance, achieve lower virulence on another. The existence of such a pattern is at the heart of a broad array of coevolutionary theory (Hamilton 1980; Lively and Apanius 1995), but despite some evidence of it in plants and invertebrates (Little 2002), we know of no examples from pathogens of vertebrates where strains virulent on one host genotype are avirulent on another and *vice versa*. We have yet to investigate this across a wide range of host and parasite genotypes in *P. chabaudi*, but in one experiment, the relative virulence of two parasite lines was never reversed on any of three mouse genotypes (Mackinnon et al. 2002).

A cost of virulence one might expect, at least following serial passage, is loss of transmission-stage production (Ebert 1998). This has sometimes been seen in malaria parasites (Day et al. 1993), but in all our experiments to date, virulence increases have been associated with increased transmission-stage production (Mackinnon and Read 1999b; Mackinnon et al. 2002).

Taking all this together, we favor the hypothesis that host death imposes significant selection against pathogen virulence, and we assume that in what follows. But exactly why malaria parasites, or any other pathogens, are as benign as they are is a key unsolved problem in parasite evolution. We know of no compelling quantitative data demonstrating, for any pathogen, the factor(s) that prevent pathogens from being more virulent.

VACCINATION AND VIRULENCE EVOLUTION: THEORY

Here we consider the consequences of widespread vaccination of a population harboring a pathogen whose virulence is determined by the balance of the cost of virulence (host death) and the benefits (less rapid clearance by the host, enhanced ability to exploit host resources for transmission). We are envisaging that the vaccine is imperfect, in that it allows at least some transmission from some vaccinated hosts. Vaccines against some acute childhood infections can generate lifelong sterilizing immunity, but vaccines against other diseases do not (Plotkin and Orenstein 1999). In general, vaccines elicit levels of protection that are the same or worse than that elicited by natural infections. Many of the vaccines currently under development are attempting to elicit protection against diseases for which natural immunity is rather poor. In the absence of technological advances that endow new-generation vaccines with protective effects better than those achieved by nature, future vaccines will reduce disease levels but will not prevent some transmission from at least some vaccinated individuals. This is certainly likely for malaria vaccines. Natural immunity seems to wane rapidly and a high degree of immunity seems to require repeated exposure to parasites (Bruce-Chwatt 1963; Day and Marsh 1991; Marsh 1992; Richie and Saul 2002). Thus, it seems probable that, at best, levels of protection as modest as those seen naturally will be achieved by malaria vaccines. Indeed, malaria vaccines that have achieved statistically significant protection have done so by delaying rather than preventing the invasion of parasites or the onset of symptoms (Bojang et al. 2001; Genton et al. 2002). For the same reason, vaccines against other childhood diseases and those still under development, such as those against HIV or hepatitis B and C, will be similarly imperfect or 'leaky.' Even some of the vaccines against childhood diseases may be very imperfect. For instance, vaccination against *Bordetella pertussis* has been staggeringly successful at reducing childhood deaths due to whooping cough, but the immunity elicited by the vaccine wanes within a few years. Consequently, the pathogen still circulates, even in populations with high levels of vaccine coverage (Ewald 1996).

Vaccines that impact some aspect of pathogen fitness without reducing it to zero have the potential to impose strong selection. The most obvious impact is selection for epitope mutants – variants that are able to at least partially evade vaccine-induced immunity. But there could also be potent selection on parasite virulence: vaccines have the potential to reduce the selection against virulent parasites. Vaccines that protect against host death also protect more-virulent pathogens from killing

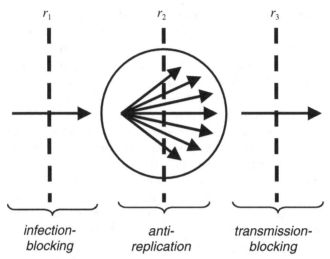

r_1 r_2 r_3

infection-   anti-   transmission-
blocking   replication   blocking

Figure 11.4. Schematic representation of the action of different types of host resistance at different stages of a parasite life cycle: r_1, infection-blocking resistance; r_2, anti-replication resistance; r_3, transmission-blocking resistance. A fourth type of resistance, r_4, anti-toxin resistance, prevents host death without impacting pathogen replication or transmission. Each of these four types of resistance is the target of various candidate malaria vaccines (Richie and Saul 2002): r_1, sporozoite and liver-stage vaccines; r_2, blood-stage vaccines; and r_3, transmission-blocking vaccines. Recently, a candidate r_4 vaccine has been proposed (Schofield et al. 2002). Figure is from Gandon et al. (2001) with permission.

their host and therefore themselves. Given the fitness benefits of virulence (transmission), more-virulent strains will then circulate in the population. Key here is the mode of action of the vaccine (Figure 11.4). In general, imperfect infection-blocking or transmission-blocking vaccines will not favor virulence increases because they do not alter the balance between the cost (mortality) and benefit (transmission) of virulence. In contrast, imperfect anti-disease vaccines (those reducing in-host replication or parasite toxicity) will always select for more-virulent pathogens: they protect both the host and the parasite from the costs of virulence. Elsewhere, we have formalized this verbal argument and we refer those interested in the mathematical details to that paper (Gandon et al. 2001). The argument is rather general and will also apply to any other medical intervention that reduces parasite replication or toxicity without eliminating the pathogens: subcurative chemotherapy and, in the veterinary context, enhanced genetic resistance through selective breeding will also select for more-virulent pathogens.

We are thus hypothesizing that anti-disease vaccines will prompt more-virulent pathogens to circulate in a population. Unvaccinated individuals, such as young children, travelers, and those unlucky enough not to get the vaccine, will therefore have a greater risk of contacting more-virulent pathogens and hence dying. If this argument is correct, it raises some interesting ethical issues. It is a quite different view of vaccine risk than that normally discussed in the tabloid press and public health circles. Normally, it is thought that, at no risk to themselves, unvaccinated individuals benefit through herd immunity from those who have been vaccinated and taken the (usually minute) risk of vaccine-induced side effects. Under our scenario, those who get the vaccine are putting at risk those who do not get the vaccine. We know of no other situation in which a medical intervention is beneficial for the recipient but makes things worse for those who do not receive it.

Our argument is rather general, though it obviously depends on a number of assumptions to which we return below. But it prompts additional questions, which can only be addressed with rather more disease-specific models. First, what is the population-wide disease burden? Even if unvaccinated individuals are at greater risk, the key question from the public health perspective is whether the population as a whole is better off after the vaccine-driven evolution has occurred. Unvaccinated individuals will be more likely to die, but the vaccine would not have been used at all if it did not reduce host death, lead to more rapid recovery, and/or make hosts less infectious. This is precisely what semi-immunity does in our mouse model (Buckling and Read 2001; Mackinnon and Read 2003). At a post-vaccine evolutionary equilibrium, how does population-wide death compare to the pre-vaccine case? Second, how rapid is this evolution likely to be? Clearly, if the time scale is measured in centuries or more, it need not be an issue for public health.

To answer these questions, we developed a malaria-specific population dynamic model in which virulence evolution was possible (Gandon et al. 2001). The model is a modified version of the standard susceptible–infected model, with two types of host (fully susceptible and semi-immune), and incorporates naturally acquired immunity, superinfection, and vector transmission. It was parameterized for a high-transmission year-round endemic *P. falciparum*. Broadly speaking, the above generalizations about the evolutionary consequences of different types of vaccine also hold for this model: infection-and transmission-blocking vaccines have minimal impact on virulence evolution, whereas vaccines targeted at malaria blood stages or parasite toxins are expect to prompt the evolution

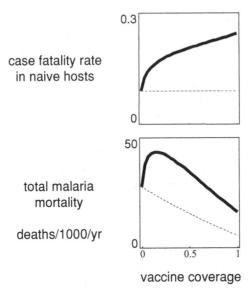

Figure 11.5. Predicted virulence in unvaccinated host and population-wide malaria mortality in a population vaccinated with an imperfect blood-stage (anti-replication, r_2) malaria vaccine. The dotted line indicates before evolution, as might be seen in a clinical trial or early in a vaccine campaign; the solid line indicates after evolution. For model details, see Gandon et al. (2001).

of more-virulent parasites. An example is shown in Figure 11.5. In the absence of pathogen evolution, pathogen virulence (here, case-fatality rate) is unaffected by vaccine coverage, but the population-wide deaths drop as vaccine coverage is increased. This is the conventional expectation, and what one would expect to see, for instance, in a clinical trial or early in the implementation of a vaccine program. However, vaccination protects virulent pathogens, so that they are expected to increase in frequency in the population. The post-vaccination equilibrium virulence is always expected to be higher, and the more people that are vaccinated, the higher it is. Consequently, case-fatality rates are higher among the unvaccinated. Among the population as a whole, the benefits of vaccination are eroded from those that would have been seen in the clinical trials or early in the implementation phase. There are even some regions of parameter space where the population as a whole is worse off than before vaccination (Figure 11.5). Moreover, once evolution has occurred, it would be difficult to halt vaccination even if death rates were similar or worse than the pre-vaccine era: withdrawal of the vaccine would put even more individuals at risk.

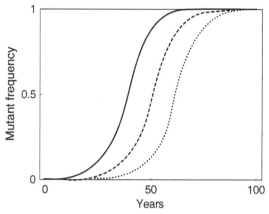

Figure 11.6. Invasion dynamics of a virulent mutant after the start of a vaccination campaign using an anti-replication vaccine ($r_2 = 0.8$ and $r_1 = r_3 = r_4 = 0$) with a vaccination coverage of 90%. In this simulation, the virulence of the resident strain is the ES virulence in the absence of vaccination ($\alpha_N = 0.153$). We present the invasion dynamics of invasion of a mutant with virulence equal to the ES level after vaccination ($\alpha^*_N = 0.418$). At the start of the vaccination campaign, we assumed that the mutant was at an initial frequency of 1% (solid line), 0.1% (dashed line), or 0.01% (dotted line). We further assumed that its distribution among naive and immune hosts was identical to the distribution of the resident strain before vaccination. For more details, see Supplementary Information in Gandon et al. (2001).

What of the time scale of this evolution? Predictions of rates of evolution are far more sensitive to specific model parameters and assumptions than are predictions about the direction of evolution or the magnitude of final equilibria. The time to fixation of variant alleles critically depends, for instance, on their initial frequency and selective advantage. In the current context, the selective advantage depends on the precise shape of the trade-off functions, and we simply know too little of the quantitative details to be sure of any estimates. Nonetheless, for plausible values, an anti-blood stage vaccine could drive virulence mutants to clinically relevant levels in a few decades (Figure 11.6). It is even possible that much more rapid evolution could occur: in some cases, epidemiological feedback on parasite evolution yields an evolutionary bistable situation where, for intermediate vaccine coverage, parasites can evolve toward high or low virulence, depending on initial conditions (Gandon et al., 2003). Clearly, there is much more work to be done on the invasion dynamics of a virulence mutant, but from our simulations we expect that, as with the evolution of vaccine epitope escape mutants and drug resistance, the evolution of virulence will occur on time scales relevant to public health (a few decades or even less).

VACCINATION AND VIRULENCE EVOLUTION: ASSUMPTIONS

Our conclusions about time scale, and the possibility that things can evolve to be worse than they were in the pre-vaccine era, are very parameter-(system) specific. However, our conclusion that anti-disease vaccines prompt evolution that, at the population level, erodes the benefits of vaccines seen in clinical trials and puts unvaccinated individuals at greater risk is a very general conclusion that flows from our assumptions. The assumptions of our model that we see as key are as follows.

1. There is genetic variation in parasite virulence.
2. Vaccination is imperfect.
3. There is a positive genetic correlation between virulence and transmission.
4. The cost of virulence is host death.

Above, we reviewed the evidence relevant to these assumptions in the malaria context. Broadly speaking, the evidence is supportive; certainly we are unaware of any compeling contradictions from either animal experiments or the more anecdotal human data. Nonetheless, there are certainly a number of outstanding empirical issues. For instance, theory requires that the virulence–transmission relationship be saturating (i.e., the marginal transmission advantages of virulence decrease as virulence increases): if it is not, selection would favor infinite virulence. We have yet to find direct evidence for saturation, although there are hints that it might be present (Mackinnon et al. 2002; Mackinnon and Read 2003) and theoretical reasons to expect it (Antia et al. 1994; Ganusov et al. 2002). There may also be other costs to virulence. For instance, extreme anemia may impact the parasites' ability to replicate. Also, genetically diverse infections are the rule rather than the exception in malaria (Day et al. 1992), and we only modeled the case of superinfection (i.e. that in which hosts can be infected by more than one clone, but if a new clone establishes then it completely replaces the previous clone, so coexistence does not occur). Superinfection and coinfection can have profound impacts on the direction of virulence evolution depending on the mechanisms of any interactions between competing genotypes and how competitive outcome within hosts translates into between-host fitness (Read and Taylor 2001; Read et al. 2002). We are pursuing these issues in our rodent model; so far generalities have proved elusive (Taylor et al. 1997; Taylor and Read 1998; Read and Taylor 2001; Mackinnon et al. 2002; Read et al. 2002; de Roode et al., 2003).

The extent to which these outstanding issues undermine our argument that blood-stage and toxin vaccines will select for more-virulent malaria parasites is unclear. We view absence of evidence as an opening for further work and understanding rather than as a fatal flaw (cf., Smith 2002, Ebert and Bull 2003). If, for instance, other costs of virulence are discovered, these will need to be explored theoretically. If anemia is a cost of virulence, vaccination may reduce this too, with the same consequence for the evolution of malaria virulence. More telling empirical tests of the idea may also be possible. For instance, it should be possible to experimentally evolve malaria parasites in semi-immune and immunologically naive mice and compare the resulting virulence.

OTHER DISEASES

Of course, the real test of our hypothesis is the outcome of real-time evolution in pathogen populations. Yet any evolutionary consequences of malaria vaccines are unlikely to become apparent during our working lives: even if the malaria vaccines currently under field trial actually work, it will be at least another 10–15 years before they could emerge from the regulatory morass of Western medicine and begin to impose selection on *Plasmodium* populations. Once they do, there will be no turning back. So what about diseases other than malaria? In many instances, vaccination has been going on for many decades, probably long enough for virulence evolution to occur. There is an urgent need to determine whether it has occurred or is occurring.

However, this will not be straightforward. Although the assumptions underpinning our model are likely to apply in many cases, they will certainly not apply to all diseases. For instance, we assume that virulence and transmission will be positively and genetically correlated. This is believed to be so in many but not all infectious diseases (Lipsitch and Moxon 1997; Weiss 2002; Ebert and Bull 2003). For instance, it is frequently stated that polio virus causes disease only when it gets into neural tissue, from which it has no transmission potential, and hence that there can be no link between virulence and transmission (Levin and Bull 1994). Actually, in this particular case we are not convinced. The reversion of the attenuated vaccine strains to wild type virulence in some parts of the world points to some sort of fitness advantage to virulence (Kew et al. 2002; Day 2003) Nevertheless, the general point must be sound: virulence and transmission need not be genetically correlated. Clearly, our analysis applies only to those situations in which they are: an absence of virulence evolution in a disease without such a correlation casts no light on our argument. We note, however, that even for *Plasmodium chabaudi* in laboratory mice, it has been·an experimental challenge to determine relevant fitness functions. Where experimental work is not possible (human diseases without an appropriate animal model), it is going to be very difficult to determine whether a particular disease is of the sort that is likely to be in accord with our model.

Even if virulence evolution has occurred, it might be a serious challenge to detect it against other changes, especially where vaccination is highly effective. For illustrative purposes, consider measles in England and Wales, where vaccination began in the late 1960s. The best test of whether contemporary measles is more virulent than measles in the pre-vaccine era is to take strains from the 1960s and current strains and compare their virulence in an experimental common garden. This is how the evolution of the virulence of myxoma virus in Australia has been elucidated (Fenner and Fantini 1999). However, in the case of a human disease, there is no ethically acceptable common garden. One is forced to compare strains using a surrogate measure of virulence (such as *in vitro* growth rates, or virulence in an animal model) or to compare case records from the 1960s with contemporary disease outcome. Such a comparison would be hard enough, given undoubted changes in societal health generally and improvements in intensive and palliative care. This comparison becomes even more problematic when vaccination has reduced serious cases to very low numbers (as with measles in England and Wales), so that sample sizes become a problem. It may also be that many of the current UK cases are from pathogens initially contracted in populations in which there has been little or no vaccination.

An altogether more worrying prospect is that very large increases in virulence are in fact about to occur but we have yet to detect the early signs. As drug resistance has demonstrated, novel alleles rising from very low frequencies can be extremely hard to detect during the early part of their expansion, but then once they are at detectable levels, they very rapidly rise toward fixation to become a public health disaster (Hastings and D'Alessandro 2000). Chloroquine resistance in malaria, for example, must have first arisen in the 1950s but was not a serious problem until it became widespread two or three decades later. The situation could be even more difficult with virulence: drug resistance is a relatively uncontroversial phenotype to assay both *in vitro* and *in vivo*. In the absence of a genetic marker of virulence, it will be very, very difficult to detect changes in pathogen virulence until the evolution has proceeded a long way.

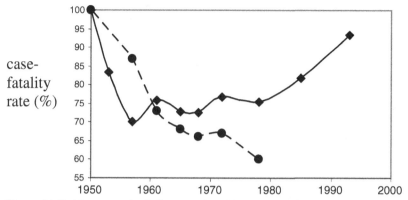

Figure 11.7. Myxomatosis virulence and rabbit resistance in Australia. Myxoma virus evolution was initially towards lower virulence, but virus virulence is increasing again (solid line, diamond), evidently in response to increases in resistance in wild rabbits (broken line, circles). Virulence of myxoma virus was estimated from mean survival times of wild isolates in a standard laboratory strain of rabbit; rabbit resistance was estimated by inoculating wild rabbits with a standard laboratory myxoma virus strain. Data are from Tables 14.5 and 14.6 of Fenner and Fantini (1999).

All of these issues make it difficult to test our hypothesis using human data in the absence of genetic markers. Nonetheless, we believe there is a strong case for trying to do so: we know of two examples of animal diseases that may be consistent with our proposed scenario. The first is myxomatosis in Australian rabbits. The dramatic drop in virulence that occurred a few years after the introduction of the myxoma virus in the 1950s is a textbook example of virulence evolution and one of the empirical examples used to support the argument that evolution favors intermediate levels of virulence that balance the benefits (high transmission and slower clearance) and costs (host death) of virulence. However, the implications of the more recent evolution of the myxoma virus have been largely overlooked (Gandon and Michalakis 2000). The virulence of isolates recovered from the field has been rising again (see Figure 11.7). In fact, more than 80% of the isolates were ranked as the most virulent strain. This percentage is the same as in the virus population introduced into Australia in the first place. The explanation for this seems to be increasing resistance in the rabbit population as a result of the strong natural selection imposed by the virus on the rabbit (Fenner and Fantini 1999). Studies of the infection process demonstrate that natural rabbit resistance has impaired the expansion of the virus population within the host (Best and Kerr 2000).

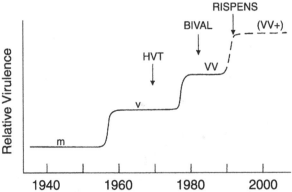

Figure 11.8. Schematic representation of Marek's disease virus virulence evolution, taken from the poultry literature. Until the 1950s, Marek's disease was caused
by strains now classified as mild (m). These were then replaced by virulent (v), very
virulent (vv), and very virulent + (vv+) strains. This evolution prompted the development of first-generation vaccines (HVT) and then second-(Bival) and third-(Rispens)
generation vaccines. The original caption to this figure reads, "There appears to
be a relationship between the introduction of new vaccines and the development
of more virulent pathotypes" (Witter 1998, p. S50). Figure is from Witter (1998)
with permission.

Thus, it is analogous to the resistance elicited by an anti-replication vaccine (r_2 in Figure 11.4). There is also evidence that more-virulent strains
transmit better in the absence of host death (May and Anderson 1983;
Best and Kerr 2000). Thus, it is most probable that increases in resistance
are responsible for increases in virus virulence.

A possible case of vaccine-driven evolution is that of Marek's disease
virus (MDV), which is a lymphoproliferative avian herpes virus that causes
substantial losses in the poultry industry. There is extremely good evidence for strain differences in virulence, and this seems to have been the
raw material that fueled substantial virulence evolution (Witter 1997a,b,
1988, 2001; Kreager 1998; Biggs 2001). From its first description in 1907
through to the middle part of the twentieth century, Marek's disease was a
mildly paralytic syndrome associated with gross enlargements of peripheral nerves and occasional lymphomas. Death was relatively rare. This suite
of symptoms was caused by strains now categorized as mild Marek's disease virus, mMDV. From the 1950s on, there has been a steady increase in
virulence (Figure 11.8), and today hyperpathogenic strains dominate the
poultry industry. These strains can kill within two weeks and are associated
with syndromes such as flaccid neck paralysis and extensive lymphomas
on most internal organs. Direct comparisons of isolates from the past 50

years are not possible, but there seems little doubt that virulence evolution has occurred: mild MDV strains can no longer be isolated in commercial operations in the U.S., and it is difficult to imagine that hyperpathogenic strains could have been present at the end of World War II without someone recording the horrendous symptoms they cause.

What drove this evolution? Initial increases in virulence, attributed to the intensification of the poultry industry, prompted the development of the first generation of live attenuated vaccines (see Figure 11.8). This was shortly followed by further increases in virulence, which prompted the development of second-generation vaccines, more virulence, and subsequently the development of a third generation of vaccines. Many in the industry consider that the vaccines last about ten years and that virulence evolution has made vaccination an extremely fragile means of MDV control (Kreager 1998; Witter 2001): the next generation of vaccines may themselves be untenably virulent.

But is there a causal link between vaccination and the virulence increases? Certainly, this is a prevalent view in the poultry literature (Witter 1997a, 1998, 2001; Kreager 1998; Biggs 2001). A number of lines of evidence show that MDV accords with our model. First, there is good experimental evidence that vaccination is less protective against more-virulent strains (Witter 1997b); Read et al. in preparation). Second, the more-virulent strains are not vaccine escape mutants: there is absolutely no evidence that they are antigenically different from the milder ancestral strains. Instead, enhanced virulence is associated with enhanced proliferative ability; it looks very much like the sort of life-history (virulence) variation we are discussing. Third, the effects of the vaccine are anti-proliferative – r_2 in our speak (Figure 11.4). The vaccines may also have a pure anti-disease effect analogous to our r_4 resistance (Biggs 2001). Fourth, vaccinated birds transmit wild type virus, so the vaccine is imperfect. However, there are no quantitative data on transmission, so although a positive genetic correlation between virulence and transmission seems likely (pathogen-induced host cell proliferation is associated with both), it has not been confirmed.

It is of course possible that changes in factors other than vaccination efficacy were responsible for the virulence increases in MDV populations. One possibility is genetically enhanced resistance in the chickens. This resistance also slows virus proliferation, which would reinforce our argument that anti-replication immunity can drive virulence increases. Another possibility is that there have been other changes in the poultry industry, not least intensification. Such hypotheses do not readily explain why more lethal strains are better able to cope with vaccination. To us, the lesson of Marek's disease is two-fold:

(1) virulence evolution on short time scales can have a huge impact on animal health, and
(2) vaccination might be the cause.

There is an urgent need to determine whether it is, and if so, whether vaccination could have the same effect in diseases of humans.

ANTI-TOXIN VACCINES

Several authors have argued that there is evidence that anti-toxin vaccines have altered the virulence of at least one human disease (Ewald 1994, 1996, 2002; Soubeyrand and Plotkin 2002; Ebert and Bull 2003). However, the orthodoxy is that vaccination prompted the evolution of *less*-virulent pathogens. The most widely cited example is the bacterial disease diphtheria, caused by *Corynebacterium diphtheriae*, although the argument has been extended to *Bordetella pertussis* (Ewald 1994, 1996, 2002; Soubeyrand and Plotkin 2002; Ebert and Bull 2003) and *Hemophilus influenzae* (Ewald 1996). The orthodox explanation for the reductions in virulence is as follows. The pathogen produces a fitness-enhancing toxin, but producing the toxin is metabolically expensive. Thus, in an unvaccinated host, the tox$^+$ variants have a fitness advantage, but in vaccinated hosts, the tox$^-$ forms have an advantage because they pay no metabolic cost for producing a toxin rendered useless by immunity. Thus, in a vaccinated population, tox$^-$ forms dominate, so that overall virulence goes down. Deliberately targeting vaccines at toxins to induce this sort of evolution (the "virulence-antigen strategy") has been advocated as a practical example of virulence management (Ewald 1994, 1996, 2002; Ebert and Bull 2003).

This argument contrasts with our view that anti-toxin vaccines will protect the host and hence the toxin-producing strains from risk of death, so that higher levels of toxin production (and hence virulence) can be sustained. The two arguments differ in what is considered the cost of virulence – host death (Gandon et al. 2001) or metabolic costs of toxin production (Ewald 1994, 1996, 2002; Soubeyrand and Plotkin 2002; Ebert and Bull 2003). In a model incorporating both costs (Gandon et al. 2002), we showed that increased or decreased virulence could result, depending on the metabolic cost of toxin production and vaccine coverage efficacy (Figure 11.9). This raises the very interesting possibility of testable predictions across a range of diseases or epidemiological conditions. It might also explain why for instance, virulence reductions in diphtheria

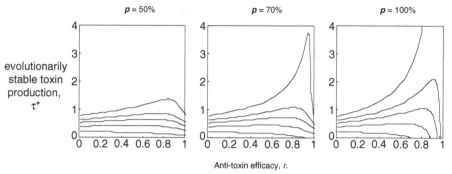

Figure 11.9. Evolutionarily stable toxin production, τ^*, against anti-toxin vaccine efficacy, r. Plotted lines are for different costs of toxin production. The upper line in each panel is when the cost of toxin production is zero; the lower lines are for progressively higher costs of toxin production, and variable levels of vaccination coverage, p. We used the same fitness function and parameter values as Gandon et al. (2002) and the epidemiological model used here is equation (6) in Gandon et al. (2001).

have apparently occurred in some contexts (e.g., Pappenheimer 1982), whereas there have apparently been increases in other populations (e.g., Mortimer and Wharton 1999). More generally, we believe our model to be the first formal attempt to model the virulence-antigen strategy; it is quite clear that it can have harmful evolutionary consequences in a fairly wide range of parameter space.

CONCLUDING REMARKS

We end with three points. First, there is absolutely no doubt that vaccination is one of the most important successes of biomedical science, and none of our data or theoretical models provides any argument against continuing to develop and implement mass vaccination. Even if we are right that *some* vaccines have the capacity to lead to the evolution of more-virulent pathogens, this is no argument for not developing them. Marek's disease is an important case in point. It may indeed be that vaccination is responsible for the hyperpathogenic strains that now exist. But it has been estimated that vaccination has saved 2 billion chickens in the U.S. since its implementation (Witter 2001). Indeed, it is difficult to envisage any situation in which transitory benefits of disease relief for even just a few decades would not be worth having. For instance, malaria has a massive economic as well as human cost (Sachs and Malaney 2002). A viable vaccine that reduced the malaria burden for even a few decades would liberate huge economic potential, given the nature of financial compounding. Economic development is a prerequisite for developing the health delivery infrastructure, implementing environmental measures, and improving housing, which could themselves have a huge impact on malaria incidence. *Thus, any effective malaria vaccine has huge potential for good.* Our strongest point is that, as with chemotherapy, we have to be aware that evolution can take the gloss off apparently magic bullets, so that we need to be vigilant and continually considering new control measures.

Second, our arguments are based on mathematical and animal models. For the most part, it is difficult to imagine any other way to investigate, in advance, the evolutionary safety of population-wide interventions in human pathogens. However, models of any sort rarely deliver certainty, and this necessarily means controversy. This is no reason to ignore the evolutionary consequences of vaccination. We suggest that more theoretical and empirical work of the sort we have been doing is called for on malaria and a range of other pathogens. Some of this work might serve to determine whether we are already unwittingly doing experiments on human populations that will have undesirable outcomes for which we should prepare. Other such studies might help us decide between possible candidate vaccines. For instance, transmission- and infection-blocking vaccines will generally lessen the risks of increased virulence and so should be incorporated in multi-component vaccines (Gandon et al. 2001). We also think that, during vaccine trials, substantially more attention should be paid to measuring transmission from vaccinated people. Finally, when vaccines go into widespread use, it is vital that extensive collections of frozen pathogen material be kept throughout, so that if evolutionary change does occur, it is possible to study it.

The Marek's disease story demonstrates the need for more work on the impact of medical and veterinary interventions on virulence evolution. Something caused MDV to become more virulent over the past fifty years. Perhaps it was something other than vaccination, but it is surely some consequence of chicken husbandry. The poultry industry deals with virulence evolution by sterilizing empty chicken sheds. Analogous options are clearly unavailable to public health managers.

ACKNOWLEDGMENTS
Our work is or has been supported by the Wellcome Trust, The Leverhulme Trust, the BBSRC, The Royal Society, the Fondation Singer Polignac, the CNRS, and the University of Edinburgh. We are grateful to Mike Gravenor for sharing the data on which Figure 11.1 is based, and to members of the Read group and Edinburgh malaria group for numerous discussions over many years.

REFERENCES

Antia, R., B. Levin, and R. M. May. 1994. Within-host population dynamics and the evolution and maintenance of microparasite virulence. *American Naturalist* **144**:457–72.

Ariey, F., D. Hommel, C. Le Scanf, J. B. Duchemin, C. Peneau, A. Hulin, J. L. Sarthou, J. M. Reynes, T. Fandeur, and O. Mercereau-Puijalon. 2001. Association of severe malaria with a specific *Plasmodium falciparum* genotype in French Guiana. *Journal of Infectious Diseases* **184**:237–41.

Best, S. M., and P. J. Kerr. 2000. Coevolution of host and virus: the pathogenesis of virulent and attenuated strains of myxoma virus in resistant and susceptible European rabbits. *Virology* **267**:36–48.

Biggs, P. M. 2001. The history and biology of Marek's disease virus. *Current Topics in Microbiology and Immunology* **255**:1–24.

Bojang, K. A., P. J. M. Milligan, M. Pinder, L. Vigneron, A. Alloueche, K. E. Kester, W. R. Ballou, D. J. Conway, W. H. H. Reece, P. Gothard, L. Yamuah, M. Delchambre, G. Voss, B. M. Greenwood, A. Hill, K. McAdam, N. Tornieporth, J. D. Cohen, and T. Doherty. 2001. Efficacy of RTS,S/ASO2 malaria vaccine against *Plasmodium falciparum* infection in semi-immune adult men in The Gambia: a randomised trial. *Lancet* **358**:1927–34.

Bruce-Chwatt, L. 1963. A longtitudinal survey of natural infection in a group of West African adults. *West African Medical Journal* **12**:199–206.

Buckling, A., and A. F. Read. 2001. The effect of partial host immunity on the transmission of malaria parasites. *Proceedings of the Royal Society of London, Series B* **268**:2325–30.

Buckling, A. G. J., L. H. Taylor, J. M. R. Carlton, and A. F. Read. 1997. Adaptive changes in Plasmodium transmission strategies following chloroquine chemotherapy. *Proceedings of the Royal Society of London, Series B* **264**:553–9.

Bull, J. J. 1994. Virulence. *Evolution* **48**:1423–37.

Bull, J. J., and I. J. Molineux. 1992. Molecular genetics of adaptation in an experimental model of cooperation. *Evolution* **46**:882–95.

Bull, J. J., I. J. Molineux, and W. R. Rice. 1991. Selection of benevolence in a host-parasite system. *Evolution* **45**:875–82.

Carlson, J., H. Helmby, A. V. S. Hill, D. Brewster, B. M. Greenwood, and M. Wahlgren. 1990. Human cerebral malaria: association with erythrocyte rosetting and lack of anti-rosetting antibodies. *Lancet* **336**:1457–60.

Chotivanich, K., R. Udomsangpetch, J. A. Simpson, P. Newton, S. Pukrittayakamee, S. Looareesuwan, and N. J. White. 2000. Parasite multiplication potential and the severity of falciparum malaria. *Journal of Infectious Diseases* **181**:1206–9.

Covell, G., and W. D. Nicol. 1951. Clinical, chemotherapeutic and immunological studies on induced malaria. *British Medical Bulletin* **8**:51–8.

Cox, F. E. G. 1988. Major animal models: rodent. In *Malaria: Principles and Practice of Malariology*, W. H. Wernsdorfer and I. McGregor, Eds., pp. 1503–43. Churchill Livingston, Edinburgh.

Day, T. 2001. Parasite transmission modes and the evolution of virulence. *Evolution* **55**:2389–400.

Day, T. 2003. Personal Commumication. Virulence evolution and the timing of disease life-history events. *Trends in Ecology & Evolution* **18**:113–118.

Day, K. P., F. Karamalis, J. Thompson, D. A. Barnes, C. Peterson, H. Brown, G. V. Brown, and D. J. Kemp. 1993. Genes necessary for expression of a virulence determinant and for transmission of *Plasmodium falciparum* are located on a 0.3-megabase region of chromosome 9. *Proceedings of the National Academy of Sciences USA* **90**:8292–6.

Day, K. P., J. C. Koella, S. Nee, S. Gupta, and A. F. Read. 1992. Population genetics and dynamics of *Plasmodium falciparum*: an ecological view. *Parasitology* **104**:S35–52.

de Roode, J. C., A. F. Read, B. H. K. Chan, and M. J. Mackinnon, 2003. Infection dynamics and virulence of three-clone infections with the rodent malaria parasite *Plasmodium chabaudi*. *Parasitology* **127**:411–418.

Dieckmann, U., J. A. J. Metz, M. W. Sabelis, and K. Sigmund, Eds. 2002. *Virulence Management: The Adaptive Dynamics of Pathogen-Host Interactions*. Cambridge University Press, Cambridge, UK.

Ebert, D. 1998. Experimental evolution of parasites. *Science* **282**:1432–5.

Ebert, D. 1999. The evolution and expression of parasite virulence. In *Evolution in Health and Disease*, S. C. Stearns, Ed., pp. 161–72. Oxford University Press, Oxford.

Ebert, D., and J. J. Bull. 2003. Challenging the tradeoff model for the evolution of virulence: is virulence management feasible? *Trends in Microbiology* **11**:15–20.

Ebert, D., and K. L. Mangin. 1997. The influence of host demography on the evolution of virulence of a microsporidian gut parasite. *Evolution* **51**:1828–37.

Engelbrecht, F., I. Felger, B. Genton, M. Alpers, and H.-P. Beck. 1995. *Plasmodium falciparum*: malaria morbidity is associated with specific Merozoite Surface Antigen 2 genotypes. *Experimental Parasitology* **81**:90–6.

Ewald, P. W. 1994. *Evolution of Infectious Diseases*. Oxford University Press, Oxford.

Ewald, P. W. 1996. Vaccines as evolutionary tools: the virulence-antigen strategy. In *Concepts in Vaccine Development*, S. H. E. Kaufmann, Ed., pp. 1–25. de Gruyter & Co., Berlin.

Ewald, P. W. 2002. Virulence management in humans. In *Adaptive Dynamics of Infectious Diseases: In Pursuit of Virulence Management*, U. Dieckmann, J. A. J. Metz, M. W. Sabelis, and K. Sigmund, Eds., pp. 399–409. Cambridge University Press, Cambridge, UK.

Fenner, F., and B. Fantini. 1999. *Biological Control of Vertebrate Pests*. CABI Publishing, Wallingford, UK.

Ferguson, H. M., M. J. Mackinnon, B. H. K. Chan, and A. F. Read, in press. Mosquito mortality and the evolution of malaria virulence. *Evolution*.

Ferguson, H. M., and A. Read. 2002a. Genetic and environmental determinants of malaria parasite virulence in mosquitoes. *Proceedings of the Royal Society of London, Series B* **269**:1217–24.

Ferguson, H. M., and A. Read. 2002b. Why is the impact of malaria parasites on mosquito survival still unresolved? *Trends in Parasitology* **18**:256–61.

Frank, S. A. 1996. Models of parasite virulence. *Quarterly Review of Biology* **71**:37–78.

Gandon, S., M. J. Mackinnon, S. Nee, and A. F. Read. 2001. Imperfect vaccines and the evolution of pathogen virulence. *Nature* **414**:751–6.

Gandon, S., M. J. Mackinnon, S. Nee, and A. F. Read. 2002. Antitoxin vaccines and pathogen virulence. *Nature* **417**:610.

Gandon, S., M. J. Mackinnon, S. Nee, and A. F. Read, 2003. Imperfect vaccination: some epidemiological and evolutionary consequences. *Proceedings of the Royal Society of London, Series B* **270**:1129–1136.

Gandon, S., and Y. Michalakis. 2000. Evolution of parasite virulence against qualitative or quantitative host resistance. *Proceedings of the Royal Society of London, Series B* **267**:985–90.

Ganusov, V., C. Bergstrom, and R. Antia. 2002. Within-host population dynamics and the evolution of microparasites in a heterogeneous host population. *Evolution* **56**:213–23.

Genton, B., I. Betuela, I. Felger, F. Al-Yaman, R. F. Anders, A. Saul, L. Rare, M. Baisor, K. Lorry, G. V. Brown, D. Pye, D. O. Irving, T. A. Smith, H. P. Beck, and M. P. Alpers. 2002. A recombinant blood-stage malaria vaccine reduces *Plasmodium falciparum* density and exerts selective pressure on parasite populations in a phase 1–2b trial in Papua New Guinea. *Journal of Infectious Diseases* **185**:820–70.

Gravenor, M. B., A. R. McLean, and D. Kwiatkowski. 1995. The regulation of malaria parasitaemia: parameter estimates for a population model. *Parasitology* **110**:115–22.

Greenwood, B., and T. Mutabingwa. 2002. Malaria in 2002. *Nature* **415**:670–1.

Hastings, I. M., and U. D'Alessandro. 2000. Modelling a predictable disaster: the rise and spread of drug resistant malaria. *Parasitology Today* **16**:340–7.

Hayward, R. E., B. Tiwari, K. P. Piper, D. I. Baruch, and K. P. Day. 1999. Virulence and transmission success of the malaria parasites *Plasmodium falciparum*. *Proceedings of the National Academy of Sciences USA* **96**:4563–8.

Herre, E. A. 1993. Population structure and the evolution of virulence in nematode parasites of fig wasps. *Science* **259**:1442–5.

James, S. P., W. D. Nicol, and P. G. Shute. 1932. A study of induced malignant tertian malaria. *Proceedings of the Royal Society of Medicine* **25**:1153–79.

James, S. P., W. D. Nicol, and P. G. Shute. 1936. Clinical and parasitological observations of induced malaria. *Proceedings of the Royal Society of Medicine* **29**:879–93.

Kew, O., V. Morris-Glasgow, M. Lanaverde, and 21 colleagues. 2002. Outbreak of poliomyelitis in Hispaniola associated with circulating Type 1 vaccine-derived poliovirus. *Science* **296**:356–9.

Kreager, K. S. 1998. Chicken industry strategies for control of tumour virus infections. *Poultry Science* **77**:1213–7.

Kun, J. F. J., R. J. Schmidt-Ott, L. G. Lehman, B. Lell, D. Luckner, B. Greve, P. Matousek, and P. G. Kremsner. 1998. Merozoite surface antigen 1 and 2 genotypes and rosetting of *Plasmodium falciparum* in severe and mild malaria in Lambaréné, Gabon. *Transactions of the Royal Society of Tropical Medicine and Hygiene* **92**:110–14.

Levin, B. R., and J. J. Bull. 1994. Short-sighted evolution and the virulence of pathogenic microbes. *Trends in Microbiology* **2**:73–7.

Lipsitch, M., and R. E. Moxon. 1997. Virulence and transmissibility of pathogens: what is the relationship? *Trends in Microbiology* **6**:31–6.

Little, T. J. 2002. The evolutionary significance of parasitism: do parasite-driven genetic dynamics occur *ex silico. Journal of Evolutionary Biology* **15**:1–9.

Lively, C. M., and V. Apanius. 1995. Genetic diversity in host-parasite interactions. In *Ecology of Infectious Diseases in Natural Populations*, B. T. Grenfell and A. P. Dobson, Eds., pp. 421–49. Cambridge University Press, Cambridge UK.

Mackinnon, M. J., D. J. Gaffney, and A. F. Read. 2002. Virulence of malaria parasites: host genotype by parasite genotype interactions. *Infection, Genetics and Evolution* 1:287–96.

Mackinnon, M. J., D. M. Gunawardena, J. Rajakaruna, S. Weerasingha, K. N. Mendis, and R. Carter. 2000. Quantifying genetic and nongenetic contributions to malarial infection in a Sri Lankan population. *Proceedings of the National Academy of Sciences USA* 97:12661–6.

Mackinnon, M. J., and A. F. Read. 1999a. Genetic relationships between parasite virulence and transmission in the rodent malaria *Plasmodium chabaudi*. *Evolution* 53:689–703.

Mackinnon, M. J., and A. F. Read. 1999b. Selection for high and low virulence in the malaria parasite *Plasmodium chabaudi*. *Proceedings of the Royal Society of London, Series B* 266:741–8.

Mackinnon, M. J., and A. F. Read. 2003. The effects of host immunity on virulence-transmissibility relationships in the rodent malaria parasite *Plasmodium chabaudi*. *Parasitology* 126:1–10.

Marsh, K. 1992. Malaria – a neglected disease? *Parasitology* 104:S53–69.

Marsh, K., and R. W. Snow. 1997. Host-parasite interaction and morbidity in malaria endemic areas. *Philosophical Transactions of the Royal Society of London* 352:1385–94.

May, R. M., and R. M. Anderson. 1983. Epidemiology and genetics in the coevolution of parasites and host. *Proceedings of the Royal Society of London, Series B* 219:281–313.

Mbogo, C. N. M., E. W. Kabiru, G. E. Glass, D. Forster, R. W. Snow, C. P. M. Khamala, J. H. Ouma, J. I. Githure, K. Marsh, and J. C. Beier. 1999. Vector-related case-control study of severe malaria in Kilifi District, Kenya. *American Journal of Tropical Medicine and Hygiene* 60:781–5.

Messenger, S. L., I. J. Molineux, and J. J. Bull. 1999. Virulence evolution in a virus obeys a trade-off. *Proceedings of the Royal Society of London, Series B* 266:397–404.

Miller, L. H., D. I. Baruch, K. Marsh, and O. Doumbo. 2002. The pathogenic basis of malaria. *Nature* 415:673–9.

Molineaux, L., H. H. Diebner, M. Eichner, W. E. Collins, G. M. Jeffrey, and K. Dietz. 2001. *Plasmodium falciparum* parasitemia described by a new mathematical model. *Parasitology* 122:379–91.

Mortimer, E. A., and M. Wharton. 1999. Diphtheria toxoid. In *Vaccines*, S. A. Plotkin and W. A. Orenstein, Eds., pp. 140–57. W. B. Saunders, Philadelphia.

Ochman, H., and N. Moran. 2001. Genes lost and genes found: Evolution of bacterial pathogenesis and symbiosis. *Science* 292:1096–8.

Pandey, B., and A. Igarashi. 2000. Severity-related molecular differences among 19 strains of dengue type 2 viruses. *Microbiology and Immunology* 44:179–88.

Pappenheimer, A. M. 1982. Diphtheria: studies on the biology of an infectious disease. *Harvey Lectures* 76:45–73.

Phillips, R. S. 2001. Current status of malaria and potential for control. *Clinical Microbiology Reviews* 14:208–26.

Plotkin, S. A., and W. A. Orenstein, Eds. 1999. *Vaccines*. W. B. Saunders, Philadelphia.

Preiser, P. R., W. Jarra, T. Capiod, and G. Snounou. 1999. A rhoptry-protein-associated mechanism of clonal phenotypic variation in rodent malaria. *Nature* **298**:618–22.

Read, A. F., P. Aaby, R. Antia, D. Ebert, P. W. Ewald, S. Gupta, E. C. Holmes, A. Sasaki, D. C. Shields, F. Taddei, and R. E. Moxon. 1999. What can evolutionary biology contribute to understanding virulence? In *Evolution in Health and Disease*, S. C. Stearns, Ed., pp. 205–16. Oxford University Press, Oxford.

Read, A. F., M. J. Mackinnon, M. A. Anwar, and L. H. Taylor. 2002. Kin selection models as evolutionary explanations of malaria. In *Adaptive Dynamics of Infectious Diseases: In Pursuit of Virulence Management*, U. Dieckmann, J. A. J. Metz, M. W. Sabelis, and K. Sigmund, Eds., pp. 165–78. Cambridge University Press, Cambridge, UK.

Read, A. F., and L. H. Taylor. 2001. The ecology of genetically diverse infections. *Science* **292**:1099–102.

Richie, T., and A. Saul. 2002. Progress and challenges for malaria vaccines. *Nature* **415**:694–701.

Robert, F., F. Ntoumi, G. Angel, D. Candito, C. Rogier, T. Fandeur, J.-L. Sarthou, and O. Mercereau-Puijalon. 1996. Extensive genetic diversity of *Plasmodium falciparum* isolates collected from patients with severe malaria in Daka, Senegal. *Transactions of the Royal Society of Tropical Medicine and Hygiene* **90**:704–11.

Rowe, J. A., J. M. Moulds, C. I. Newbold, and L. H. Miller. 1997. *P. falciparum* rosetting mediated by a parasite-variant erythrocyte membrane protein and complement-receptor 1. *Nature* **388**:292–5.

Sachs, J., and P. Malaney. 2002. The economic and social burden of malaria. *Nature* **415**:680–5.

Schofield, C. J., M. C. Hewitt, K. Evans, M.-A. Siomos, and P. H. Seeberger. 2002. Synthetic GPI as a candidate anti-toxic vaccine in a model of malaria. *Nature* **418**:785–9.

Sibley, L. D., and J. C. Boothroyd. 1992. Virulent strains of *Toxoplasma gondii* comprise a single clonal lineage. *Nature* **359**:82–5.

Simpson, J. A., K. Silamut, K. Chotivanich, S. Pukrittayakamee, and N. J. White. 1999. Red cell selectivity in malaria: a study of multiple-infected erythrocytes. *Transactions of the Royal Society of Tropical Medicine and Hygiene* **93**:165–8.

Smith, T. 2002. Imperfect vaccines and imperfect models. *Trends in Ecology and Evolution* **17**:154–156.

Snow, R. W., M. Craig, U. Deichmann, and K. Marsh. 1999. Estimating mortality, morbidity and disability due to malaria among Africa's non-pregnant population. *Bulletin of the World Health Organization* **77**:624–40.

Soubeyrand, B., and S. A. Plotkin. 2002. Antitoxin vaccines and pathogen virulence. *Nature* **417**:609–10.

Taylor, L. H., M. J. Mackinnon, and A. F. Read. 1998. Virulence of mixed-clone and single-clone infections of the rodent malaria *Plasmodium chabaudi*. *Evolution* **52**:583–91.

Taylor, L. H., and A. F. Read. 1998. Determinants of transmission success of individual clones from mixed-clone infections of the rodent malaria parasite, *Plasmodium chabaudi*. *International Journal of Parasitology* **28**:719–25.

Taylor, L. H., D. Walliker, and A. F. Read. 1997. Mixed genotype infections of malaria parasites: within-host dynamics and transmission success of competing clones. *Proceedings of the Royal Society of London, Series B* **264**:927–35.

Timms, R. 2001. *The Evolution and Ecology of Virulence in Mixed Infections of Malaria Parasites.* Ph.D. thesis, University of Edinburgh, Edinburgh.

Timms, R., N. Colegrave, B. H. K. Chan, and A. F. Read. 2001. The effect of parasite dose on disease severity in the rodent malaria *Plasmodium chabaudi. Parasitology* **123**:1–11.

Trape, J. F., G. Pison, A. Spiegel, C. Enel, and C. Rogier. 2002. Combating malaria in Africa. *Trends in Parasitology* **18**:224–30.

Weiss, R. A. 2002. Virulence and pathogenesis. *Trends in Microbiology* **10**:314–17.

Witter, R. L. 1997a. Avian tumour virsuses: persistent and evolving pathogens. *Acta Veterinaria Hungarica* **45**:251–66.

Witter, R. L. 1997b. Increased virulence of Marek's disease virus field isolates. *Avian Diseases* **41**:149–63.

Witter, R. L. 1998. The changing landscape of Marek's disease. *Avian Pathology* **27**:S46–53.

Witter, R. L. 2001. Protective efficacy of Marek's disease: Vaccines. *Current Topics in Microbiology and Immunology* **255**:57–90.

Yoeli, M., B. Hargreaves, R. Carter, and D. Walliker. 1975. Sudden increase in virulence in a strain of *Plasmodium berghei yoelii. Annals of Tropical Medicine and Parasitology* **69**:173–8.

TWELVE

Infection and the Diversity of Regulatory DNA

Lindsay G. Cowell, N. Avrion Mitchison, and Brigitte Muller

INTRODUCTION

The importance of regulatory DNA in long-term evolution is well recognized. Major evolutionary advances such as the origin of the bacteria, vertebrates, and mammals can be interpreted largely in terms of regulation of gene expression (1,2). Short-term evolution also is mediated substantially by changes in gene expression, and polymorphism in regulatory DNA provides a major component of genetic variation in natural populations (3–5). Such variation has been particularly well studied in respect to disease susceptibility in human populations, including susceptibility to infection. The immune system is a rich source of regulatory genetic variation, notable in the following four areas:

1. Inflammation, where variation in the expression of pro- and anti-inflammatory cytokines is conspicuous.
2. Th1/Th2/Treg balance, where differentiation into distinct T-cell subsets is regulated by the timing and level of gene expression, as can be modeled by Hopff bifurcation (6).
3. In the constitutive immune system, where multiple copies of, e.g., interferon genes occur.
4. In the generation of diverse antigen receptor repertoires, where variability among signal sequences mediating V(D)J recombination may regulate ordered assembly and allelic exclusion of the H, β, and δ loci and may bias the preselection repertoire.

In short, the immune system does not know what it will need to cope with next and uses regulatory DNA to provide some of the flexibility needed to function effectively. The same means are also used elsewhere to provide balance and flexibility, e.g., in the cardiovascular system (7).

The problem is that variation is harder to interpret in regulatory than in coding DNA. For instance, EMBOSS (the European Molecular Biology Open Software Suite) contains 43 programs for interpreting protein structure, and just one for regulatory DNA. We understand little about the key event, in which transcription factors (TFs) use a scaffold of regulatory DNA to build an "enhanceosome" assembly around the transcription bubble (8). That structure will be far harder to solve than the structure of any single protein. Hitherto, the approach to this problem has been largely through biochemistry, using such tools as EMSA (electrophoresis mobility shift assay), foot printing, and reporter constructs. Furthermore, biophysical studies of DNA ligation by TFs are beginning to contribute recently. Our thesis here is that genome informatics could provide important new insights. The problem is how regulatory DNA evolves in response to selective pressure during microevolution. To solve that problem we need to answer the following five key questions:

1. Can regulatory DNA be identified from its sequence properties, without recourse to experimental testing? Current informatics methods have been reviewed elsewhere (9–13). The variability that provides the fodder for evolution makes regulatory DNA extremely difficult to characterize. Much progress has been made by means of these methods, but further methods that utilize the patterns of variability are still greatly needed (14).

2. Can we detect a pattern in the appearance of nucleotide substitutions in the promoter region, or do they occur at random? If patterned, how does their distribution relate to the recognition elements that bind TFs?

3. Can we extrapolate from this patterning into anatomy? Will the patterning around particular recognition elements tell us which cell types and parts of the body are under selective pressure?

4. Nucleotide substitutions in the coding region can tell us about the evolutionary time scale. Can we in this way obtain a kinetic picture of evolution in the regulatory region?

5. Can we predict which genes will be most revealing in this way? For instance, do highly variable (or highly replicated) coding regions predict highly variable promoters?

Although final answers are not available, we are increasingly confident that they can be found.

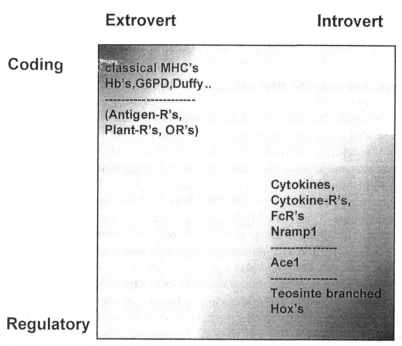

Figure 12.1. The distinction between extrovert and introvert genes. Extrovert genes include the classical MHC (types I and II) and genes whose products are recognized by malarial parasites, such as hemoglobin (Hb), glucose-6-phosphate dehydrogenase (G6PD), and the Duffy blood group. Also included are those where diversity has evolved through serial duplication, e.g., antigen receptor genes (Ig, TCR), plant resistance-to-infection genes (R), and olfactory receptors' genes (OR). Introvert genes include Fc receptors (FcR), transporters (Nramp1), and those involved in cardiovascular disease, such as the angiotensin cleaving enzyme (Ace-1), and in morphogenesis (Teosinte branched, Hox genes).

THE DISTRIBUTION OF DIVERSITY BETWEEN REGULATORY AND CODING DNA

Figure 12.1 summarizes our view of genetic diversity. Polymorphism in coding sequences occurs most conspicuously in the "extrovert" genes, which encode proteins that bind parasite and other foreign molecules (3). "Introvert" genes vary predominantly in their regulatory sequences, whereas extrovert genes tend to vary in their coding sequences and are therefore largely irrelevant to the present discussion (apart from the hitch-hiking effect mentioned in How Regulatory DNA Influences the Immunopathology of Infection). As the shading in Figure 12.1 indicates, we suppose that the introvert genes predominate. Overlaps occur, e.g., when a virus learns to recognize a cytokine receptor (15). Coding variation in extrovert genes may dominate susceptibility to infection, but in the major consequence of infection – immunopathology mediated by hypersensitivity – the reverse may well apply, with regulatory variation predominating.

HOW REGULATORY DNA INFLUENCES THE IMMUNOPATHOLOGY OF INFECTION

In an imaginary world in which humans lack inflammation, rather as Amphibia do at present, we would die more frequently of infection but with less chronic morbidity. Benefits thus balance drawbacks, so that genetic variance is to be expected and is in fact conspicuous. The pro- and anti-inflammatory cytokines (IL-1, IL1RA, IL-6, IL-10, TNFα, and TGFβ) all have polymorphisms in their regulatory DNA (16), some of which are known to affect susceptibility to chronic infection, notably in periodontal disease (17). These effects are particularly well documented in organ transplantation (18) and autoimmunity (19). This view predicts that variability in regulatory DNA should play a major role in the genetics of long-term morbidity post infection (hospitalization, long-term survival). It will be of particular interest to determine whether heterozygosity (e.g., in cytokine gene regulatory DNA) helps long-term survival in, for instance, HTLV-1 infection.

Studies on the major histocompatibility complex (MHC) contribute relevant information. In both man and mouse, MHC alleles and loci

(i) have different expression (20–22) and
(ii) modulate the immune response differentially, independently it seems of epitope selection (23–25).

It has now become clear that differential expression can regulate T-cell subsets (e.g., high expression favors Th1 differentiation) and that this effect can account for at least part of the modulation (22,26). Sequence analysis of murine MHC promoters reveals a high level of diversity that results from two effects working in parallel (27,28). One is hitchhiking (linkage disequilibrium with polymorphic coding sequences) (29), which seems to be essential for this diversity, as shown by the contrasting evolution of the CIITA (class II transactivator) gene. This gene strongly modulates MHC II expression; but presumably because its coding sequence has not diversified, its promoter has also not done so. The other effect is direct selection operating on the MHC promoter, as shown by the predominant distribution of single-nucleotide polymorphisms (SNPs) in and around the TF recognition elements. Furthermore, an MHC class I gene that had approximately equal coding diversity but lacked immunomodulatory function also had a nondiversified promoter.

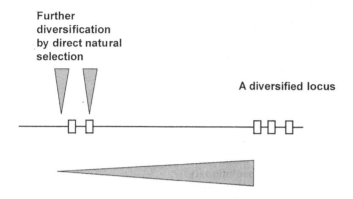

Further
diversification
by direct natural
selection

A diversified locus

Upstream diversifying by hitchhiking

Figure 1 2.2. Hitchhiking (linkage diseqilibrium upstream from variable coding sequences) has a diversifying effect upstream, which allows further diversification by natural selection. The effect is prominent in the MHC.

We also examined human MHC II promoters. For this purpose we took a biophysical approach to examining the ability of varying promoter sequences to bind transcription factors. First we obtained synthetic samples of naturally occurring oligonucleotides that varied slightly in sequence. Each oligonucleotide was then tested for its ability to bind nuclear extracts, by means of plasmon resonance. The only variation that affected the level of binding occurred in sequences adjacent to the TATA box. Furthermore, this effect could be verified by showing that the oligonucleotides detected in this way also yielded different values by EMSA (the electrophoresis mobility shift assay mentioned above (46)).

This view of the origin of diversity in MHC promoters is summarized in Figure 12.2. It indicates that in the pursuit of promoter polymorphism, genes with extrovert function should not be neglected, thus qualifying the message from Figure 12.1.

PROMOTER DIMORPHISM

An unexpected feature of the polymorphism emerged during the above study of mouse MHC II promoters. To our surprise, each of the three promoters under study turned out to have two basic polymorphic types, each made up of a linked series of alternative nucleotide substitutions. The separation into two haplotypes was not complete, as limited recombination between different sites within each promoter had evidently taken place. The haplotypes were not in linkage disequilibrium between the three loci. Thus we were confronted with three instances of promoter dimorphism, for which we had no explanation. Enlarging on this limited data, we

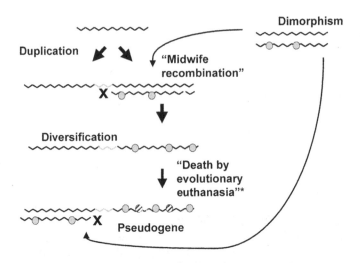

*suggested: as in loss of H2E, common in wild mice

Figure 12.3. How gene dimorphism could contribute to evolution.

speculated that dimorphism (in either regulatory or coding sequences) may play a part in the birth and death of genes as shown in Figure 12.3. Conceivably, we proposed, it might assist newly duplicated genes to diversify at the early time when one of the pair is at greatest risk of extinction, thus serving a "midwife" function. A dimorphic gene might duplicate and subsequently recombine with the old, nonduplicated gene of the opposite dimorphic type, thus generating a new chromosome in which the two dimorphic types are both in tandem.

A second possibility, we proposed, is that dimorphism could assist the loss of function of a duplicated gene, thus serving an "evolutionary euthanasia" function. One member of a pair of duplicated genes (that have undergone differentiation) might develop a dimorphism, in which one of the haplotypes mimics the function of the other duplicated gene. This would make the other gene redundant and leaves it prone to extinction. We proposed that this could explain why the *H2E* gene so often looses function, because its own function is mimicked (up to a point) by one of the dimorphic haplotypes at *H2A*.

These two possibilities are opposed to one another, in a way that might be expected of genomes that seem to vary habitually in size. This is all highly speculative and hardly worth mentioning except that promoter dimorphism does not seem to have been encountered previously. It is worth keeping an eye out for in the future.

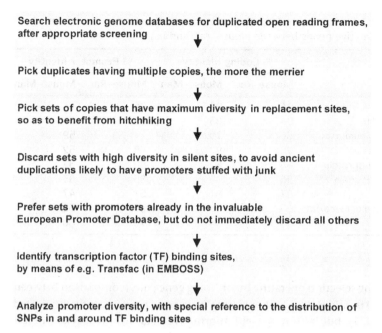

Figure 12.4. A sketch map for genome-wide promoter tracking.

FISHING FOR PROMOTER DIVERSIFICATION USING THE WHOLE GENOME

Far more data will be required if we are ever to tackle seriously the questions about diversity in regulatory DNA posed previously. The work that we have so far cited, including our own, has been conducted on a case-by-case basis, allowing information gradually to accrue from a variety of *ad hoc* genetical studies. We think it possible that this cottage industry is ripe for modernization. We draw inspiration from recent success in tracking the fate of duplicated genes. Open reading frame (ORF) duplicates can be identified by pair-wise comparison of all ORFs in a genome (30,31). Thus far the main effort has been to analyze these duplicates for nucleotide substitution at replacement (R) and silent (S) codon sites. The R/S ratio is a useful parameter, because silent substitutions provide a "biological clock," and replacements measure the force of natural selection. A straightforward path leads from there towards tracing the evolutionary divergence of upstream promoter sequences, as outlined in Figure 12.4. We suppose most the direction most likely to yield with least effort the maximum information about promoter evolution, although variant strategies are also applicable (13). Furthermore, as related genomes become available (5), the same approach could be pursued in relation to interspecific evolution. The

Table 12.1. Divergence in coding and promoter nucleotide sequences of representative genes between mouse, rat, and man

	% Coding Identity		% Promoter Identity	
Gene	Mouse–Rat	Mouse–Man	Mouse–Rat	Mouse–Man
Histone 4	94	90	86	44
Nucleolin	94	83	82	65
Serum albumin	89	79	74	59
Prealbumin	98	92	58	52
Alpha-fetoprotein	86	82	80	62
Ferritin H chain	96	90	85	53
Renin	87	81	51	47
Proopiomelanocortin	93	86	86	68
c-myc	95	94	78	60

purifying selection operating on ortholog gene pairs (comparison between species) is milder than that operating on paralog pairs (recently duplicated genes) (31), but both may yield informative data. It is of special interest to the present volume that sequences predictive of secreted and membrane proteins emerge as overrepresented among paralogs and as diverging relatively rapidly (31). Such proteins are relatively likely to have extrovert functions, as shown in Figure 12.1, and to mediate host–parasite interactions in infection.

We present a small-scale model experiment that uses the European Promoter Database (EPD). EPD is an annotated nonredundant collection of eukaryotic POL II promoters, for which the transcription start site has been determined experimentally. It is structured in a way that facilitates extraction of biologically meaningful promoter subsets for comparative sequence analysis, and at the time of use it contained 119 rat promoters and larger numbers of mouse and human ones. We selected the nine genes that were represented in all three species collections, and compared for each one the coding and promoter sequences (the latter running from −1 to −500) of mouse, rat, and man, with the results shown in Table 12.1. The promoters were compared by means of the STRETCHER program in EMBOSS (The European Molecular Biology Open Software Suite), and the coding sequences were compared by means of BLAST. As expected, the data show higher identity between mouse and rat than between mouse and man, in both coding and promoter sequences. The important point is that the divergence in promoter sequences was consistently larger than that in coding sequences. The divergence was highest in the more distal part of the promoters (−200 to −500) presumably because of lesser need there to conserve transcription factor binding sites. This divergence presumably reflects the accumulation of junk DNA as mentioned in Figure 12.4 rather than selection, at least for the most part. Over their full 500 nucleotides, the promoters diverged too far to allow application of BLAST (or MATCHER) rather than STRETCHER. It is for this reason that we recommend above the indirect strategy of using coding sequences to identify duplicate genes within a genome rather than going directly to the promoters. Furthermore, EPD, valuable as it is in its present state, does not provide full coverage. We suspect that this state of affairs is familiar, although we are not aware of it having been spelled out previously.

FISHING FOR PROMOTER DIVERSIFICATION IN SMALLER POOLS: POACHING BY VIRUSES

Viruses poach host genes and incorporate them into their own genomes. Notably, of the ~100 genes per herpes virus genome, 10–15% have been copied from host RNA (32,33). Little is known about the regulatory DNA sequences used by these viruses, and the possibility of something akin to host regulation has not been excluded. For an example of reverse poaching, see "Recombination Signal Sequence Variability below.

FISHING FOR PROMOTER DIVERSIFICATION IN SMALLER POOLS: CANCER

Cancer cells come under selective pressure to inactivate unnecessary genes, to acquire drug resistance, and possibly to acquire resistance to immune surveillance. The extent of the resulting changes is being extensively explored by means of DNA microarrays. It is already known that the mechanisms of change include promoter crippling (34,35), and it is likely that regulatory DNA will in the future move up in priority on the agenda of cancer research from the lowly position that it occupies at present.

HOW DO TFs BIND TO DNA, AND HOW DOES THAT AFFECT DNA STRUCTURE?

We refer above to the predominant distribution in regulatory DNA of SNPs in and around TF recognition elements. Indeed, when a disease-associated SNP is discovered at a distance from known recognition elements, this often precipitates a search for a novel one near by. So where do we stand in understanding how an SNP could affect recognition? We take two well-worked biophysical examples. One is the binding of cyclic-AMP response element (CRE) to CRE-binding protein (CREB) (36), and the other is of the TATA box to TATA-box binding protein (TBP)(37). Both of these excellent studies place their work in the context of ongoing efforts to press ahead with the biophysical approach.

Of the two, the TATA box study is the more relevant to our purpose, because it pays attention to sequence variants. Many of these variants have been cocrystallized with TBP, their structure solved, and their transcriptional activity (ranging from a nominal 100% down to 1%) measured. The factors that combine to produce high TBP activity include minor groove widening, as well as roll, rise, and shift at the end of the box, untwisting within the box, and lowered water density around the DNA. Binding makes the TATA DNA sequence bend and kink at its ends. Interestingly, binding also bends the CRE nucleotide in the other study.

The studies covered sequences of 14 nucleotides for CRE and 8 for TATA. This may be compared with the functionally important substitution found at position -174 in the IL-6 promoter, 14 nucleotides away from a CRE center 38. The majority of common substitutions in murine MHC II promoters occur within this distance from the CRE center (38). Whether or not substitutions so far away could influence DNA bending within the canonical CRE sequence is at present unclear.

RECOMBINATION SIGNAL SEQUENCE VARIABILITY

We finish with an example from non-promoter regulatory DNA, the recombination signal sequences (RSSs) that are the binding sites for the V(D)J recombinase. Extensive gene duplication has generated hundreds of antigen receptor gene segments; each one is associated with at least one RSS, providing a tremendous source of regulatory genetic variation. RSSs contain a conserved heptamer (consensus: CACAGTG) and nonamer (consensus: ACAAAAACC) separated by a less conserved spacer of $12 \pm$ pm1 or $23 \pm$ pm1 bp (12-RSSs and 23-RSSs, respectively) (39). The inter- and intraspecific variability is high for all RSS positions but the first three (39,40). The efficiency with which each RSS mediates recombination depends on its sequences, so this variability has important implications for immune function and may be maintained by selection.

Antigen receptor genes are assembled by the somatic rearrangement of V, D, and J gene segments. At each locus, there are multiple gene segments of each type with diverse coding and signal sequences, but only one of each type of gene segment will participate in physiologic rearrangement. The relative frequencies with which the gene segments participate in rearrangement are nonrandom, and there is evidence that the biases may result in part from differences in RSS efficiency, thus "... a bias in the use of gene segments during recombination could prove beneficial if it enhanced the production of selectable receptors" (41, p. 1053). Biased recombination efficiencies could also be advantageous if receptor specificities for microorganisms that are highly pathogenic or are encountered early in life are favored.

The B-cell receptor (Bcr) H chain genes and the Tcr β and δ chain genes are composed of V, D, and J gene segments (others are composed of only V and J gene segments). Rearrangement at these loci is ordered: D to J (Vδ to Dδ) joining precedes V to DJ (VDδ to Jδ) joining, and direct V to J rearrangements are not observed. The ordered rearrangement of β chain gene segments is regulated at least in part by the Dβ- and Jβ-associated 12-RSSs (42). 23-RSSs are more variable than 12-RSSs (43). However, we hypothesize that this may reflect the presence of two distinct 23-RSS subsets: those associated with D-to-J type gene segments and those associated with V-to-DJ type gene segments. The specificity of the signal may differ between the two sets of 23-RSS, or the V-to-DJ type RSS may simply be less efficient. Recent data from the Schlissel laboratory (44) show that the non-core region of the RAG2 enzyme is required for normally efficient V-to-DJ type rearrangement and distinguishes between the two groups of 23-RSS. This provides an RSS-mediated mechanism for the regulation of ordered assembly and allelic exclusion at the H, β, and δ loci.

A final note on the host–parasite relationship: RSS provide an interesting example of reverse poaching in that the host genome is hypothesized to have done the poaching. The RAG enzymes may have descended from an ancient transposon, and the RSS are assumed to be relics of the transposon's inverted repeats (45). The capture of this system for site-specific DNA cleavage and relegation provided the means for evolution of adaptive immunity in jawed vertebrates.

FROM INFORMATICS TO EXPERIMENT, AND BACK AGAIN

Improvement of procedures for the automated screening of anonymous DNA heads the list of important tasks for informatics. Useful models of regulatory DNA, however, will not only provide descriptors for use in identifying sites of interest but also will help elucidate the sequence properties determining function and predict the functional level of putative sites. We believe that successful models of variable sites must account for the patterns underlying the variability (see, e.g., 40). The scope outlined here for future informatics would lead to the gathering of further data, particularly in the areas of

(i) population genetics,
(ii) biochemistry with reporter constructs, EMSA, and plasmon resonance, and
(iii) reverse genetics with promoter reshuffling.

In population genetics, we expect increasing use will be made of the DNA banks, such as the one from inbred wild-derived mice at the Jackson

304   L. G. Cowell, N. A. Mitchison, and B. Muller

Laboratory. These data will in turn demand further informatics. The trade
will be two-way and will require scientists equipped to handle both sides.

REFERENCES

1. Gehring, W. J., Qian, Y. Q., Billeter, M., Furukubo-Tokunaga, K., Schier, A. F.,
   Resendez-Perez, D., Affolter, M., Otting, G., and Wuthrich, K., 1994. Home-
   odomain – DNA recognition. *Cell* 78:211–23.
2. Davidson, E. H., 2001. *Genomic Regulatory Systems: Development and Evolution.*
   Academic Press, San Diego.
3. Mitchison, N. A., 1997. Partitioning of genetic variation between regulatory
   and coding gene segments: the predominance of software variation in genes
   encoding introvert proteins. *Immunogenetics* 46:46–52.
4. Carroll, S. B., 2000. Endless forms: the evolution of gene regulation and morphological diversity. *Cell* 101:577–80.
5. Enard, W., Khaitovich, P., Klose, J., Zollner, S., Heissig, F., Giavalisco, P., Nieselt-
   Struwe, K., Muchmore, E., Varki, A., Ravid, R., Doxiadis, G. M., Bontrop, R. E.,
   and Paabo, S., 2002. Intra- and interspecific variation in primate gene expression
   patterns. *Science* 296:340–3.
6. Hofer, T., Nathansen, H., Lohning, M., Radbruch, A., and Heinrich, R., 2002.
   GATA-3 transcriptional imprinting in Th2 lymphocytes: a mathematical model.
   Proc. Natl. Acad. Sci. USA 26:26.
7. Paillard, F., Chansel, D., Brand, E., Benetos, A., Thomas, F., Czekalski, S.,
   Ardaillou, R., and Soubrier, F., 1999. Genotype-phenotype relationships for
   the renin angiotensin-aldosterone system in a normal population. *Hypertension*
   34:423–9.
8. Maniatis, T., Falvo, J. V., Kim, T. H., Kim, T. K., Lin, C. H., Parekh, B. S., and
   Wathelet, M. G., 1998. Structure and function of the interferon-beta enhanceosome. *Cold Spring Harb. Symp. Quant. Biol.* 63:609–20.
9. Gelfand, M. S., 1995. Prediction of function in DNA sequence analysis. *J. Com-
   put. Biol.* 2:87–115.
10. Bucher, P., 1999. Regulatory elements and expression profiles. *Curr. Opin. Struct.
    Biol.* 9:400–7.
11. Vanet, A., Marsan, L., and Sagot, M. F., 1999. Promoter sequences and algorith-
    mical methods for identifying them. *Res. Microbiol.* 150:779–99.
12. Stormo, G. D., 2000. DNA binding sites: representation and discovery. *Bioinfor-
    matics* 16:16–23.
13. Coleman, S. L., Buckland, P. R., Hoogendoorn, B., Guy, C., Smith, K., and
    O'Donovan, M. C., 2002. Experimental analysis of the annotation of promoters
    in the public database. *Hum. Mol. Genet.* 11:1817–21.
14. Burge, C. B., 1998. Modeling dependencies in pre-mRNA splicing signals. In
    Computational Methods in Molecular Biology, S. L. Salzberg, D. B. S., S. Kasif, Eds.
    Elsevier Science, Amsterdam.
15. Carrington, M., Dean, M., Martin, M. P., and O'Brien, S. J., 1999. Genetics
    of HIV-1 infection: chemokine receptor CCR5 polymorphism and its conse-
    quences. *Hum. Mol. Genet.* 8:1939–45.

16. Mitchison, N. A., 2001. Polymorphism in regulatory gene sequences. *Genome Biol.* 2:1–6.
17. Kornman, K. S., and Duff, G. W., 2001. Candidate genes as potential links between periodontal and cardiovascular diseases. *Ann. Periodontol.* 6:48–57.
18. Asderakis, A., Sankaran, D., Dyer, P., Johnson, R. W., Pravica, V., Sinnott, P. J., Roberts, I., and Hutchinson, I. V., 2001. Association of polymorphisms in the human interferon gamma and interleukin-10 gene with acute and chronic kidney transplant outcome: the cytokine effect on transplantation. *Transplantation* 71:674–7.
19. Becker, K. G., Simon, R. M., Bailey-Wilson, J. E., Freidlin, B., Biddison, W. E., McFarland, H. F., and Trent, J. M., 1998. Clustering of non-major histocompatibility complex susceptibility candidate loci in human autoimmune diseases. *Proc. Natl. Acad. Sci. USA* 95:9979–84.
20. Louis, P., Eliaou, J. F., Kerlan-Candon, S., Pinet, V., Vincent, R., and Clot, J., 1993. Polymorphism in the regulatory region of HLA-DRB genes correlating with haplotype evolution. *Immunogenetics* 38:21–6.
21. Louis, P., Vincent, R., Cavadore, P., Clot, J., and Eliaou, J. F., 1994. Differential transcriptional activities of HLA-DR genes in the various haplotypes. *J. Immunol.* 153:5059–67.
22. Baumgart, M., Moos, V., Schuhbauer, D., and Muller, B., 1998. Differential expression of major histocompatibility complex class II genes on murine macrophages associated with T cell cytokine profile and protective/suppressive effects. *Proc. Natl. Acad. Sci. USA* 95:6936–40.
23. Nishimura, Y., Kamikawaji, N., Fujisawa, K., Yoshizumi, H., Yasunami, M., Kimura, A., and Sasazuki, T., 1991. Genetic control of immune response and disease susceptibility by the HLA-DQ gene. *Res. Immunol.* 142:459–66.
24. Ottenhoff, T. H., Walford, C., Nishimura, Y., Reddy, N. B., and Sasazuki, T., 1990. Natural variation in immune responsiveness, with special reference to immunodeficiency and promoter polymorphism in class II MHC genes. *Eur. J. Immunol.* 20:2347–50.
25. Oliveira, D. B., and Mitchison, N. A., 1989. HLA-DQ molecules and the control of *Mycobacterium leprae*-specific T cell nonresponsiveness in lepromatous leprosy patients. *Clin. Exp. Immunol.* 75:167–77.
26. Mitchison, N. A., Muller, B., and Segal, R. M., 2000. Natural variation in immune responsiveness, with special reference to immunodeficiency and promoter polymorphism in class II MHC genes. *Hum. Immunol.* 61:177–81.
27. Cowell, L. G., Kepler, T. B., Janitz, M., Lauster, R., and Mitchison, N. A., 1998. The distribution of variation in regulatory gene segments, as present in MHC class II promoters. *Genome Res.* 8:124–34.
28. Mitchison, N. A., and Roes, J., 2002. Patterned variation in murine MHC promoters. *Proc. Natl. Acad. Sci. USA* 19:19.
29. Beck, S., and Trowsdale, J., 2000. The human major histocompatibility complex: lessons from the DNA sequence. *Annu. Rev. Genomics Hum. Genet.* 1:117–37.
30. Lynch, M., and Conery, J. S., 2000. The evolutionary fate and consequences of duplicate genes. *Science* 290:1151–5.

31. Kondrashov, F. A., Rogozin, I. B., Wolf, Y. I., and Koonin, E. V., 2002. Selection in the evolution of gene duplications. *Genome Biol.* 3:1–8.
32. Alba, M. M., Das, R., Orengo, C. A., and Kellam, P., 2001. Genomewide function conservation and phylogeny in the Herpesviridae. *Genome Res.* 11:43–54.
33. Alba, M. M., Lee, D., Pearl, F. M., Shepherd, A. J., Martin, N., Orengo, C. A., and Kellam, P., 2001. VIDA: a virus database system for the organization of animal virus genome open reading frames. *Nucl. Acids Res.* 29:133–6.
34. Muschen, M., Re, D., Brauninger, A., Wolf, J., Hansmann, M. L., Diehl, V., Kuppers, R., and Rajewsky, K., 2000. Somatic mutations of the CD95 gene in Hodgkin and Reed-Sternberg cells. *Cancer Res.* 60:5640–3.
35. Lee, T. J., Kim, S. J., and Park, J. H., 2000. Influence of the sequence variations of the HLA-DR promoters derived from human melanoma cell lines on nuclear protein binding and promoter activity. *Yonsei Med. J.* 41:593–9.
36. Derreumaux, S., and Fermandjian, S., 2000. Bending and adaptibility to proteins of the cAMP DNA-responsive element: molecular dynamics contrasted with NMR. *Biophys. J.* 79:656–69.
37. Qians, X., Strahs, D., and Schlick, T., 2001. Dynamic simulations of 13 TATA variants refine kinetic hypotheses of sequence/activity relationships. *J. Mol. Biol.* 308:681–703.
38. Humphries, S. E., Luong, L. A., Ogg, M. S., Hawe, E., and Miller, G. J., 2001. The interleukin-6-174 G/C promoter polymorphism is associated with risk of coronary heart disease and systolic blood pressure in healthy men. *Eur. Heart J.* 22:2243–52.
39. Ramsden, D. A., Baetz, K., and Wu, G. E., 1994. Conservation of sequence in recombination signal sequence spacers. *Nucl. Acids Res.* 22:1785–96.
40. Cowell, L. G., Davila, M., Yang, K. Y., Kepler, T. B., and Kelsoe, G., 2002. Prospective estimation of recombination signal efficiency and identification of functional cryptic signals in the genome by statistical modeling. *J. Exp. Med.* 197:207–20.
41. Hesse, J. E., Lieber, M. E., Mizuuchi, K., and Gellert, M., 1989. V(D)J recombination: a functional definition of the joining signals. *Genes and Development* 3:1053–61.
42. Livak, F., and Petrie, H. T., 2001. Somatic generation of antigen-receptor diversity: a reprise. *Trends Immunol.* 22:608–12.
43. Bassing, C. H., Alt, F. W., Hughes, M. M., D'Auteuil, M., Wehrly, T. D., Woodman, B. B., Gartner, F., White, J. M., Davidson, L., and Sleckman, B. P., 2000. Recombination signal sequences restrict chromosomal V(D)J recombination beyond the 12/13 rule. *Nature* 405:583–6.
44. Liang, H. E., Hsu, L. Y., Cado, D., Cowell, L. G., Kelsoe, G., and Schlissel, M. S., 2002. The "dispensable" portion of RAG-2 is necessary for efficient V-to-DJ rearrangement during B and T cell development. *Immunity* 17:639–51.
45. Hansen, J. D., and McBlane, J. F., 2000. Recombination-activating genes, transposition, and the lymphoid-specific combinatorial immune system: a common evolutionary connection. *Curr. Top. Microbiol. Immunol.* 248:111–35.
46. Heldt, C., Listing, J., Sozeri, O., Blasing, F., Frischbutter, S., Muller, B., 2003. Differential expression of HLA class II genes associated with disease susceptibility and progression in rheumatoid arthritis. *Arthritis Rheuma.* 48:2779–87.

THIRTEEN

Genetic Epidemiology of Infectious Diseases: The First Half-Century

Newton E. Morton

INTRODUCTION

The discovery of blood groups by Landsteiner (1900) coincided with the rediscovery of Mendelism. During the next generation, immunology and genetics developed together, but the ideas and even the vocabulary that are essential today were slow to evolve. The term *population genetics* was not used by Fisher, Wright, or Haldane but was introduced by C. C. Li (1948) in his magisterial synthesis that anticipated division of the field into what we now call evolutionary genetics and genetic epidemiology. Neel and Schull (1954) devoted a chapter of their book to *epidemiological genetics*, which evolved into *genetic epidemiology* to reflect the dual origins of disease (Morton et al., 1967). A characteristic of the time was that definitions appeared first in books rather than in articles. However, the pace quickened as human genetics rose from the ashes of eugenics. Haldane (1949) suggested that thalassemia minor might be protective against malaria. The young regional societies of human genetics and the first international congress of human genetics were excited by critical evidence for the link between hemoglobin S and *falciparum* malaria (Allison, 1954). At the same time, Watson and Crick (1953) created molecular biology, which provided the techniques and markers for genetic epidemiology of infectious disease. The essential vocabulary was slowly invented: *linkage disequilibrium* (Ohta and Kimura, 1969), *haplotype* (Ceppellini et al., 1967), *diplotype* (Morton, 1983), and linkage disequilibrium unit or LDU (Maniatis et al., 2002).

ABO BLOOD GROUPS

Fifty years ago polymorphisms were generally thought to be rare and maintained by selection. Now DNA technology has revealed so many polymorphisms that most must be effectively neutral and useful only to localize functionally selective markers. Among the classical markers, the ABO blood groups have the strongest claim to selection by infectious disease, because each ABO phenotype produces antibodies against the epitopes it lacks, perhaps in response to provocation by microorganisms carrying those epitopes (Springer et al., 1959).

Association with noninfectious disease, such as cancers of the digestive system, may be secondary to the role of ABH-labeled microorganisms as risk factors for cancer. Microbiological characterization of antigens shared with humans has lagged behind development of fluorescent probes and strain-specific markers, and therefore understanding of mechanisms is indirect and fragmentary. Two decades of research were prompted by Aird et al. (1953), who noted that group O and gastric cancer were more common in the north of the United Kingdom than in the south. Their careful study of cases and controls in various cities confirmed an association, but with group A. This observation was repeated and extended to other cancers of the digestive tract in many populations, taking advantage of preoperation blood typing, and confirmed in some studies by transmission disequilibrium in sibships. Therefore, population stratification and linkage disequilibrium were effectively ruled out. Group O and nonsecretors are at increased risk of duodenal ulcers, for unknown reasons that may involve microorganisms. Among the many studies of infectious disease (Mourant et al., 1978), six associations in multiple samples stand out (Table 13.1). Of these, bilharzial hepatic failure is estimated to have the highest relative risk for groups A and B and the greatest χ^2 when evidence on relative risk is pooled over samples, but there is significant heterogeneity (Woolf, 1955). Rheumatic heart disease is also strongly associated with groups A and B, with weakly significant heterogeneity among studies. The other reports have highly significant main effects and no evidence of heterogeneity.

Smallpox (not listed in Table 13.1) has received the greatest attention and provides a cautionary tale. Vogel et al. (1960) predicted an association with group A from two facts: the frequency of group B is elevated in Central Asia at the expense of group A, and they found A activity in a cowpox–chick culture. Their prediction was apparently verified by Vogel and Chakravartti (1966) among fatal cases in two Indian epidemics. However, no other group was able to confirm an association (Azevedo et al., 1964). Springer and Weiner (1962) attacked the immunological work and emphasized the evolutionary potential of pathogens to evade host defenses, and Harris et al. (1963) found A activity in chicks but not in

Table 13.1. Possible ABO and SE associations in multiple samples

Disease	Phenotype	N	RR	χ^2_1	χ^2_{n-1}
Influenza A$_2$	A/O	701	0.67	22.87	ns
Bilharzial hepatic failure	A/O	353	4.05	65.98	**
	B/O	353	2.19	16.41	ns
Tropical eosinophilia	A/O	780	2.22	43.01	ns
	B/O	780	1.75	21.98	ns
Rheumatic heart disease	A/O	5103	1.23	36.53	*
	B/O	5103	1.28	24.65	ns
Cerebrospinal syphilis	A/O	341	1.57	10.59	ns
Rheumatic fever	SE+/SE−	1308	0.78	11.04	ns

Note: *, P < .05; **, P < .01.

cowpox. Mourant et al. (1978) concluded that no association with small-pox had been demonstrated and that eradication of the disease made further research impractical. However, the past year has raised the specter of smallpox and other infectious diseases as a terrorist weapon. If countered by present vaccines, a small proportion of recipients will die of postvaccinal encephalitis. The study of genetic factors in resistance to infectious disease and the sequelae of vaccination may return to the limelight, more fruitfully this time through advances in molecular biology. It is no longer necessary to infer the epitopes of pathogens indirectly, for example that the canine tapeworm must have P-like antigens because high-quality anti-P is produced by P-individuals with echinococcus infection (Cameron and Staveley, 1957), although cyst fluid alone is not antigenic (Levine et al., 1958).

HYPERNORMAL CONTROLS

One of the most useful tricks of epidemiology is to replace random controls by hypernormals who may be defined in various ways (Morton and Collins, 1998). The gain in power is easily achieved in case-control studies but can be realized in family studies only when the initial sample

Table 13.2. Venereal disease in Chinese prostitutes

Infection Group	Number	Mean Age	Exposure Time (years)
Repeated gonorrhea	56	35.8	5.5
Latent syphilis	31	39.3	10.5
Double infection	31	37.3	6.7
Resistant	30	32.5	4.3

Table 13.3. HLA in Chinese prostitutes

Group	Number	per cent Aw19 B17	per cent A11 B15
Double infection	31	26[a]	3
Normal controls	238	7	13
Resistant	30	7	30
Resistant > 2 years	15	0	47[a]

[a] $P < .001$.

is very large and a small subsample is chosen for genetic analysis. One design is especially useful for infectious disease: affected cases and normal controls with heavy environmental exposure. Its use was pioneered by Chan and Rajan (1982) in a study of HLA-resistance to venereal disease among Chinese prostitutes in an extremely high-risk area of Singapore. The four infection groups were similar in mean age and exposure (Table 13.2). However, the HLA distribution was markedly different when the highly susceptible and highly resistant groups were compared (Table 13.3). This old study has not been repeated in Singapore or elsewhere with more precise characterization of the immune response and the HLA genotypes. However, a similar design has been used to identify CCR5 as the HIV-1 receptor, comparing heavily exposed controls who remain seronegative with seropositive cases (Samson et al., 1996). The protective effect of the homozygous deletion Δccr5 is striking (Table 13.4), and it formed the basis for more detailed characterization of genetic risk factors. The deletion cannot make the HIV-1 receptor, recalling the protective effect of the Duffy (a−b−) allele that cannot form the receptor for *vivax* malaria, to which Africans have long been known to be resistant (Miller et al., 1975). Other papers in this volume describe the exciting developments in understanding of malaria-dependent polymorphisms that began half a century ago.

Table 13.4. CCR5 and HIV-1 response

Allele	Seronegative	Seropositive	Total
CCR5+	1278	1368	2646
Δccr5	130	78	208
Total	1408	1446	2854

Note: $\chi_1^2 = 15.56$, $P < 10^{-4}$.

Table 13.5. Race by schistosomiasis in Caatinga do Moura

| | Severity | | | | |
Race	Negative 0	Mild 1	Moderate 2	Severe 3	Total
Black 0	156	101	12	2	271
Mulatto 1	340	248	84	19	691
White 2	132	100	50	12	294
Total	628	449	146	33	1256

Note: Regression of infection on race: $\chi_1^2 = 8.98$, P = .003. Regression of severity on race: $\chi_1^2 = 23.66$, P $< 10^{-5}$. Residual: $\chi_5^2 = 7.14$.

GROUP DIFFERENCES

In the absence of strict randomization of the environment there is no infallible method to control environmental differences between groups. However, several methods have been useful. Covariance adjustment minimizes measurable environmental differences, but errors of the independent variables leave residual differences that cannot be interpreted (Rao et al., 1977). Linkage disequilibrium is a powerful tool, especially in children of interracial crosses. Infectious disease tempts us to push group differences to the limit by suggesting that genetic factors for pathogen resistance will accumulate in populations that have been exposed to intense pathogen pressure for a long time. This prediction clearly works for malaria-dependent polymorphisms but is ambiguous for other infectious diseases. For example, CCR5 is polymorphic in Eurasian populations that have only recently been exposed to HIV-1, suggesting that they have been exposed to another pathogen that uses the CCR5 receptor. *Schistosoma mansoni* was introduced to the New World by slaves. At least four Brazilian studies have concluded that Caucasian ancestry is a risk factor for severity of schistosomiasis in populations of mixed origin classified by racial phenotype (Cardoso, 1953; Prata and Schroeder, 1967; Nunesmaia et al., 1975; Bina et al., 1978). Table 13.5 gives results from the last of these studies, taken from the village of Caatinga do Moura. The authors considered that the environment was homogenous, with whites not especially exposed to schistosomiasis. The regression of severity on a rough measure of white ancestry is highly significant ($\chi_1^2 = 23.66$, P $< 10^{-5}$), with no significant residual ($\chi_5^2 = 7.14$). This village has also been used for marker studies of schistosomiasis that neglected ethnicity (Marquet et al., 1999). Combination of the two approaches is desirable if susceptibility factors are to be not only localized but identified. Many other infectious diseases have more or less convincing evidence of geographic origin or exposure intensity that could orient genetic studies.

POSITIONAL CLONING

For half a century, linkage has been the main tool to localize susceptibility factors, and the methods are fairly stable (Morton, 2002). On the contrary, linkage disequilibrium became popular only recently (Morton and Collins, 2002). There is an infinite number of LD measures, one of which (ρ) has been shown to be optimal in two respects – by obeying the Malecot theory that describes LD both in time and with distance on the genetic map and by being least sensitive to allele frequencies. Other measures do not have the same parameters and are seriously inefficient (Morton, 2000). Therefore, use of diverse measures of association results in a loss of much of the value of LD studies. This is assuming special importance as the LD map is developed with capability to extend the resolution provided by linkage, specify efficient spacing of SNPs for positional cloning, and detect selective sweeps characteristic of particular haplotypes and populations. Properly constructed LD maps promise to be as useful in genetics as linkage maps have been but with the much greater resolution required to detect hot and cold spots at the kb level.

There are several obstacles to these developments. A recent paper argues for use of the correlation metric, for which there is no evolutionary theory and the distribution is highly skewed and heteroscedastic, highly sensitive to allele frequencies, and has poor goodness of fit (Teare et al., 2002). The principal compensation is that the asymptote is smaller, but nearby values are no more informative than with any other asymptote.

A serious problem is that populations differ much more in LD than in linkage. Given adequate sample size and dense markers, an LD map is best in the population for which it was created. A large effort is being made to construct LD maps from three small samples representing the major ethnic groups (Couzin, 2002). This material cannot usefully be pooled, and a composite map is incorrect for each constituent. These problems will not be dispelled by replacing LD with some measure of haplotype diversity. Efficient use of candidate regions inferred by linkage, LD, haplotypes, function, and chromosomal abnormalities will require more judgment than a genome scan. The next half-century will be very different from the last, but the old problems should be remembered until they are solved. Despite many established associations, the roles of genes in infectious diseases have hardly begun to be elucidated.

REFERENCES

Aird L, Bentall HH, and Roberts JAF (1953). A relationship between cancer of the stomach and the ABO blood groups. *Brit. Med. J.* 1:799–801.

Allison AC (1954). The distribution of the sickle-cell trait in East Africa and elsewhere, and its apparent relationship to subtertian malaria. *Trans. Soc. Trop. Med. Hyg.* 48:312–18.

Azevedo E, Kreiger H, and Morton NE (1964). Smallpox and the ABO blood groups in Brazil. *Am. J. Hum. Genet.* 16:451–4.

Bina JC, Tavares-Neto J, Prata A, and Azevedo ES (1978). Greater resistance to development of severe schistosomiasis in Brazilian negroes. *Hum. Biol.* 50:41–9.

Cameron GL, and Staveley JM (1957). Blood group P-substance in hydatid cyst fluids. *Nature* 179:147–8.

Cardoso W (1953). A esquistossomose mansonica no negro. *Med. Cir. Farm.* 202:89–93.

Ceppellini R, Curtoni ES, Mattiuz PL, Miggiano V, Scudeller G, and Serra A (1967). Genetics of leucocyte antigens. A family study of segregation and linkage. In *Histocompatability Testing, 1967* (Curtoni ES, Mattiuz PL, and Tosi RM, Eds.), pp. 149–85. Munksgaard, Copenhagen.

Chan SH, and Rajan VS (1982). HLA – resistance and receptibility to venereal disease. In *Immunogenetics in Rheumatology* (Dawkins RL, Christensen FT, and Zilko PJ, Eds.), pp. 20–3. Excerpta Med. Elsevier Co., Amsterdam.

Couzin J (2002). New mapping project splits the community. *Science* 296:1391–3.

Haldane JBS (1949). Disease and evolution. *Ricerca Sci.* 19(Suppl.):68–76.

Harris R, Harrison GA, and Rondle CJM (1963). Vaccinia virus and human blood group-A substance. *Acta Genet.* 13:44–57.

Landsteiner K (1900). Zur Kenntnis der antifermentativen, lytischen und agglutinierenden Wirkungen des Blutserums and der Lymphe. *Zbl. Bakt. Abt.* 1(27): 357–62. Translation in: UIS Army Med. Res. Lab. (1970). Selected contribution to the literature of blood groups and immunology, 1. Fort Knox, Kentucky.

Levine P, Celano M, and Staveley JM (1958). The antigenicity of P substance in Echinococcus cyst fluid coated on to tanned red cells. *Vox Sang.* 3:434–8.

Li CC (1948). *An Introduction to Population Genetics.* Peking University Press, Peking.

Maniatis N, Collins A, Xu C-F, McCarthy LC, Hewitt DR, Tapper W, Ennis S, Ke X, and Morton NE (2002). The first linkage disequilibrium (LD) maps: Delineation of hot and cold blocks by diplotype analysis. *Proc. Natl. Acad. Sci. USA* 99:2228–33.

Marquet S, Abel L, Hillaire D, and Dessein A (1999). Full results of the genome-wide scan which localises a locus controlling the intensity of infection by *Schistosoma mansoni* on chromosome 5q31-q33. *Eur. J. Hum. Genet.* 7:88–97.

Miller LH, Mason ST, Dvorak JA, McGiness MH, and Rothman IK (1975). Erythocyte receptors for (*Plasmodium knowlesi*) malaria: Duffy blood group determinants. *Science* 189:561–3.

Morton NE (1983). Factor-union phenotype systems. In *Methods in Genetic Epidemiology* (Morton NE, Rao DC, and Labuel J-M, Eds.), pp. 14–16. S. Karger, Basel.

Morton NE (2000). LODs past and present. In *Perspectives on Genetics* (Crow J and Dove WJ, Eds.), pp. 453–8, 705. University of Wisconsin Press, Madison, WI.

Morton NE, and Collins A (1998). Tests and estimates of allelic association in complex inheritance. *Proc. Natl. Acad. Sci. USA* 95:11389–93.

Morton NE, and Collins A (2002). Toward positional cloning with SNPs. *Curr. Opin. Mol. Ther.* 4:259–64.

Morton NE, Chung CS, and Mi M-P (1967). *Genetics of Interracial Crosses in Hawaii.* S. Karger, New York.

Mourant AE, Kopec AC, and Domaniewska-Sobczak K (1978). *Blood Groups and Diseases.* Oxford University Press, Oxford.

Neel JV, and Schull WJ. (1954). *Human Heredity.* University of Chicago Press, Chicago.

Nunesmaia GN, Azevedo ES, Arandas EA, and Widmer CG (1975). Composicao racial e anaptoglobinemia em portadores de equistossomose mansonica forma hepatoesplenica. *Rev. Inst. Med. Trop. Sao Paulo* 17(3):160–3.

Ohta T, and Kimura M (1969). Linkage disequilibrium due to random genetic drift. *Genet. Res.* 13:47–55.

Prata A, and Schroeder S (1967). A comparison of whites and negroes infected with *Schistosoma mansoni* in a hyperendemic area. *Gaz. Med. Bahia* 67:93–8.

Rao DC, Morton NE, Elston RC, and Yee S (1977). Causal analysis of academic performance. *Behav. Genet.* 7:147–59.

Samson M, Libert F, Donanz BJ, Rucker T, Liesnard C, Farber CM, Saragosti S, Lapoumeroulie C, Cognaux T, Forceille C, and 12 colleagues (1996). Resistance to HIV-1 infection in Caucasian individuals bearing mutant alleles of the CCR-5 chemokine receptor gene. *Nature* 382:251–3.

Springer GF, and Weiner AS (1962). Alleged causes of the present day world distribution of the human ABO blood groups. *Nature* 193:444–6.

Springer GF, Horton RE, and Forbes M (1959). Origin of anti-human blood group B agglutinins in White Leghorn chicks. *J. Exp. Med.* 110:221–44.

Teare MD, Dunning AM, Durocher F, Rennant G, and Easton DF (2002). Sampling distribution of summary linkage disequilibrium measures. *Ann. Hum. Genet.* 66:223–33.

Vogel F, and Chakravartti MR (1966). ABO blood groups and smallpox in a rural population of West Bengal and Bihar (India). *Humangenetik* 3:166–80.

Vogel F, Pettenkoffen HJ, and Helmbold W (1960). Uber die populationsgenetik der ABO-blutgruppen. *Acta Genet.* 10:267–94.

Watson JD, and Crick FHC (1953). Molecular structure of nucleic acids: A structure for deoxyribose nucleic acid. *Nature* 171:737–8.

Woolf B (1955). On estimating the relation between blood groups and disease. *Ann. Hum. Genet.* 19:251–3.

FOURTEEN

The Impact of Human Genetic Diversity on the Transmission and Severity of Infectious Diseases

Michel Tibayrenc

THE INFECTIOUS THREAT

This opening century will appear as both "the golden age of genetics and the dark age of infectious diseases" (Tibayrenc, 2001a). On the battlefront of infectious diseases, the situation is more than just a concern, due to the threat of emerging and reemerging infectious diseases (ERID). In developing countries, infectious diseases still are the main demographic regulating factor. In particular, Africa is more than ever afflicted with sleeping sickness, malaria, bilharziosis, and other major parasitoses. The three "diseases of poverty," namely malaria, tuberculosis, and AIDS, have become the top priority of the World Health Organization. The industrial world has not been spared. In France, 12,000 people die every year of nosocomial infections. In New York City, 25% of the *Mycobacterium tuberculosis* strains are resistant to antibiotics.

ENVIRONMENTAL AND BIOLOGICAL FACTORS

Transmission and severity of infectious diseases are the result of a complex interplay between environmental and biological (built-in) parameters. There is no doubt that environmental factors play a major role in the present resurgence of infectious diseases, through climatic changes, massive migrations, economic inequalities, and political instability. However, even in acting on these environmental factors, control is more efficient when sophisticated knowledge of the biology of the disease under survey is available. For example, in Latin America, Chagas disease is a parasitic disease caused by the flagellate *Trypanosoma cruzi* and transmitted by triatomine bugs (hematophagous tree bugs). Efficient control of their transmission has been achieved in some advanced countries, such as Brazil,

315

through improvement of human habitats (triatomine bugs preferentially colonize poor habitats). However, this control has been greatly improved by a thorough understanding of the biology of the many vector species, their population structures, and the genetic mechanisms of their insecticide resistance (information available at http://eclat.fcien.edu.uy/).

THE GOLDEN AGE OF GENETICS?

The good news in this dark situation is the immense progress of genetic science. Unfortunately, however, this progress concerns technology more than it does theory (Tibayrenc, 2001b). We are flooded with an ocean of data generated by powerful technologies (automatic sequencing, DNA chips, bioinformatics), and we count on megacomputers to sort it all out. On the other hand, genetic and evolutionary theories have not changed much since the time I was a student. Detailed knowledge of the structure and functioning of the genomes provides a hope that evolutionary theory can be taken out of its speculative status into a "neoPasteurian era" based on straightforward observation of crude data and reliable facts. However, to make this enterprise successful, it will be crucial to rely on a specific population and holistic thought (see below). Moreover, the genome of a host is shaped by the action of a pathogen, and vice versa. In the case of vector-borne diseases, the genome of the vector interferes too. It is therefore highly desirable to study the genetics of the host, the pathogen, and the vector both at the individual and population levels and even more desirable to focus on the interactions between them, for it is a unique biological phenomenon – coevolution (Tibayrenc, 1998).

THE SEARCH FOR SUSCEPTIBILITY GENES FOR INFECTIOUS DISEASES

Progress in the field of genetics provides us with powerful tools to look for the genes responsible for susceptibility to infectious diseases. This work relies on two main approaches, finding gene candidates and linkage studies.

In the gene candidate approach, a gene is suspected a priori to be involved in susceptibility to a given disease, either because it is associated with biological processes that make it a plausible candidate (for example, immunological mechanisms) or because it has been identified by animal experiments.

In the linkage approach, a given region of the genome is located through whole genome screening by microsatellite markers and association analyses relying on twin/sibpair/family/pedigree studies. The null hypothesis is a total lack of linkage (free recombination between the marker used and the hypothetical gene). If the recombination rate is significantly lower than 0.5, the linkage hypothesis is retained. Of course, all methods of gene identification are more successful when the genetic component of susceptibility is strong, when few genes are involved, and when precise hypotheses on the Mendelian inheritance of the involved genes are available. However, such favorable situations are probably the exception rather than the rule. Nonparametric tests do not require a working hypothesis on inheritance. However, they are less powerful than parametric tests. The more classical parametric linkage test is the lod (logarithm of the odds) score (Morton, 1955): Lod $= \log_{10}$ [(probability of data if disease and marker are linked)/(probability of data if disease and marker recombine freely)].

Thanks to new powerful technologies, systematic screening is becoming easier and the number of microsatellite markers is now considerable.

INDIVIDUALS VS POPULATIONS

Medical approaches developed in the Western world still tend to favor individual-level treatment to the detriment of population-level treatment. Few medical doctors think in terms of population/evolution. Now the two sides are complementary. Curative medicine is designed to take care of the individual, whereas preventive medicine and disease control are conceivably more efficient when they consider the population level. In the human species, not only individuals but also populations are targets of natural selection and evolution through multilevel and **group selection** (Wilson, 1997). This must be especially true in the field of infectious diseases. At the two levels, those of individuals and populations, there could be differences in genetic susceptibility, not only to the diseases themselves, but also to the drugs, antibiotics, and vaccines designed to control them.

THE DISTRIBUTION OF OVERALL DIVERSITY IN
THE HUMAN SPECIES

Our species exhibits considerable phenotypic polymorphism contrasting with relatively low genetic diversity. In fact, when usual population genetic markers such as isoenzymes or microsatellites are used, the main human ethnic groups are separated by **genetic distances** that are comparable to the distances observed between local populations of *Drosophila*, and the intrapopulation diversity is higher than the interpopulation diversity (Nei and Roychoudhury, 1974, 1993). However, the phylogenetic trees designed after these markers match rather well the classical ethnic subdivisions of the human species (Cavalli-Sforza et al., 1994; Nei, 1978; Nei and Roychoudhury, 1993). Moreover, the nodes of the main branches of such trees are quite robust, which indicates that genetic exchanges between

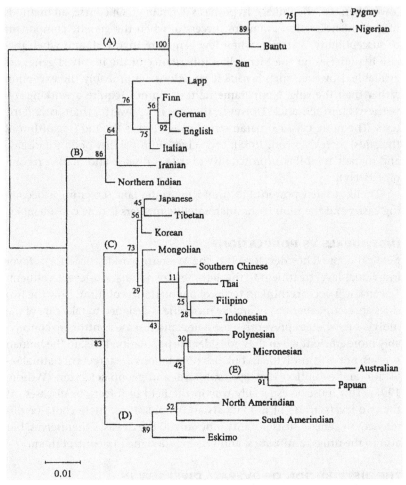

Figure 14.1. Neighbor-joining phylogenetic tree of 26 human populations built with genetic markers. Lengths of the branches represent the genetic distances (Nei, 1972) that separate the different groups. (A) = Africans; (B) = Caucasians; (C) = Greater Asians; (D) = Amerindians; and (E) = Australopapuans (after Nei and Roychoudhury, 1993).

these groups have been rather limited, at least until recently (see Figure 14.1). Lastly, it is important to emphasize that the markers used to design these trees are considered selectively neutral. They are therefore no more than relative time markers and say nothing about the relevant properties of the organisms under survey. Obviously human morphological diversity is not randomly distributed throughout the species but is linked to geographical distance. Ethnic diversity is the result of local selective pressures (climatic among others) acting strongly at the population level. It is probable that this phenotypic diversification occurred in a relatively short time, because it is currently accepted that all present human populations diverged from a common ancestral population about 100,000 years ago (Solignac, 1998). It is striking that human phenotypic diversity is not limited to morphology or skin color and extends to unexpected physiological parameters. Medical doctors trained for tropical medicine (as in my case) are aware that Africans (and also African Americans) have frequent abnormalities in their electrocardiograms (repolarization) that have no pathological significance (Gentillini, 1993).

INFECTIOUS DISEASES AND HUMAN EVOLUTION
Haldane (1949) was the first to propose that infectious diseases have been the strongest selective force for human evolution for the past 5,000 years. Actually, the limitation to the last 5,000 years is at best gratuitous. A strong indication of the impact of infectious diseases on human evolution is the molecular evolution pattern of the HLA system. Some HLA alleles have a very ancient **coalescence time**, prior to the divergence among Catarrhinians, and are genetically more similar to some *Macacus* alleles than to other human HLA alleles. The explanation proposed (Ayala and Escalante, 1996) is balanced selection by infectious agents, mainly malaria. As mentioned in The Infectious Threat section, it would be misleading to maintain that infectious diseases are no longer a selective force at present. Even in industrial countries, the mortality rate due to them is still considerable. The situation is, of course, even worse in southern countries.

DIFFERENCES AMONG INDIVIDUALS AND AMONG HUMAN POPULATIONS
The selective pressure caused by infectious diseases differs drastically among geographical places and continents. It is therefore not astonishing that different human populations do not have the same genetic susceptibility to given infectious agents. It is well known that native Amerindian populations have been decimated by the pathogenic agents brought by the Europeans. Tuberculosis is a classical case of this (Poulet, 1994). The differences in AIDS epidemiology between Africa, on the one hand, and North America and Western Europe, on the other hand, have obvious environmental explanations. However, it is now clear that genetic factors are also involved (Dean et al., 1996; Samson et al., 1996), which makes Africans more susceptible to AIDS than Europeans due to allelic frequency differences for the CCR-5 **chemokine** receptor gene. Another viral disease,

hepatitis C, shows a different genetic susceptibility distribution among human groups, too (Thio et al., 2001).

A telling example of differential ethnic impact of infectious diseases is the case of malaria in Africans, African Americans, and Melanesians (Labie, 1994; Miller, 1994). Balanced polymorphism due to malaria has been fairly well demonstrated for hemoglobine S (**sickle cell anemia**). Homozygous state S/S is 100% lethal in Africa, whereas the heterozygous individuals (S/A) have a selective advantage by comparison with children who do not have the mutant allele (A/A) genotype. Heterozygous individuals (S/A) are resistant to *Plasmodium falciparum*, the agent of the most malignant form of malaria. This persistence of a deleterious gene by selective pressure of a disease is a case of **balanced polymorphism**. A very similar case of balanced polymorphism is Melanesian ovalocytosis, which is due to a mutation in erythrocyte band 3. Such cases of balanced polymorphism suggest a very high mortality rate from malaria in Africa and Papua New Guinea. Although their genetic links with malaria are not clear, it has been postulated (Miller, 1994) that the high frequency of some other diseases in African Americans and Africans are due to a survival advantage they confer in malarious areas. This is the case for hypertension and iron overload with cirrhosis. It is worth noting that although African Americans underwent much genetic admixture from Caucasian populations, they still exhibit very specific genetic polymorphisms linked to malaria selective pressure. Statistical differences between African Americans and Caucasian Americans are also reflected in their HLA polymorphism (see below), which has been most probably shaped by the selective pressure of infectious diseases. It is probable that major transmissible diseases other than malaria have played a major selective role in sub-Saharan Africa. Human African trypanosomiasis (HAT) is a possible example.

The few examples cited above suggest that ethnic diversity is a parameter to be taken into account in the study of human genetic diversity and susceptibility to infectious diseases, even at a fine level, because, for example, susceptibility to malaria is different among West African ethnicities (Modiano et al., 1996, 1999).

Differences between ethnic groups, even at a fine level, are important to take into account for surveillance and control of infectious diseases and for easier identification of susceptible individuals. However, one has to keep in mind that in the same population there can be considerable differences in genetic susceptibility to infectious diseases among individuals (Abel and Dessein, 1997). Both levels, that of the population and that of individuals, must therefore be considered in attempts to control infectious diseases.

A SORELY NEEDED ENTERPRISE: THE HUMAN GENOME DIVERSITY PROJECT (HGDP)

In the euphoria following the launch of the Human Genome Program (HGP), the simple idea came that with only one individual for each part of the genome to be sequenced, one would miss an important parameter, which is the genetic diversity of our species. This was all the more concerning, because the individuals selected for the HGP were all Caucasians. Now European populations exhibit a notable genetic homogeneity, and much of the overall genetic variability of our species is found in African populations (Solignac, 1998). It was quickly proposed (Cavalli-Sforza et al., 1991) that a program complementary to the HGP be established, with a survey of a representative sample of human diversity through the analysis of diversified gene sequences. The project focused on those human populations who had remained genetically isolated for a long time, because they are considered to be more informative than urban populations, which underwent much admixture. The project generated much opposition from many organizations representing indigenous people, mainly the World Council of Indigenous People (WCIP) (Kahn, 1994). Although the concern of these organizations is understandable, from a scientific and medical point of view, having a clear idea of human gene sequence diversity remains highly desirable. Since the HGDP itself was stopped, other valuable initiatives have been launched. The African-American Diversity Project has been designed (Kahn, 1994). A specific interest of it is that HLA polymorphism, which has been strongly selected by infectious diseases, including malaria (Ayala and Escalante, 1996), presents notable differences between African Americans and Caucasian Americans, which leads to a lower success rate of organ transplantations in the first of these two communities. The European Union has launched a diversity program in Europe through a network of laboratories stretching from Barcelona to Budapest (Kahn, 1994). Lastly, the Chinese Human Genome Diverstity Project has already yielded valuable results (Chu et al., 1998).

CONCLUDING REMARKS

Identifying candidate genes of susceptibility to infectious diseases is an indispensable step. However, monogenic susceptibility mechanisms are probably the exception rather than the rule. Even if all genes involved in a polygenic disease are identified, however, it does not mean that the complete mechanism leading to the disease is elucidated. Gene interactions may be extremely complex. The genetic background of the simplest phenotypic traits at the individual level (shape of the nose, for example) is totally unknown, although such traits are obviously inherited. If we consider much more complicated phenotypes such as susceptibility to infectious diseases at the population level, or even more complex, host/pathogen coevolution, our ignorance is even greater (Tibayrenc, 2001b). This means that a precise knowledge of genetic susceptibility to infectious diseases (and to antibiotics/drugs/vaccines) can be hoped for only when we know much better the structure and function of the human genome, at the individual *and* the population level. The human genome project is only a first step toward reaching this goal. As stressed above, this project should not be developed to the detriment of population studies. It is desirable that the HGDP be designed so that clinical records of the individuals surveyed are carefully selected, and if possible, the individuals are selected with this need in mind.

Pure genetic approaches are not the only avenues to explore. Very simple morphological traits, such as symmetry, seem to have a strong link with susceptibility to infectious disease (Thornhill and Moller, 1997), possibly because symmetry is an indicator of hormonal equilibrium and/or fair immunological defenses. Such nonconventional approaches must not be neglected.

GLOSSARY

Balanced polymorphism   genetic polymorphism that persists in a population because the heterozygotes for the alleles concerned have a higher fitness than either homozygote.

Chemokine   cytokines that stimulate leucocyte chemotaxis.

Coalescence time   time elapsed between the common ancestral copy (one gene in one individual) and two or more copies of a given gene at the present time.

Genetic distance   various statistical quantities inferred from genetic data, estimating the genetic dissimilarities among individuals or populations. The most widely used are Nei's standard genetic distance (1972) and Jaccard distance (1908). Although the statistics differ, most genetic distances start from an estimation of the percentage of band mismatch on electrophoresis gels.

Group selection   natural selection acting upon a group of two or more individuals by which positively selected traits are more beneficial to the group than to the individual.

HLA   human leucocyte antigens involved in the rejection or acceptance of tissue and organ grafts and transplants or in the destruction of pathogens.

Sickle cell anemia   a generally lethal form of hemolytic anemia observed in individuals homozygous for the autosomal, codominant gene H^s. The red cells of these individuals contain an abnormal hemoglobin, Hb^s. These red cells exhibit a reversible shape alteration when the oxygen concentration in the plasma falls slightly and they get a sickle-like form. These pathological red cells have a shortened lifetime. About 0.2% of African American babies suffer from sickle cell anemia.

REFERENCES

Abel, L., and Dessein, A. J. 1997. The impact of host genetics on susceptibility to human infectious diseases. *Curr. Opin. Immunol.* 9:509–16.

Ayala, F. J., and Escalante, A. 1996. The evolution of human populations: a molecular perspective. *Mol. Phylogenet. Evol.* 5:188–201.

Cavalli-Sforza, L. L., Menozzi, P., and Piazza, A. 1994. *The History and Geography of Human Genes*. Princeton University Press, Princeton, NJ.

Cavalli-Sforza, L. L., Wilson, A. C., Cantor, C. R., Cook-Deegan, R. M., and King, M. C. 1991. Call for a world-wide survey of human genetic diversity: a vanishing opportunity for the human genome project. *Genomics* 11:490–1.

Chu, J. Y., Huang, W., Kuang, S. Q., Wang, J. M., Xu, J. J., Chu, Z. T., Yanga, Z. Q., Lina, K. Q., Li, P., Wu, M., Geng, Z. C., Tang, C. C., Du, R. F., and Jing, L. 1998. Genetic relationship of populations in China. *Proc. Natl. Acad. Sci. USA* 20:11763–8.

Dean, M., Carrington, M. Winckler, C., Huttley, G. A., Smith, M. W., Allikmets, R., Goedert, J. J., Buchbinder, S. P., Vittinghoff, E., Gomperts, E., Donfield, S., Vlahov, D., Kaslow, R., Saah, A., Rinaldo, C., Detels, R., Hemophilia Growth and Development Study, Multicenter AIDS Cohort Study, Multicenter Hemophilia Cohort Study, San Francisco City Cohort, ALIVE Study, Stephen O'Brien. 1996. Genetic restriction of HIV-1 infection and progression to AIDS by a deletion allele of the CKR5 structural gene. *Science* 273:1856–62.

Gentillini, M. 1993. *Médecine Tropicale*. Flammarion Editions, Paris.

Haldane, J. B. S. 1949. Disease and evolution. *La ricerca scientifica* Suppl. 19:68–76.

Hill, A. V. S. 2001. The genomics and genetics of human infectious disease susceptibility. *Annu. Rev. Hum. Genet.* 2:373–400.

Jaccard, P. 1908. Nouvelles recherches sur la distribution florale. *Bull. Soc. Vaudoise Sci. Nat.* 44:223–70.

Jarolim, P., Palek, J., Amato, D., Hassan, K., Sapak, P., Nurse, G. T., Rubin, H. L., Zhai, S., Sahr, K. E., and Liu, S. C. 1991. Deletion in erythrocyte band 3 gene in malaria-resistant Southeast Asian ovalocytosis. *Proc. Natl. Acad. Sci. USA* 88:11022–6.

Kahn, P. 1994. Genetic diversity project tries again. *Science* 266:720–2.

Labie, D. 1994. Polymorphismes génétiques et développement du paludisme: au delà du cas de la drépanocytose. *Médecine/Science* 10:905–6.

Miller, L. H. 1994. Impact of malaria on genetic polymorphism and genetic diseases in Africans and African Americans. *Proc. Natl. Acad. Sci. USA* 91:2415–19.

Modiano, D., Petrarca, V., Sirima, B. S., Nebie, I., Diallo, D., Esposito, F., and Coluzzi, M. 1996. Different response to *Plasmodium falciparum* malaria in west African sympatric ethnic groups. *Proc. Natl. Acad. Sci. USA* 93:13206–11.

Modiano D., Chiucchiuini, A., Petrarca, V., Sirima, B. S., Luoni, G., Roggero, M.A., Corradin, G., Coluzzi, M., and Esposito, F. 1999. Interethnic differences in the humoral response to non-repetitive regions of the *Plasmodium falciparum* circum-sporozoite protein. *Am. J. Trop. Med. Hyg.* 61:663–7.

Morton, N. E. 1955. Sequential tests for the detection of linkage. *Am. J. Hum. Genet.* 7:277–318.

Nei, M. 1972. Genetic distance between populations. *Am. Nat.* 106:283–92.

Nei, M. 1978. The theory of genetic distance and evolution of human races. *Jap. J. Hum. Genet.* 23:341–69.

Nei, M., and Roychoudhury, A. K. 1974. Genetic variation within and between the three major races of man, Caucasoids, Negroids and Mongoloids. *Am. J. Hum. Genet.* 26:421–43.

Nei, M., and Roychoudhury, A. K. 1993. Evolutionary relationships of human populations on a global scale. *Mol. Biol. Evol.* 10:927–43.

Poulet, S. 1994. *Organisation Génomique de* Mycobacterium tuberculosis *et Épidémiologie Moléculaire de la Tuberculose.* Ph.D. dissertation, University of Paris 6, Paris.

Samson, M., Libert, F., Doranz, B. J., Rucker, J., Liesnard, C., Farber, C. M., Saragosti, S., Lapoumeroulie, C., Cognaux, J., Forceille, C., Muyldermans, G., Verhofstede, C., Burtonboy, G., Georges, M., Imai, T., Rana, S., Yi, Y., Smyth, R. J., Collman, R. G., Doms, R. W., Vassart, G., and Parmentier, M. 1996. Resistance to HIV-1 infection in caucasian individuals bearing mutant alleles of the CCR-5 chemokine receptor gene. *Nature* 382:722–5.

Solignac, M. 1998. Génétique, population et evolution. In *Principes de Génétique Humaine* (Feingold, J., Fellous, M., and Solignac, M., Eds.), pp. 511–58. Hermann Editions, Paris.

Thio, C. L., Thomas, D. L., Goedert, J. J., Vlahov, D., Nelson, K. E., Hilgartner, M. W., O'Brien, S. J., Karacki, P., Marti, D., Astemborski, J., and Carrington, M. 2001. Racial differences in HLA class II associations with hepatitis C virus outcomes. *J. Infect. Dis.* 184:16–21.

Thornhill, R., and Moller, A. P. 1997. Developmental stability, disease and medicine. *Biol. Rev. Camb. Phil. Soc.* 72:497–548.

Tibayrenc, M. 1998. Beyond strain typing and molecular epidemiology: integrated genetic epidemiology of infectious diseases. *Parasitol. Today* 14:323–9.

Tibayrenc, M. 2001a. The golden age of genetics and the dark age of infectious diseases. *Infect., Genet., Evol.* 1(1):1–2.

Tibayrenc, M. 2001b. The golden age of genetics? *Infect., Genet., Evol.* 1(2):83–4.

Wilson, D. S. 1997. Human groups as units of selection. *Science* 276:1816–17.

FIFTEEN

Evolution and the Etiology of Diabetes Mellitus

Kyle D. Cochran and Gregory M. Cochran

INTRODUCTION

Over the course of the past century, medical science has produced rapid advances in human health. At first, the germ theory and Koch's postulates framed the search for pathogens causing major infectious diseases, and later, Mendelian genetics and modern molecular techniques were used to identify the genes responsible for inherited syndromes. A number of major human diseases are of unknown origin, including, but not limited to atherosclerosis, multiple sclerosis, rheumatoid arthritis, schizophrenia, manic depression, ulcerative colitis, many cancers, and diabetes mellitus.

In some cases, these diseases have plagued humanity for evolutionarily long periods. These diseases present a puzzle to evolutionary biologists. What selective pressures can maintain a high frequency for these deleterious syndromes over evolutionarily long periods? Using an adaptationist point of view, we have suggested that if a syndrome has been common in a well adapted population for an evolutionarily long period and maintains a large negative selective effect on the population, then the syndrome is likely to be caused, directly or indirectly, by infection (Cochran et al., 2000).

Diabetes may be one such disease. It is ancient, has a severe selective effect, and is relatively common; this combination of factors leads us to suspect that diabetes may be caused by infection. In this study, using an evolutionary perspective and with an eye on infection, we examine the etiology of diabetes mellitus.

EVOLUTIONARY PRESSURE, EPIDEMIOLOGY, AND FITNESS

What are the evolutionary pressures driving human disease syndromes? As a consequence of the different evolutionary calculus of a pathogen and its host, virulence can be maintained by pathogens indefinitely. Because of this, pathogens need not tend toward mutualism or symbiosis, and the frequency of severe infectious disease can be maintained perpetually (Ewald, 1994). This situation produces an evolutionary arms race, leading to coevolution of the human and the pathogen. This genetic diversity is reflected in the variation in susceptibility and protective genes contained in human populations (Hamilton, 1982). The human leukocyte antigen (HLA) system, also called the major histocompatibility complex, plays an important role in disease susceptibility. Proteins encoded in the HLA system are responsible for the recognition and presentation of foreign antigens to the immune system. Many important human diseases have strong associations with particular HLA alleles, including diabetes mellitus, a subject to which we return.

The fact that infectious disease has molded the human genome is, of course, no surprise. Some have suggested that infectious disease is responsible for sex itself (the Parasite Red Queen theory) and that infection is one of the only possible selective pressures able to overcome the two-fold loss in fitness that sexual reproduction incurs relative to clonal reproduction (Hamilton, 1982; Hamilton *et al.*, 1990).

In any case, it seems certain that the single largest selective force in human evolution after the agricultural revolution is infectious disease. Examples of the selective force of infection abound, ranging from the Black Death of the Middle Ages, *Plasmodium falciparum* malaria in sub-Saharan Africa, and the epidemics in Native Americans during the European conquest to the current AIDS pandemic. This selective force caused several notable cases of heterozygote advantage, e.g., sickle cell and cystic fibrosis and many cases of susceptibility genes, including HLA variants associated with rheumatoid arthritis, multiple sclerosis, lupus, narcolepsy, and the juvenile form of diabetes mellitus (Vogel and Motulsky, 1997).

As a consequence of the introduction of vaccination, the germ theory of disease, and antiseptic technique, the selective pressure of infectious disease in the industrialized world has decreased immensely in the past 250 years. In particular, mortality data in western populations shows the enormous advances that have taken place in human life span. These advances in life span are mostly due to improvements in childhood mortality rates. For example, in Prussia in the middle of the eighteenth century, only about 50% of children survived to age twenty. Strikingly, about 25% of newborns died in the first year of life alone. In contrast, in 1955 Berlin, approximately 95% of children survived to age twenty (Vogel and Motulsky, 1997).

Historically, the recognition of infection as the cause of disease syndromes was based upon several key indicators. In diseases in which infectious symptoms, e.g., sneezing, fever, etc., were obvious, and acute infection was quickly followed by symptoms, an infectious etiology was quickly investigated. When animal models were available, the application of Koch's postulates made the diagnosis of an infectious agent certain. However, when symptoms were subtle or internal, the classification of etiology became more difficult. If the onset of symptoms occurred much later than the time of infection or if the method of infection was subtle, then ascription of infectious etiology could also be delayed. Water-borne, vector-borne, and sexually transmitted diseases such as cholera, yellow fever, and gonorrhea provide examples of disease syndromes in which subtle chains of transmission made the recognition of infectious etiology more difficult. Furthermore, when the infectious agent was small, for example a virus, the classification of infectious etiology sometimes awaited the introduction of more precise methods of observation or detection (Cochran *et al.*, 2000).

As the organisms causing infectious diseases have become more difficult to detect, the epidemiology of these disease syndromes has become more important in determining etiology. Some of the key epidemiological indicators for infectious causation are: geographical variation in disease incidence, epidemic waves of disease (particularly in island populations), and temporal and spatial clustering of disease. As we describe, the juvenile form of diabetes mellitus exhibits all of these characteristics (see p. 330).

Environmental factors can also cause long-term, damaging, common human diseases, e.g., rickets, scurvy, goiter, beriberi, and pellagra. From an adaptationist viewpoint, given enough time and sufficient variation at the proper loci, as humans adapt to a new environment the frequency of environmentally caused human disease should decrease. There are certainly exceptions to this adaptationist view, particularly in the case of dietary deficiencies for which humans are unable to compensate. However, other than the possible link between spina bifida and folic-acid deficiency in pregnancy, there are virtually no diseases of unknown etiology for which vitamin or nutrient deficiency is strongly suspected as a causal factor.

The modern era has witnessed a great increase in the prevalence of some disease syndromes. In principle, these diseases could have a new environmental cause, a cause to which humans are not yet well adapted. Some evolutionary biologists have concluded that humans are not yet adapted to an agricultural diet (Nesse and Williams, 1994), yet it is well known that significant evolution in vertebrates can occur in as little as ten to twenty generations if genetic variations at the proper loci are available. Most humans have had hundreds of generations to adapt to an agricultural diet, and modern molecular genetics is providing evidence of an enormous amount of genetic variation in the human genome. In general, basic metabolic functions are expected to be robust. We do not expect such functions to fail due to moderate perturbations, because our evolutionary past was full of such perturbations.

In modern times, many of the most important chronic human diseases have also been ascribed to a combination of genetic susceptibility and environmental causes. However, since their initial classification as multifactorial, many diseases have been found to have infectious causes. Peptic ulcer, for example, was once thought to be caused by stress, smoking, alcohol consumption, genetic susceptibility, and stomach acid. Now, it is known that the vast majority of peptic ulcers are caused by *Helicobacter pylori*. Over the past two decades, other subtle, chronic diseases have been linked causally to infectious disease, including many cancers (Zur Hausen, 1991). Atherosclerosis, like diabetes, is considered a multifactorial disease. As with peptic ulcer, it is said to be caused by stress, smoking, cholesterol, high fat diets, and genetic predispositions, and yet there are indications that infection may cause coronary heart disease (Saikku *et al.*, 1988), although these infectious indicators are not, as yet, generally accepted.

In the case of simple Mendelian genetic diseases, statistics on monozygotic (MZ) and dizygotic (DZ) twins can aid us in determining etiology. For Mendelian genetic diseases, the MZ twin concordance should be approximately 100%, whereas the DZ twin concordance should occur in Mendelian ratios, depending on the type of inheritance of the syndrome. However, great care must be taken in relying on twin concordance as an indicator for genetic causation. Despite a high concordance rate, which might at first glance suggest a genetic cause, there are several infectious diseases that have shown quite high MZ twin concordance. For example, leprosy had a 60 to 80% MZ twin concordance rate, and tuberculosis had a 50% MZ twin concordance rate, and dizygotic twin concordances were far lower (Vogel and Motulsky, 1997).

What evolutionary pressures drive the frequency of genetic causes of human disease? For simple Mendelian syndromes, the frequency of disease corresponds to the frequency of deleterious genes. At equilibrium, the loss of a disease-producing allele must be balanced by the reintroduction of that allele. If a genetic disease generates no other compensating fitness benefit, is caused by a dominant mutation, and is at equilibrium, then the frequency per generation is approximately equal to the per-generation mutation rate. The average mutation rate for humans is approximately 1 mutation in 10^9 nucleotides per generation in humans (Vogel and Motulsky, 1997), and the most common, mutation-driven genetic diseases of the germ-line occur at a frequency of approximately 1 mutation per 10^4 persons per generation. An example of this is neurofibromatosis. Neurofibromatosis is the most common, damaging autosomal dominant genetic disease known. Its mutation rate is between 4.0 and 6.5* 10^{-5}. This high rate of mutation is related to the extreme length of the *NFI* gene, some 350 kilo-base-pairs (Online *Mendelian Inheritance In Man*, 2002).

How high must the negative selective effect be to rule out genetic causes? To address that question, we turn to the concept of fitness. *Fitness* refers to Hamilton's concept of inclusive fitness (Hamilton, 1964) that accounts not only for direct effects of reproduction and mortality but also indirect effects due to the rearing of the young and aid to near-relatives. In order to estimate the relative impact of fitness-reducing syndromes, we use the concept of fitness load to characterize the burden of any syndrome or disease on a population. Closely related to the concept of genetic load (Muller, 1950), fitness load estimates the load on a population induced by the increased mortality or reduced fertility of sufferers of a fitness-reducing syndrome. Fitness load can be used to indicate if a fitness-reducing syndrome could reasonably be ascribed to genetic or infectious causation. It is calculated by taking the product of the prevalence of the syndrome and an estimate of its average selective disadvantage. From our estimates of the most common simple Mendelian diseases, the evolutionary pressure for purely genetic causation of disease is limited to those diseases that have a fitness load of less than 0.001 (Cochran *et al.*, 2000).

Several diseases of unknown etiology, including diabetes mellitus, have much higher fitness loads than the fitness loads for the most common, damaging mutation-driven Mendelian disease. The MZ twin concordance for many of these multifactorial diseases is only moderately high. Failing to find single-gene causes for such multifactorial syndromes, many geneticists have turned to more complex gene interactions to explain the prevalence of certain diseases. As we describe, the adult form of diabetes is one of those syndromes (see p. 331). Despite a great deal of effort, these models have met with little success.

In certain cases, fitness load alone can be used to predict the etiology of common fitness-reducing syndromes. Using this metric, we would have predicted that infectious agents caused all disease syndromes with fitness loads above 0.01; these diseases include malaria, bubonic plague, cholera, smallpox, AIDS, tuberculosis, schistosomiasis, and a host of others. This threshold (0.01) is about two orders of magnitude above the estimated fitness load for the most common known cases of fatal Mendelian genetic disease, thus the threshold is conservative. We suggest that fitness loads above 0.0001 are suspiciously high, and for diseases with unknown etiology, the history and epidemiology of these syndromes should be carefully investigated for indications of infectious causation.

Accordingly, let us turn now to a case study of diabetes mellitus. The etiology of diabetes mellitus is unknown; it is one of the major concerns in health science research worldwide. Its fitness load (for both its juvenile and adult forms) is suspiciously high. In the following sections, we discuss the history of human diabetes, its epidemiology and evolutionary impact, and some of the proposed causes of diabetes, including the evidence for an environmental or infectious etiology.

DIABETES MELLITUS: A CASE STUDY

Diabetes mellitus (DM) has been known for several thousand years. By 1500 BC, the Ayurveda recorded that insects and flies were attracted to the urine of some people, that the urine tasted sweet, and that it was associated with certain diseases. The ancient Greeks also described diabetes. The Greek term "diabetes," literally "to go through," describes the excessive urination associated with extreme forms of diabetes. The term "mellitus," which literally means honey, indicates the presence of sugar in the urine, one of its classic symptoms. Successful treatment of DM did not occur until the twentieth century, when Macleod, Banting, Best, and Collip successfully refined pancreatic extracts to produce insulin (MacCracken and Hoel, 1997).

Diabetes mellitus is a metabolic disorder characterized by hyperglycemia resulting from impairment in insulin production and/or insulin response. According to the National Institute of Diabetes and Digestive and Kidney Diseases, the prevalence of DM in the United States population was estimated at 6.2% in 2000, with 11.1 million diagnosed cases and an estimated 5.9 million undiagnosed cases. The complications associated with DM include blindness caused by diabetic retinopathy, renal failure, diabetic neuropathy, blood vessel damage, and increased risk of heart disease. Diabetes is one of the top ten leading causes of death in the United States (National Institute of Diabetes and Digestive and Kidney Diseases, 2002).

In general, DM is divided into two types. Although the symptoms of type 1 and type 2 diabetes mellitus are similar, the pathology of the two types differs markedly, suggesting very different causes for the two types of diabetes mellitus. Type 1 diabetes mellitus (T1DM) occurs primarily in children and adolescents and is due to the destruction of the insulin-producing beta cells in the pancreas; it is also called juvenile-onset diabetes or insulin-dependent diabetes mellitus (IDDM). In clinical T1DM, more than 90% of the insulin-producing beta cells of the pancreas are destroyed, resulting in near total loss of insulin production (Merck, 1999).

Type 2 diabetes mellitus (T2DM) is characterized by impaired use of insulin in the body. T2DM is also called adult-onset or non-insulin-dependent diabetes mellitus (NIDDM). The age at onset for T2DM is normally from 40 to 60, but this is highly variable. Unlike T1DM, T2DM is not usually characterized by destruction of insulin-producing cells in the pancreas. In healthy individuals, skeletal muscle glucose uptake and hepatic glucose output is balanced by insulin production in the pancreas. In T2DM, hyperglycemia results from a disturbance in this balance; pancreatic insulin production ultimately fails to compensate for insulin resistance and glucose production in other tissues. Over all age groups, approximately 10% of all diabetes cases are classified as type 1, and 90% are classified as type 2 (Merck, 1999). Individuals diagnosed with diabetes after age 45 are nearly all type 2 diabetics (Kenny *et al.*, 1995).

The classic symptoms of type 1 DM are polyuria, polydipsia, weight loss, and fatigue. In type 2, and in some cases of type 1, many of these symptoms may be less pronounced or absent. About 40% of diabetics take oral agents to manage their blood glucose levels, and 40% require insulin injections. The remaining 20% of diabetics treat DM with exercise and control of their diet. Type 1 diabetics normally require daily insulin injections (World Health Organization, 2002).

THE EPIDEMIOLOGY OF T1DM

European frequencies of T1DM differ markedly from region to region – approximately 35 new cases per 100,000 inhabitants are reported annually in Finland, whereas 2 to 3 per 100,000 inhabitants are reported in Macedonia. Diagnosticians have seen a steady increase in type 1 frequency in Europe in the past few decades. Overall, the prevalence of T1DM is about 0.3% in Caucasians. The incidence of T1DM in China is dramatically lower, reaching as low as 0.1/100,000 per year (Karvonen *et al.*, 2000), reflecting · broad ethnic differences in T1DM rates. Seasonal trends in the incidence of T1DM onset have also been found, including island epidemics of T1DM and increased incidence of T1DM cases after epidemics (LaPorte *et al.*, 1995).

The monozygotic (MZ) twin concordance for T1DM is about 30–50%, implying that nongenetic factors have a large etiologic role in T1DM. In spite of the low MZ twin concordance, some HLA types are strongly associated with T1DM. In particular, several studies have shown a slightly increased risk for persons with both HLA B8 and B15 antigens. HLA DR3, DR4, and DR3/DR4 heterozygotes also have somewhat increased risk of T1DM, whereas HLA DRB1*1401 and DQA1*0102-DQB1*0602 haplotypes seem protective (Online *Mendelian Inheritance In Man*, 2002).

Several studies have found that single amino acid changes on the HLA DQ alpha and beta chains (DQA1*Arg-52 and DQB1*non-Asp-57) greatly increase the risk of T1DM; these DQ mutations represent the strongest HLA/T1DM associations found to date. The Arg-52 (in Japan) and non-Asp-57 sequences may explain most of the difference in T1DM frequencies between ethnic groups (Bao *et al.*, 1989). Using genotype-specific T1DM age-adjusted incidence rates from U.S. whites, Dorman and colleagues made predictions for T1DM incidence rates in four groups with known non-Asp-57 gene frequencies: U.S. blacks, Chinese, Norwegians, and Sardinians. Each of these predictions fell within the 95% confidence limits of the observed rates, indicating that the geographic variation in 1DDM risk is intimately related, in these groups, to the HLA DQB1*non-Asp-57 genotype (Dorman *et al.*, 1990).

THE EVOLUTIONARY IMPACT OF T1DM

Understanding the evolutionary impact of T1DM is best attempted by examining those groups in which it is most common, as long as it is really the same disease, because the evolutionary anomaly will be most pronounced. Thus, for T1DM, the Finns are of special interest. In the case of T1DM, for the Finns the modern prevalence is about 0.3% whereas the average selective disadvantage, prior to modern medical intervention, was at least 50%. From these figures, the estimated fitness load is between 0.001 and 0.01 (Cochran *et al.*, 2000). This estimate assumes that the historical prevalence of T1DM in this at-risk population was about the same as its modern levels. The obvious DM diagnosis, its youthful age at onset, and some limited historical data suggest that this approximation holds. For example, in the United States, there was a 0.37% prevalence rate for DM in 1935 (Kenny *et al.*, 1995).

This level of fitness load is relatively high, so high, in fact, that from theoretical calculations, the worldwide distribution of T1DM, its epidemiological indicators of infection, and an understanding of the fitness load of the most common genetic diseases, we would hypothesize that T1DM's load could only be maintained over evolutionary time by an infectious cause. There is considerable evidence suggesting that this is the case.

THE ETIOLOGY OF T1DM

Because its MZ twin concordance is low, nutrition and pathogens are potential factors in the etiology of T1DM in genetically susceptible individuals. Several studies have linked early exposure to bovine serum albumin in cow's milk to T1DM (Dorman *et al.*, 1995), although this association seems weak. Viral infections have also been strongly associated with the onset of T1DM, including infections caused by coxsackie B viruses, human cytomegalovirus (CMV), mumps, and rubella.

INFECTION AND T1DM

There is a great deal of evidence suggesting that coxsackie B viruses can cause T1DM in susceptible individuals. For instance, recent studies have found a high incidence of enterovirus RNA in T1DM patients and their siblings at T1DM diagnosis, as well as evidence of diabetogenic strains of coxsackie virus (Yin *et al.*, 2002). In addition to seasonality and waves of infection in island populations, there is also evidence of an increase in the incidence of T1DM cases after epidemics caused by enteroviruses (LaPorte *et al.*, 1995; Yin *et al.*, 2002).

In T1DM, the beta cells in the pancreas are attacked and destroyed either by direct viral action or by the immune system. It appears that this may be a case of molecular mimicry. Glutamic acid decarboxylase 65 (*GAD65*), a 65-kilodalton islet-cell antigen, appears in many T1DM patients and is a suspected autoantigen. GAD65 shares amino acid homology with the P2-C protein of the coxsackie B4 virus, suggesting that it is the source of the autoimmune response (Online *Mendelian Inheritance In Man*, 2002). Coxsackie B viruses have also been shown to induce diabetes in animal models (Fairweather and Rose, 2002).

Human CMV gene segments have been significantly associated with recently diagnosed T1DM cases. Particular mumps variants and congenital rubella are also associated with T1DM. In the case of mumps, diabetes may occur immediately after infection, whereas in congenital rubella, infection occurs *in utero*, and diabetes occurs many years later (Dorman *et al.*, 1995). The diabetes cases caused by mumps and rubella currently tend to be rare in the industrialized world due to widespread MMR vaccination programs, but in the past, mumps and rubella could have contributed significantly to the incidence of T1DM.

THE EPIDEMIOLOGY OF T2DM

Let us turn now to T2DM. T2DM comprises the vast majority of diabetes cases, particularly in the western world. The majority of T2DM cases seem to be related to insulin resistance and pancreatic beta-cell failure, but not autoimmunity. A smaller, though not insignificant fraction (10 to 33%) of T2DM cases exhibits beta-cell autoimmunity (Rewers and Hamman, 1995).

Rates of T2DM are moderate in populations of European descent, with a prevalence of between 3 and 10%. The prevalence of T2DM is higher in

Asian Indians and Africans, about twice the rate of European Caucasians, and the highest prevalence of T2DM is found in Australian aboriginal communities, the Pima Native Americans, and urban Micronesians living on the island of Nauru, where the prevalence of T2DM can reach 40 to 50% of the total population (Karvonen *et al.*, 2000). T2DM also occurs much earlier than normal in the Pima Indians. In some traditional societies, such as the Machupe Indians in Chile and the Bantu in Tanzania, T2DM is virtually unknown (Gohdes, 1995; Rewers and Hamman, 1995).

Familial aggregation of T2DM is exhibited in all populations. In addition, obesity is strongly linked to T2DM; 60 to 90% of diabetes occurs in obese individuals (Fajans, 1981). The prevalence of T2DM has increased dramatically over the past few decades as western populations have become both more sedentary and more obese. However, studies suggest that other factors besides obesity must be involved. As the Fajans study indicates, not all T2DM diabetics are obese, and furthermore, not all obese individuals become type 2 diabetics. In one study, in male twins discordant for T2DM, no difference in dietary input was found (Rewers and Hamman, 1995). Therefore, it seems that obesity is a risk factor but not a causative agent of T2DM.

The build-up of islet amyloid polypeptide (IAPP), also known as amylin, has been strongly associated with T2DM patients. Amylin is cosecreted with insulin by the pancreatic beta cells. The association of amylin build-up with diabetes was first noted in 1901, when Opie first described a hyaline staining substance that was associated with diabetes mellitus. We now know this as islet amyloid (Hayden and Tyagi, 2001). More than 70% of T2DM patients are suspected to have pancreatic amylin build-up. In a study of Pima Indians, 77% of T2DM Pima had IAPP pancreatic build-up at autopsy versus only 7% of controls. IAPP DNA sequences for T2DM and non-T2DM patients do not differ, so mutations in IAPP do not account for this association. Amylin build-up has been proposed as a possible cause for both insulin resistance and T2DM, but so far, it is not known if it is a cause or an effect of T2DM (Rewers and Hamman, 1995).

There is an urban/rural split in the frequency of T2DM in at-risk populations; in one study from Polynesia, urban Samoans had a three times higher frequency of T2DM than rural inhabitants. In addition, migrants to western countries have shown elevated frequencies of T2DM relative to their homelands (Rewers and Hamman, 1995).

Some studies have suggested an extremely high MZ twin concordance of 80 to 90% for T2DM, leading to the search for diabetogenic genes (Online *Mendelian Inheritance In Man*, 2002). There are, however, a wide range of MZ twin concordances reported in the literature. Ascertainment bias has affected some twin studies, particularly those studies showing the highest MZ twin concordance. For two twin studies without suspected ascertainment bias, the MZ twin concordance was only 55 to 58%, and the DZ twin concordance was 17% (Rewers and Hamman, 1995).

A small percentage of T2DM cases have been successfully associated with genetic mutations. These cases include mitochondrial mutations that account for perhaps 0.5% of the total T2DM prevalence; these cases are often associated with hearing loss. Mutations in the *NIDDM1* and *NIDDM2* nuclear genes are factors increasing the risk of T2DM in Mexican American and Finnish subpopulations, respectively (Online *Mendelian Inheritance In Man*, 2002). Over sixty rare genetic diseases are associated with glucose intolerance (Rewers and Hamman, 1995).

Because Mexican Americans contain an admixture of Native American and European genes, and because Amerindians often have a high frequency of T2DM, several studies have suggested that the high prevalence of T2DM in Mexican Americans is due to the admixture of Native American genes (Hanis *et al.*, 1991). The presence of a particular allele of the super-gene *GM* responsible for the constant part of the heavy-chain of four immunoglobulins has been found to be negatively associated with T2DM in the Pima and Papago Indian tribes. This allele, *Gm3;5,13,14*, is also a sign of Caucasian admixture (Cavalli-Sforza *et al.*, 1996).

Mutations in at least six genes are responsible for a distinct, though rare form of T2DM. This syndrome, called maturity-onset diabetes of the young (MODY), is comprised of six subtypes, all of which involve the alteration of insulin secretion (Elbein, 2002). These subtypes are caused by mutations in the hepatocyte nuclear factor-4-alpha, glucokinase, hepatic transcription factor-1, insulin promoter factor-1, hepatic transcription factor-2, and the *NEURODI* genes. MODY is an autosomal dominant genetic disease with age at onset less than age 25 (Online *Mendelian Inheritance In Man*, 2002).

In spite of a great deal of research and a high MZ twin concordance, the genes that are associated with T2DM seem limited to small groups in at-risk populations. So far at least, widespread genes "for" the most common form of T2DM have not been found (Elbein, 2002).

THE EVOLUTIONARY IMPACT OF T2DM

The estimation of fitness load for T2DM is considerably more complicated than it is for T1DM. Although diabetes has been known for literally millennia, subtypes of diabetes have only been categorized in recent times. In the Pima Indians, the population with the highest known rate of T2DM, the modern prevalence is 40 to 50%. Using historical demographical data prior to modern medicine (Hamilton, 1966), we estimate that the average selective disadvantage for a T2DM-like disease with similar age at onset would have been about 0.15. Therefore, the calculated fitness load for this level of selective disadvantage and prevalence is 0.075. Due to its extremely high fitness load, if this level of T2DM had been present for an evolutionarily long time period, then we would be confident that an infectious disease caused T2DM.

However, until recently, T2DM was virtually unknown in many, if not all, of the subpopulations where its prevalence is currently highest. What might have changed to account for this increase? Our evolutionary framework would point to a change in the environment and/or the spread of a new pathogen in at-risk populations. For many years, researchers have recognized that nutrition and exercise factors are strongly associated with T2DM, but there are also several recent lines of evidence that suggest infectious disease plays a significant part in T2DM.

THE ETIOLOGY OF T2DM
Because T2DM is a heterogeneous disease, there have been several etiological hypotheses suggested for it. Some studies have shown high MZ twin concordance, leading to the search for diabetogenic genes. Despite years of search, the percentage of T2DM caused by known genetic mutations is probably less than 5% in most populations (Elbein, 2002).

Other hypotheses suggest that T2DM was evolutionarily adaptive. Perhaps the most well known adaptive hypothesis for T2DM is the thrifty genotype hypothesis (first suggested in 1962 and reviewed in Neel *et al.*, 1998). Neel suggested that populations enduring periods of feast and famine would favor genes that efficiently use and store carbohydrates. In the thrifty genotype model, Neel theorized that energy could be stored in the adipose tissues during times of feast and then used during times of famine. In this model, obesity, as long as exercise levels were high, would rarely lead to diabetes, and individuals with the thrifty genotype would be favored.

The thrifty genotype theory fails to account for the worldwide distribution of the subpopulations most at risk for T2DM. If the thrifty genotype had a single genetic source, then the common group of ancestors for Native Americans, Polynesians, and Australian aborigines lived 50,000 to 70,000 years ago (Cavalli-Sforza *et al.*, 1996). If Africans and Asian Indians were also included in the group of at-risk populations, then a common ancestor for this thrifty genotype would have come out of Africa some 100,000 years ago. If that were the case, then all of modern humanity would have a thrifty genotype, and yet ethnic differences reflect two- to eight-fold differences in T2DM rates between populations.

Some of the time, and for some people, there was more than enough to eat in humanity's evolutionary past. Tribal chieftains, for example, usually had plenty of food and had significantly more children than average. This implies that there was at least some selective pressure favoring non-thrifty alleles. Neel's hypothesis fails to provide support for its claim that the Amerinds, for example, needed to be especially thrifty. For instance, there is evidence that hunter–gatherers had more stable food sources than agriculturalists (Cordain *et al.*, 1999). Because Amerinds have not farmed as long as Europeans, and Australian aborigines never farmed at all, recurrent famine may have had its strongest impact on Europeans.

In addition, the subpopulations most at risk for T2DM are all characterized by their isolation and their lack of HLA diversity. In the Old World, even small populations have a very large number of HLA alleles, with each allele occurring at a relatively low frequency. However, in Native Americans, Polynesians, and the Australian aborigines, only a small number of HLA alleles are present, and their frequencies are often quite large; in some Native American tribes, the dominant HLA allele frequency can exceed 50% (Cavalli-Sforza *et al.*, 1996). The lack of HLA diversity in Native Americans has been shown to indicate a lack of strong overdominant selective pressures among these populations (Slatkin and Muirhead, 2000).

These populations were particularly vulnerable to infectious disease because they have spent several thousand years under weak disease pressure. HLA homogeneity is a result and a sign of this. Might these populations also be especially prone to subtle, chronic infectious diseases? Other diseases associated with high-risk T2DM groups might provide a clue. "New World syndrome" (Weiss *et al.*, 1984) consists of the constellation of gallbladder disease, gallbladder cancer, T2DM, and obesity. This disease constellation occurs at extremely high rates in Native Americans and populations with Native American admixture. Intriguingly, recent research has produced evidence strongly associating chronic viral infection of the liver in a significant fraction of type 2 diabetics.

INFECTION AND T2DM

There are several lines of evidence that implicate infection as a cause for T2DM. Temporally, there has been an epidemic of T2DM in the past few decades in the United States, accompanied by a decrease in its age at onset (National Institute of Diabetes and Digestive and Kidney Diseases, 2002). Recently, several lines of evidence have shown that inflammatory markers predict the development of T2DM. These indicators include the presence of C-reactive protein and gamma globulin, as well as higher fibrinogen and white-cell counts (Schmidt *et al.*, 1999; Festa *et al.*, 2000; Lindsay *et al.*, 2001). These markers are indicative of an acute inflammatory response that is ongoing several years before the diagnosis of T2DM.

Perhaps the most crucial line of evidence is the finding that chronic hepatitis C virus (HCV) is strongly linked to T2DM. In 1994, Allison and colleagues first showed a link between hepatitis C viral infection and type 2 diabetics in a cirrhotic population (Allison *et al.*, 1994). Recently, a flurry of reports have indicated that patients with chronic HCV are two to five times more likely to have T2DM than those without HCV; patients with HCV and liver cirrhosis are over three times more likely to have T2DM. The association with HCV is significant, even when controlling for obesity. HCV is also a predictor of T2DM after liver transplantation (Mason *et al.*, 1999; Alexander, 2000; Mehta *et al.*, 2000).

The significance of amylin build-up may also represent a response to infection. Other amyloidoses, such as atherosclerosis and Alzheimer's disease, cause amyloid build-up, and although the evidence for infectious causation is controversial, several studies have associated infectious organisms (*Chlamydia pneumoniae*) with these amyloid plaques (Saikku *et al.*, 1988; Balin *et al.*, 1998).

Interestingly, weight loss has been connected to improvements in liver abnormalities in HCV patients (Hickman *et al.*, 2002), leading to the suggestion that weight and HCV could be cofactors in some T2DM cases. Diabetogenic strains also seem to exist. In one study from Israel, strain HCV lb was found to be significantly diabetogenic (Knobler *et al.*, 2000), and in a study from the United States, strain HCV 2a was significantly associated with patients having HCV and T2DM (Mason *et al.*, 1999).

HCV is not likely to have been a common cause of T2DM prior to the twentieth century. It is transmitted from person to person by sexual contact, blood transfusion, and intravenous drug use. The frequency of HCV is also not high enough to explain the vast majority of T2DM cases. However, it may be the single biggest known cause of T2DM (Caronia *et al.*, 1999).

CONCLUSIONS

As we have discussed, the evidence of infectious causation of T1DM is strong. From the standpoint of evolutionary theory, its fitness load is suspiciously high. Several viruses are associated with T1DM; the strongest association is with the coxsackie B virus. If an autoimmune reaction can be avoided, we suggest that a vaccine immunizing against diabetogenic strains of coxsackie B virus should be developed with an emphasis toward inoculating those populations at highest risk for T1DM. As we discussed, individuals with particular HLA DQ alleles are particularly vulnerable to T1DM, and we suggest that populations with a high frequency of those individuals should be inoculated in early infancy.

The etiology of T2DM is more of a mystery. Although the MZ twin concordance is relatively high, molecular geneticists have found genetic causes for less than 5% of all T2DM cases. Using estimates of fitness load, genetic causation cannot be ruled out, and obesity is clearly an important risk factor. The latest evidence suggests, however, that infection may play a very important role.

There is strong evidence suggesting that hepatitis C virus can cause a significant fraction of T2DM. In addition, a number of signs typically associated with infectious disease predict the onset of T2DM, which in no way fits with conventional theories of T2DM causation. The highest rates of T2DM infection are in historically isolated subpopulations lacking HLA diversity. This pattern of isolation and genetic homogeneity is typical of populations especially prone to infectious disease. The observed ethnic differences in T2DM susceptibility are also typical of an infectious disease, as is the enhanced prevalence of T2DM in urban, congested populations.

In T2DM, genes, behavior, and diet are cited as risk factors, i.e., T2DM is a multifactorial disease. Multifactorial diseases are difficult to treat, particularly when human behavior modification is involved, as it is, given the association between diabetes and obesity. It is our view that infectious diseases are preferable in at least one respect to multifactorial diseases: once a disease has been shown to be infectious, medical science has been able to devise successful countermeasures.

Over the past thirty years, most research on the cause of T2DM has focused on genetics and diet, and infection was thought to have no role in T2DM. HCV serves as an existence proof; viruses can cause T2DM. Recent evidence suggests that it is time for researchers to focus on the possibility of infectious causation for T2DM.

REFERENCES

Alexander, G. J. (2000). An association between hepatitis C virus infection and type 2 diabetes mellitus: what is the connection? *Annals of Internal Medicine*, **133**(8), 650–2.

Allison, M. E. D., Wreghitt, T., Palmer, C. R., and Alexander, G. J. M. (1994). Evidence for a link between hepatitis C virus infection and diabetes mellitus in a cirrhotic population. *Journal of Hepatology*, **21**, 1135–39.

Balin, B. J., Gerard, H. C., Arking, E. J., and 5 colleagues. (1998). Identification and calization of *Chlamydia pneumoniae* in the Alzheimer's brain. *Medical Microbiology and Immunology (Berlin)*, **187**, 23–42.

Bao, M.-Z., Wang, J.-X., Dorman, J. S., and Trucco, M. (1989). HLA-DQ-beta non-asp-57 allele and incidence of diabetes in China and the USA. (Letter) *Lancet*, **II**, 497–8.

Caronia, S., Taylor, K., Pagliaro, L., Carr, C., Palazzo, U., Petrik J., O'Rahilly, S., Shore, S., Tom, B. D., and Alexander, G. J. (1999). Further evidence for an association between non-insulin-dependent diabetes mellitus and chronic hepatitis C virus infection. *Hepatology*, **30(4)**, 1059–63.

Cavalli-Sforza, L. L., Menozzi, P., and Piazza, A. (1996). *The History and Geography of Human Genes* (abridged paperback edition). Princeton University Press, Princeton, NJ.

Cochran, G. M., Ewald, P. W., and Cochran, K. D. (2000). Infectious causation of disease: an evolutionary perspective. *Perspectives in Biology and Medicine*, **43(3)**, 406–48.

Cordain, L., Miller, J., and Mann, N. (1999). Scant evidence of periodic starvation among hunter-gatherers (Letter). *Diabetologia*, **42(3)**, 383–4.

Dorman, J. S., LaPorte, R. E., Stone, R. A., and Trucco, M. (1990). Worldwide differences in the incidence of type I diabetes are associated with amino acid variation at position 57 of the HLA-DQ beta chain. *Proceedings of the National Academy of Sciences USA*, **87**, 7370–4.

Dorman, J. S., McCarthy, B. J., O'Leary, L. A., and Koehler, A. N. (1995). Risk factors for insulin-dependent diabetes. In *Diabetes in America* (Harris, M. I., Cowie, C. C., Stern, M. P., Boyko, E. J., Reiber, G. E., and Bennett, P. H., Eds.), 2nd ed., pp. 165–78. National Diabetes Data Group, National Institutes of Health, National Institute of Diabetes and Digestive and Kidney Diseases, NIH Publication No. 95–1468.

Elbein, S. C. (2002). Perspective: the search for genes for type 2 diabetes in the postgenome era. *Endocrinology*, **143(6)**, 2012–8.

Ewald, P. W. (1994). *Evolution of Infectious Disease*. Oxford University Press, New York.

Fairweather, D., and Rose N. R. (2002). Type 1 diabetes: virus infection or autoimmune disease? *Nature Immunology*, **3(4)**, 338–40.

Fajans, S. (1981). Etiologic aspects of types of diabetes. *Diabetes Care*, **4**, 69–75.

Festa, A., D'Agostino, R., Jr., Howard, G., Mykkanen, L., Tracy, R. P., and Haffner, S. M. (2000). Chronic subclinical inflammation as part of the insulin resistance syndrome: the Insulin Resistance Atherosclerosis Study (IRAS). *Circulation*, **102(1)**, 42–7.

Gohdes, D. (1995). Diabetes in North American Indians and Alaskan natives. In *Diabetes in America* (Harris, M. I., Cowie, C. C., Stern, M. P., Boyko, E. J., Reiber, G. E., and Bennett, P. H., Eds.), 2nd ed., pp. 683–702. National Diabetes Data Group, National Institutes of Health, National Institute of Diabetes and Digestive and Kidney Diseases, NIH Publication No. 95–1468.

Hamilton, W. D. (1964). The genetical evolution of social behavior. *Journal of Theoretical Biology*, **7**, 1–52.

Hamilton, W. D. (1966). On the moulding of senescence by natural selection. *Journal of Theoretical Biology*, **12**, 12–45.

Hamilton, W. D. (1982). Pathogens as causes of genetic diversity in their host populations. In *Population Biology of Infectious Diseases: Report of the Dahlem Workshop* on *Population Biology of Infectious Disease Agents, Berlin, 1982, March 14–19* (Anderson, R. M., and May, R. M., Eds.), pp. 269–96. Springer Verlag, Berlin.

Hamilton, W. D., Axelrod, R., and Tanese, R. (1990). Sexual reproduction as an adaptation to resist parasites (a review). *Proceedings of the National Academy of Sciences USA*, **87**, 3566–73.

Hanis, C. L., Hewett-Emmett, D., Bertin, T. K., and Schull, W. J. (1991). Origins of U. S. Hispanics. Implications for diabetes. *Diabetes Care*, **14**(7), 618–27.

Hayden, M. R., and Tyagi, S. C. (2001). "A" is for amylin and amyloid in type 2 diabetes mellitus. *Journal of the Pancreas* (online), **2**(4), 124–39.

Hickman, I. J., Clouston, A. D., Macdonald, G. A., Purdie, D. M., Prins, J. B., Ash, S., Jonsson, J. R., and Powell, E. E. (2002). Effect of weight reduction on liver histology and biochemistry in patients with chronic hepatitis C. *Gut*, **51**(1), 89–94.

Karvonen, M., Viik-Kajander, M., Moltchanova, E., Libman, I., LaPorte, R., and Tuomilehto, J. (2000). Incidence of childhood type 1 diabetes worldwide. Diabetes Mondiale (DiaMond) Project Group. *Diabetes Care*, **23**(10), 1516–26.

Kenny, S. J., Aubert, R. E., and Geiss, L. S. (1995). Prevalence and incidence of non-insulin-dependent diabetes. In *Diabetes in America* (Harris, M. I., Cowie, C. C., Stern, M. P., Boyko, E. J., Reiber, G. E., and Bennett, P. H., Eds.), 2nd ed., pp. 47–67. National Diabetes Data Group, National Institutes of Health, National Institute of Diabetes and Digestive and Kidney Diseases, NIH Publication No. 95–1468.

Knobler, H., Schihmanter, R., Zifroni, A., Fenakel, G., and Schattner, A. (2000). Increased risk of type 2 diabetes in noncirrhotic patients with chronic hepatitis C virus infection. *Mayo Clinic Proceedings*, **75**(4), 355–9.

LaPorte, R. E., Matsushima, M., and Chang, Y.-F. (1995). Prevalence and incidence of insulin-dependent diabetes. In *Diabetes in America* (Harris, M. I., Cowie, C. C., Stern, M. P., Boyko, E. J., Reiber, G. E., and Bennett, P. H., Eds.), 2nd ed., pp. 37–46. National Diabetes Data Group, National Institutes of Health, National Institute of Diabetes and Digestive and Kidney Disease, NIH Publication No. 95–1468.

Lindsay, R. S., Krakoff, J., Hanson, R. L., Bennett, P. H., and Knowler, W. C. (2001). Gamma globulin levels predict type 2 diabetes in the Pima Indian population. *Diabetes*, **50**(7), 1598–603.

MacCracken, J., and Hoel, D. (1997). From ants to analogues. Puzzles and promises in diabetes management. *Postgraduate Medicine*, **101**(4), 138–40, 143–5, 149–50.

Mason, A. L., Lau, J. Y., Hoang, N., Qian, K., Alexander, G. J., Xu, L., Guo, L., Jacob, S., Regenstein, F. G., Zimmerman, R., Everhart, J. E., Wasserfall, C., Maclaren, N. K., and Perrillo, R. P. (1999). Association of diabetes mellitus and chronic hepatitis C virus infection. *Hepatology*, **29**(2), 328–33.

Mehta, S. H., Brancati, F. L., Sulkowski, M. S., Strathdee, S. A., Szklo, M., and Thomas, D. L. (2000). Prevalence of type 2 diabetes mellitus among persons with hepatitis C virus infection in the United States. *Annals of Internal Medicine*, **133**(8), 592–9.

Merck (1999). Disorders of carbohydrate metabolism. In *The Merck Manual of Diagnosis and Therapy*, 17th ed., (M. H. Beers and R. Berkow, Eds.), Ch. 13. Merck & Co., New York. (http://www.merck.com/pubs/mmanual/section2/chapter13/13a.htm).

Muller, H. J. (1950). Our load of mutations. *American Journal of Human Genetics*, **2**, 111–76.

National Institute of Diabetes and Digestive and Kidney Diseases (2002). National Diabetes Statistics fact sheet: general information and national estimates on diabetes in the United States, 2000. U.S. Department of Health and Human Services, National Institutes of Health, Bethesda, MD.

Neel, J. V., Weder, A. B., and Julius, S. (1998). Type II diabetes, essential hypertension, and obesity as "syndromes of impaired genetic homeostasis": the "thrifty genotype" hypothesis enters the 21st century. *Perspectives in Biology and Medicine*, **42**(1), 44–74.

Nesse, R. M., and Williams, G. C. (1994). *Why We Get Sick: The New Science of Darwinian Medicine*. Times Books, New York.

Online *Mendelian Inheritance In Man* (2002), Online *Mendelian Inheritance in Man*, OMIM^{TM}. Center for Medical Genetics, Johns Hopkins Univ. (Baltimore, MD) and National Center for Biotechnology Information, National Library of Medicine (Bethesda, MD). http://www3.ncbi.nlm.nih.gov/omim/.

Rewers, M., and Hamman, R. F. (1995). Risk factors for non-insulin dependent diabetes. In *Diabetes in America* (Harris, M. I., Cowie, C. C., Stern, M. P., Boyko, E. J., Reiber, G. E., and Bennett, P. H., Eds.), 2nd ed., pp. 179–220. National Diabetes Data Group, National Institutes of Health, National Institute of Diabetes and Digestive and Kidney Diseases, NIH Publication No. 95–1468.

Saikku, P., Leinonen, M., Mattila, K., and 3 colleagues. (1988). Serological evidence of an association of a novel *Chlamydia*, TWAR, with chronic coronary heart disease and acute myocardial infarction. *Lancet*, **2**, 983–86.

Schmidt, M. I., Duncan, B. B., Sharrett, A. R., Lindberg, G., Savage, P. J., Offenbacher, S., Azambuja, M. I., Tracy, R. P., and Heiss, G. (1999). Markers of inflammation and prediction of diabetes mellitus in adults (Atherosclerosis Risk in Communities study): a cohort study. *Lancet*, **353**(9165), 1649–52.

Slatkin, M., and Muirhead, C. A. (2000). A method for estimating the intensity of overdominant selection from the distribution of allele frequencies. *Genetics*, **156**(4), 2119–26.

Vogel, F., and Motulsky, A. G. (1997). *Human Genetics*, 3rd ed. Springer Verlag, Berlin.

Weiss, K. M., Ferrell, R. E., Hanis, C. L., and Styne, P. N. (1984). Genetics and epidemiology of gallbladder disease in New World native peoples. *American Journal of Human Genetics*, **36**(6), 1259–78.

World Health Organization (2002). World Health Organization Fact Sheet No. 138. Diabetes Mellitus. Revised April, 2002. http://www.who.int/inf-fs/en/fact138.html.

Yin, H., Berg, A. K., Tuvemo, T., and Frisk, G. (2002). Enterovirus RNA is found in peripheral blood mononuclear cells in a majority of type 1 diabetic children at onset. *Diabetes*, **51**(6), 1964–71.

Zur Hausen, H. (1991). Viruses in human cancers. *Science*, **254**, 1167–73.

SIXTEEN

The Future of Human Evolution

Luca Cavalli-Sforza

CULTURAL VS. GENETIC EVOLUTION

Human evolution is unique, having developed to the utmost extent a new survival tool: culture. Culture is found in many other animals, but the enormous power of communication made possible by language is not found in any other organism. What do I mean by culture? I like one definition I found in *Webster's International Dictionary* that I will paraphrase succinctly: the ensemble of knowledge, including customs and technologies, transmitted and accumulated through the generations, that played and continue to play an essential role in the evolution of our behavior. The mechanism of evolution of anything that can be reproduced and transmitted, be it genes or ideas, is dominated by three factors, which were first understood and defined in the study of genetic evolution: mutation, natural selection, and chance (random genetic drift). Although the factors underlying cultural and genetic evolution are very different, the same or similar principles and models are useful in understanding both.

Mutation is transmissible change, of genes or DNA in genetic evolution, and of ideas or brain circuits in cultural evolution. The second and third factors have been loosely described in genetic evolution as survival of the fittest and survival of the luckiest. When we speak of cultural evolution it may seem superficially that natural selection is not involved, i.e., that it is not the environment around us that decides our fate but rather that we take our destiny into our own hands by accepting or rejecting ideas, customs, and habits. The word "cultural selection" has been suggested (Campbell, 1976). But it would be wrong to forget that, over and above cultural selection, natural selection always remains in control. Our Darwinian fitness, the measurement of selective value in terms of survival

343

and reproduction of individuals is affected both by physical adaptedness to the environment and by the cultural choices we make, under the effect of genes or of learning. Choosing Coca-Cola or Pepsi-Cola when we are thirsty may be trivial from a Darwinian fitness perspective, but looking right rather than left when one leaves the curb to cross the road is obviously very important, especially if one is a foreigner who has just arrived in the UK. This decision may be influenced by our genes, or by our training, and both affect absent-mindedness, but knowledge of traffic rules in a new country, a culturally acquired trait, clearly has an important influence on our destiny. More obvious examples are the use of drugs, sexual habits such as promiscuity or use of condoms – cultural choices whose survival values are being automatically tested these days by natural selection. The advent of culture has also reinforced the effect of drift, because some very effective cultural transmitters endowed with authority, power, prestige, or persuasion skills – kings, political leaders, demagogues, popes, saints, entertainers, artists, entrepreneurs, impostors – may have enormous cultural influence.

When speaking of genetic evolution, one tends to forget to mention another basic mechanism, transmission, because in genetics we are so used to almost perfect genetic transmission, and most exceptions are listed separately as mutations. But genetic (Mendelian) and cultural transmission are very different. Genetic transmission in higher organisms is almost exclusively from parent to child (vertical transmission) and is very highly conservative. By contrast, much cultural transmission is among unrelated people (horizontal transmission), and it can be much more rapid. But there is also vertical cultural transmission, which is often more conservative than we think. In fact, there is even a frequent clash between slow cultural change for vertically transmitted traits and rapid cultural horizontal change: we witness it as part of the clash between generations. Models of cultural transmission have been developed by Cavalli-Sforza and Feldman (1972). The terms "vertical" and "cultural" in this context were borrowed from parasitology.

Mutation in cultural evolution (a new idea or an innovation) is probably more frequent than it is in genetic evolution. We don't know the physical basis of new ideas, but most likely it is not DNA change. Rather, it must be a new brain circuit. Moreover, cultural mutation is not random like genetic mutation. It is usually a response to a specific need. It may not necessarily take care of the need, or it may have a higher cost than the benefit it produces. It may be designed more for the good of the inventor than for that of society, and the two benefits may be incompatible.

Why have humans developed enormous power? The answers seem unequivocal: technology and communication. The first need not be emphasized. Communication is a special technology that helps spread other technologies, and it has had considerable evolution itself. Here we intend it as transfer of knowledge, for cooperative purposes. Language is the major tool of communication: to the extent that it is used for cooperative purposes, which is of course not always the case, it has been the most important propeller of human evolution. Its evolution must have gone through many steps that must have been the major cause of development of the human brain, leading to its present size, which was reached some 300,000 years ago, along with other changes that made us human and contributed to the rapid evolution of the brain. It may seem, superficially, that communication is not enough. What two idiots can communicate to each other may be of no help to either of them. But for communication by a structured language to develop and be helpful there is a need of a deep understanding of the meaning of words. Thus, to be able to speak usefully, one must be able to use to a non-trivial extent what we call reason or intelligence. It seems fairly obvious that human intelligence and language developed together.

RECENT HUMAN EVOLUTION: OUT OF AFRICA 2
The rate of cultural evolution is now undergoing another major increase because of the very recent globalization of communication. To better understand this point, it is useful to rapidly review the history of the past 100,000 years, in which we can witness how cultural evolution has gained control over genetic evolution. The standard model of the evolution of modern humans is sometimes labeled "Out of Africa 2" because there was an earlier expansion from Africa about two million years ago, limited to the Old World. According to this model, there was a very recent expansion about 100 kya (kiloyears ago) of a rather small group – we can call it a small tribe – that lived in East Africa. All modern humans living today are descendants of this tribe. We may in the future find exceptions to this rule, but none has yet been found. This tribe went through three major expansions (population growth and more or less simultaneous geographic expansion). The first was to Africa beginning about 100 kya. The two later ones were to the rest of the world: one was about 50–60 kya along the southern coast of Asia to Southeast Asia, Oceania, and then to East Asia; the other was about 40 kya to the center of Asia, Europe, northeast Asia, and America. By 12 kya, the tip of South America had been reached. Polynesia was settled a few thousand years later.

Probably the major propeller of this cultural evolution was some important refinement in the use of language, connected with a more advanced use of logical mechanisms. These are in a sense cultural changes, but they are also biological ones because they may have been favored or made possible by specific genetic changes in the brain. There were other, non-linguistic, important assists from true, purely cultural innovations: navigation was necessary for the settlement of Oceania but may have started earlier, and a set of new tools, called Aurignacian tools, probably developed somewhere in Eurasia.

Language evolution must have gone through many steps over the 2 million years of evolution of the genus Homo. The first may have been the creation of a very large vocabulary, under continuous enlargement. Increase in brain size must have been parallel to the development of language. Primates can manage relatively correct use of a few hundred words after intensive training, but humans have developed many thousands as a minimum, and most recently even hundreds of thousands. Pronunciation of words required the development of a complex organ, including the larynx, and correct use of words required the development of a specific drive to learn the mother tongue, which lasts only for the first four or five years. Among the most advanced steps in language evolution, a major one was perhaps the development of grammar + syntax, which made it possible to express very complex thoughts (ideas). It may have been the last step prior to human expansion to the whole world, because all modern languages share equally complex, rather similar syntaxes. In general, however, syntax has more constraints and seems to evolve more slowly than the other parts of language. There are indications that the great progress in language was complete only when the major expansions from Africa began. Art, and to a large extent religion, also exploded in the past 50,000 years and is found in all places settled by modern humans, but was not practiced, at least certainly not to the same degree, by lateral, extinct human branches like Neandertal.

PROBABLE SINGLE ORIGIN OF MODERN HUMAN LANGUAGE

The rate of language evolution is rapid because the cultural mutation rate is high, especially for sounds and words. Today there exist more than 5,000 mutually incomprehensible languages (Cavalli-Sforza, 1999). It takes 1,000 years for two closely related languages to lose understandability. It becomes very difficult to recognize that two languages separated for more than 10 ky have a common origin. It is not surprising, therefore, that many linguists, who are not used to comparing remote languages, refuse to discuss whether there was a single origin of language. It is easier to believe that modern languages had a single origin, if one knows about the genetic evolution of modern humans. Genetic markers inherited through one parent only, such as the Y chromosome and mitochondrial DNA (inherited, respectively, by the father and by the mother), are very informative for reconstructing the ancestry of modern humans. They all show that all modern humans tested so far descend from a very small population that lived in East Africa (the size of a hunter–gatherer tribe). Hence, there must have been a single language at the beginning of the last expansion.

There is a simple theoretical explanation that helps in understanding the kinetics of a mutation causing greater communication skills. For any communication mechanism, there must be at least two people involved, a transmitter and a receiver, with the capacity to interchange roles. If the increased skill is transmitted genetically, the first two individuals carrying the advantageous gene must be a parent and a child, or two siblings, etc. Otherwise, the same mutation should appear independently in two unrelated individuals, born a short distance from each other in time and space, and this is a very improbable event. Later, after developing in a small section of a population, the gene responsible for the same mechanism of communication can expand fairly rapidly to the whole population. Genetic transformation takes place quickly if the population is small, especially if the new genetic trait confers considerable advantage to its carriers and therefore is rapidly selected (Eschel and Cavalli-Sforza, 1982). Thus a genetic mutation allowing substantial progress in communication can invade a small population in a few centuries.

The advantage gained by increased communication may be of inestimable value to members of a tribe that will become better able to communicate and thus avoid dangers. A hunter–gatherer tribe has a prereproductive mortality rate of approximately 50% (Cavalli-Sforza, 1986). Even a small reduction in this mortality rate can increase the net reproduction rate enough to justify the increase from a few thousand individuals to 1,000 times more at the beginning of agriculture, 10 kya (in some places earlier).

MEDICINE AND NATURAL SELECTION

Mortality has been greatly changed in the past 150 years by the advent of medicine. Prereproductive mortality has evolved in developing countries in the past 25–50 or more years, mostly under cultural acceptance of hygienic rules learnt from developed countries, and has gone down 50–60% to rates that vary greatly from country to country but are on average 10–20%. In developed countries, where the mortality rate was already at this level two centuries ago, it went down and is now less than 1%. This means that natural selection due to differential mortality after birth has practically stopped in developed countries.

Modern medicine is changing the fitness of phenotypes that were seriously pathological in premedical times. Many people who would have been dead from infectious diseases a few centuries ago have now reached the age at which cancer is common, or at least that of the ever-present cardiac/circulatory failures. Thus there is an increasing chance that we will, if not reach, at least get near the age of 120 that some demographers have extrapolated as the limit of the human life span from classical mathematical models. Both of these usually terminal classes of disease are being treated with increasing success. But inevitably, medicine is increasing tremendously the cost of staying alive. Medicine is probably also largely dysgenic: the clearest example is the increased frequency of genetically sterile people who can now be properly treated and then transmit their sterility, which will inevitably increase the number of those who are sterile. It is likely that many of the high and increasing costs of medicine will be with us for a long time.

In most cultures, however, fertility is now largely a custom rather than a physically determined trait. There are enormous changes in gene frequencies due to variation in net population growth rates in different parts of the world, determined by culturally transmitted customs. Thompson (1972) calculated the rate of change of ABO blood group gene frequencies in the world due to the differential growth rate of populations with different cultures, and he found that it equaled or topped rates of natural selection considered very strong. Repeating the same calculation today would most probably give even more surprising rates. Another example is the browning of the human species, due to the low reproduction rate of European and North American people. This is culturally driven evolution, which will last as long as culturally controlled fertility differences remain.

The world was almost fully settled 10–12 kya, when population density was high enough that collection of food produced in nature became insufficient. Food production by agriculture and animal breeding started at that time in several different temperate areas of the world. This promoted a demographic transition to higher numbers, which caused a numerical increase in the global population of about 1,000-fold in the past 10,000 years. We are still growing, at the highest rate ever reached. In the past ten centuries, Europe, India, and China grew considerably in numbers because of a decrease in mortality due to new inventions that increased wealth and food production. China stopped its extramural expansion by suspending interest in its otherwise excellent transoceanic fleet, the consequence of a political decision made about 500 years ago, but continued to grow within its bounds (Diamond, 1997). Europeans kept growing and also started expanding outside of Europe thanks to transoceanic navigation.

As European growth became excessive, in spite of the chances of extramural expansion, in the nineteenth century birth rates started going down among Europeans. With the exception of France, where birth rates started to decrease before death rates, at least in some social classes, in the European demographic transition, the fall of the birth rate started to follow with a lag of about one generation the fall of the death rate. The population growth rate of the continent is now only slightly above zero, much of it due to immigration, with countries such as Italy and Germany having negative growth rates. Very recently the economically underdeveloped world made a demographic transition in the opposite direction and has started growing at an unprecedented rate, around 3% per year (a doubling rate close to 25 years). This is probably due to the introduction of modest and inexpensive, but effective, rules of hygiene and medical help, which have lowered the death rate by about five times. With the exception of China, there has been no dramatic change in birth rates, and it may still take more time before the world starts slowing down its present race toward disaster.

The few populations of hunter–gatherers still in existence, who need large areas of unspoiled environment, have experienced strong restrictions on their capacity to survive from the serious, long-time competition of neighboring groups. They have universally responded, probably beginning a long time ago, with measures that limited their birth rates. Examples from African Pygmies are a sex taboo for three years following the birth of a child, and "cultural menopause:" a woman whose daughter has had a child avoids having further children. The expressed aim of these customs is not malthusian, but in practice they generate zero – if not negative – population growth.

Excessive population growth is counteracted in nature by three types of brakes: epidemics, wars, and famines. The world currently experiences all three, and one epidemic, AIDS, is taking on terrifying proportions. There have been many major epidemics in past centuries, and the plagues in Europe have caused tremendous losses, up to two-thirds and more of whole countries. Our species will survive AIDS, thanks to preexisting genetic resistance mechanisms, but considering the duration of the disease and the rate at which new infections increase, we seem to be ready for major catastrophes, unless a really effective therapy is found.

The best weapons of modern medicine are often used ineffectively. Who will win the war against bacteria or other microorganisms, started in the twentieth century with the help of antibiotics and chemotherapeutics: resistant microbes or humans inventing more drugs? There may be a finite limit to the number of new drugs that are possible. The knowledge of bacterial genomes may give us new weapons but not an infinite number of them. Shall we resort to living in quarantine space stations to avoid infectious diseases by multiply resistant organisms? Or can we become more cautious in the use of antibiotics and chemotherapeutics?

Communication is not always good. Physical communication – transportation – is the main cause of the rapid spread of epidemics, as in the classical examples of plague epidemics in Europe. It is causing the spread of AIDS through Africa and probably other parts of the world. Trucks drive all the routes in Africa in great numbers, carrying various commodities, and drivers usually have sexual contacts at every stop. Prostitutes flock to truck stops and contribute to the spread of venereal disease. And the disease is reaching all parts of the world. It is difficult to abandon age-old habits. Even several Eskimos apparently find it difficult to abandon the embarrassing old custom of lending their wives to guests, according to a modern travel guide book (*Lonely Planet*).

We can conclude that in humans, cultural evolution has largely replaced genetic evolution. Medicine is a product of cultural evolution that has almost completely suppressed preproductive mortality and has therefore reduced considerably the impact of natural selection. It would be very difficult to give up medicine, which is perhaps the least objectionable aspect of progress and is abhorred only by a few cultures (more exactly, cults). But cultural selection is far from rational, as it is often the expression of the will of a few people, be they entrepreneurs or politicians who have strong influence or power and act in their own interest rather than that of the social group. It must also be acknowledged that even with the best of intentions, it may be very difficult to find good solutions to hard social problems. Fortunately, cultural decisions remain under the constant rule of natural selection, which may act ruthlessly but is geared towards the maintenance of the species. Eventually, under the higher control of natural selection, there is a reasonable probability that human matters will stay on course, though with largely inevitable local or global catastrophes.

REFERENCES

Campbell, D. T., 1976. On the conflicts between biological and social evolution and between psychology and moral tradition. *American Psychologist*, 31:1103–26.

Cavalli-Sforza, L. L., Ed., 1986. *African Pygmies*. Academic Press, Orlando, FL.

Cavalli-Sforza, L. L., 1999. *Genes, Peoples and Languages*. Penguin Press, London.

Cavalli-Sforza, L. L., and M. W. Feldman, 1972. *Cultural Transmission and Evolution.* Princeton University Press, Princeton NJ.

Diamond, J., 1997. *Guns, Germs and Steel.* W. W. Norton, New York, NY.

Eshel, I., and L. L. Cavalli-Sforza, 1982. Assortment of encounters and evolution of cooperativeness. *Proc. Natl. Acad. Sci. USA* 79:1331–5.

Lonely Planet, 2003. Lonely Planet Publication, London

Thompson, E., 1972. Rates of change of World ABO blood group frequencies. *Ann. Hum. Genet.* 35:357–61.

Index